NEUROMETHODS

Series Editor
Wolfgang Walz
University of Saskatchewan
Saskatoon, SK, Canada

For further volumes:
http://www.springer.com/series/7657

Optical Imaging of Neocortical Dynamics

Edited by

Bruno Weber

Institute of Pharmacology and Toxicology, University of Zurich, Zurich, Switzerland

Fritjof Helmchen

Brain Research Institute, University of Zurich, Zurich, Switzerland

Humana Press

Editors
Bruno Weber
Institute of Pharmacology and Toxicology
University of Zurich
Zurich, Switzerland

Fritjof Helmchen
Brain Research Institute
University of Zurich
Zurich, Switzerland

ISSN 0893-2336 ISSN 1940-6045 (electronic)
ISBN 978-1-4939-5970-9 ISBN 978-1-62703-785-3 (eBook)
DOI 10.1007/978-1-62703-785-3
Springer New York Heidelberg Dordrecht London

Printed on acid-free paper

Humana Press is a brand of Springer
Springer is part of Springer Science+Business Media (www.springer.com)

Series Preface

Under the guidance of its founders Alan Boulton and Glen Baker, the Neuromethods series by Humana Press has been very successful since the first volume appeared in 1985. In about 17 years, 37 volumes have been published. In 2006, Springer Science + Business Media made a renewed commitment to this series. The new program will focus on methods that are either unique to the nervous system and excitable cells or which need special consideration to be applied to the neurosciences. The program will strike a balance between recent and exciting developments like those concerning new animal models of disease, imaging, in vivo methods, and more established techniques. These include immunocytochemistry and electrophysiological technologies. New trainees in neurosciences still need a sound footing in these older methods in order to apply a critical approach to their results. The careful application of methods is probably the most important step in the process of scientific inquiry. In the past, new methodologies led the way in developing new disciplines in the biological and medical sciences. For example, Physiology emerged out of Anatomy in the nineteenth century by harnessing new methods based on the newly discovered phenomenon of electricity. Nowadays, the relationships between disciplines and methods are more complex. Methods are now widely shared between disciplines and research areas. New developments in electronic publishing also make it possible for scientists to download chapters or protocols selectively within a very short time of encountering them. This new approach has been taken into account in the design of individual volumes and chapters in this series.

Saskatoon, SK, Canada ***Wolfgang Walz***

Preface

Optical methods for functional imaging are indispensible tools for modern neuroscience due to their exceptional spatial and temporal sampling options and the large variety of possible functional read-outs ranging from voltage changes to metabolic states. As such, state-of-the-art optical imaging techniques have emerged as pivotal approaches for advancing our understanding of information processing in the mammalian neocortex, and many of the most imperative issues have now become technically accessible. The main goal of the present volume is to provide a collection of the most relevant optical methods currently used for investigating neocortical circuit dynamics. We believe that we were able to share our enthusiasm for this endeavor with a group of outstanding authors who have been instrumental in advancing optical methods in the past years and who at the same time are very successful users of those methods for interrogating the neocortex. We hope that the volume is useful for students as well as neuroscientists interested in applying optical approaches to answer their specific questions. The wealth of optical imaging methods that are currently applied in the neurosciences has forced us to limit the range to methods that (1) are applied in experiments in vivo, mainly in the rodent brain, and (2) measure the functional properties on the spatial scale of cortical circuits. We have divided the volume into three parts:

Part I is an introductory section that covers the physical fundamentals of optical imaging and introduces the molecular tools and imaging devices used for in vivo optical imaging. The first chapter outlines the methodological concepts of using light to address current questions in neuroscience. Next, two chapters treat the physical principles of imaging and the optical properties of brain tissue, respectively, followed by two further chapters that introduce functional indicators as well as light-controllable molecules. These five chapters thus provide a background for all the chapters in *Parts II and III.*

Part II covers the most relevant methods and their applications to investigate neuronal activity in the neocortex across a wide range of spatial and temporal scales. Several chapters present the advantages of two-photon microscopy to study neocortical dynamics, largely using in vivo calcium imaging from individual neurons and their dendrites or from large neuronal populations. Wide-field voltage-sensitive dye imaging is demonstrated as a complementary technique to reveal large-scale dynamics. In addition, the opportunities to probe causal relationship between cellular and circuit mechanisms and behavior with optogenetic tools are explained.

Part III focuses on optical imaging methods used to probe signals that do not directly reflect neuronal activity but provide information on the metabolic state of the cortex. The first two chapters address methods for imaging signals that originate from astrocytes and microglia. The remaining three chapters introduce techniques for measuring hemodynamics as well as oxygenation of blood and tissue.

Optical imaging is one of the most dynamic technological fields in neuroscience. Novel techniques, such as new optical probes or imaging instruments, are continually emerging and published in the most prestigious journals. In a way, this ever-changing field is thus very difficult to capture in a methods book, because the moment it is produced, the technological landscape may have changed again significantly. We nevertheless hope that our collection is useful even in the light of rapid changes, in part because some of the fundamental concepts remain unchanged.

Our main thanks go to all the authors for their outstanding contributions to this volume. We would also like to thank Michaela Thallmair and Ladan Egolf for their editorial assistance.

Zürich, Switzerland *Bruno Weber*
Fritjof Helmchen

Contents

Part III Elucidating Metabolic States of Cortical Circuits

Contributors

WALTHER AKEMANN • *Knöpfel Laboratory for Neuronal Circuit Dynamics, RIKEN Brain Science Institute, Wako, Saitama, Japan*
DAVID A. BOAS • *Martinos Center for Biomedical Imaging, Massachusetts General Hospital, Harvard Medical School, Charlestown, MA, USA*
BIANCA BRAWEK • *Institute of Physiology II, Eberhard Karls University of Tübingen, Tuebingen, Germany*
ANDERS M. DALE • *Department of Neurosciences and Radiology, University of California, San Diego, La Jolla, CA, USA*
KARL DEISSEROTH • *Department of Bioengineering, Department of Psychiatry and Behavioral Sciences, Howard Hughes Medical Institute, Stanford University, Stanford, CA, USA*
ANNA DEVOR • *Department of Neurosciences and Radiology, University of California, San Diego, La Jolla, CA, USA*
JONATHAN D. DRISCOLL • *Department of Physics, Division of Physical Sciences, University of California, San Diego, La Jolla, CA, USA*
ANDREW K. DUNN • *Department of Biomedical Engineering, University of Texas at Austin, Austin, TX, USA*
LIEF E. FENNO • *Department of Bioengineering, Department of Psychiatry and Behavioral Sciences, Howard Hughes Medical Institute, Stanford University, Stanford, CA, USA*
OLGA GARASCHUK • *Institute of Physiology II, Eberhard Karls University of Tübingen, Tuebingen, Germany*
BENJAMIN F. GREWE • *Department of Biological Sciences, Department of Applied Physics, Stanford University, James H. Clark Center for Biomedical Engineering and Sciences, Stanford, CA, USA*
FLORENT HAISS • *Institute of Pharmacology and Toxicology, University of Zurich, Zurich, Switzerland*
FRITJOF HELMCHEN • *Brain Research Institute, University of Zurich, Zurich, Switzerland*
HAJIME HIRASE • *RIKEN Brain Science Institute, Wako, Saitama, Japan*
HONGBO JIA • *Institute of Neuroscience and Institute for Advanced Study, Technical University Munich, Munich, Germany*
BJÖRN M. KAMPA • *Brain Research Institute, University of Zurich, Zurich, Switzerland*
DAVID KLEINFELD • *Department of Physics, Division of Physical Sciences, University of California, San Diego, La Jolla, CA, USA*
THOMAS KNÖPFEL • *RIKEN Brain Science Institute, Wako, Saitama, Japan; Brain Sciences Division, Imperial College, London, United Kingdom*
ARTHUR KONNERTH • *Institute of Neuroscience and Institute for Advanced Study, Technical University Munich, Munich, Germany*
MATTHEW E. LARKUM • *Neuroscience Research Center (NWFZ), Charité—Universitätsmedizin Berlin, Berlin, Germany*

HENRY LÜTCKE • *Brain Research Institute, University of Zurich, Zurich, Switzerland*
DAVID J. MARGOLIS • *Brain Research Institute, University of Zurich, Zurich, Switzerland*
TSUNEKO MISHIMA • *RIKEN Brain Science Institute, Wako, Saitama, Japan*
MASANORI MURAYAMA • *Laboratory for Behavioral Neurophysiology, RIKEN Brain Science Institute, Wako, Saitama, Japan*
HIROKI MUTOH • *Knöpfel Laboratory for Neuronal Circuit Dynamics, RIKEN Brain Science Institute, Wako, Saitama, Japan*
JUNICHI NAKAI • *Saitama University Brain Science Institute, Saitama, Wako, Saitama, Japan*
MASAMICHI OHKURA • *Saitama University, Brain Science Institute Saitama, Saitama, Japan*
MICHAEL J. PESAVENTO • *Department of Physics, Division of Physical Sciences, University of California, San Diego, La Jolla, CA, USA*
CARL C.H. PETERSEN • *Laboratory of Sensory Processing, Brain Mind Institute, École Polytechnique Fédérale de Lausanne (EPFL), Lausanne, Switzerland*
EMMANUEL ROUSSAKIS • *Department of Biochemistry and Biophysics, University of Pennsylvania, Philadelphia, PA, USA*
PAYAM A. SAISAN • *Department of Neurosciences, University of California, San Diego, La Jolla, CA, USA*
SAVA SAKADŽIĆ • *Martinos Center for Biomedical Imaging, MGH, Harvard Medical School, Charlestown, MA, USA*
FRANK SCHEFFOLD • *Department of Physics and Fribourg Center for Nanomaterials, University of Fribourg, Fribourg, Switzerland*
ANDY Y. SHIH • *Department of Physics, Division of Physical Sciences, University of California, San Diego, La Jolla, CA, USA*
YOSHIAKI SHINOHARA • *RIKEN Brain Science Institute, Wako, Saitama, Japan*
LOUISE E. SINKS • *Department of Biochemistry and Biophysics, University of Pennsylvania, Philadelphia, PA, USA*
HAMUTAL SLOVIN • *Gonda Brain Research Center, Bar-Ilan University, Ramat Gan, Israel*
NORIO TAKATA • *RIKEN Brain Science Institute, Wako, Saitama, Japan*
IVAN TENG • *Department of Neurosciences, University of California, San Diego, La Jolla, CA, USA*
PEIFANG TIAN • *Department of Physics, John Carroll University, University Heights, OH, USA; Department of Neurosciences, University of California, San Diego, La Jolla, CA, USA*
IVO VANZETTA • *Institut de Neurosciences de la Timone, UMR 7289 CNRS, Aix-Marseille Université, Marseille, France*
SERGEI A. VINOGRADOV • *Department of Biochemistry and Biophysics, University of Pennsylvania, Philadelphia, PA, USA*
BRUNO WEBER • *Institute of Pharmacology and Toxicology, University of Zurich, Zurich, Switzerland*
MOHAMMAD A. YASEEN • *Martinos Center for Biomedical Imaging, MGH, Harvard Medical School, Charlestown, MA, USA*
PAVEL ZAKHAROV • *Physics Institute, University of Fribourg, Fribourg, Switzerland*

Part I

Fundamentals of Using Light to Study Neocortical Dynamics

Chapter 1

Neocortex in the Spotlight: Concepts, Questions, and Methods

Fritjof Helmchen and Bruno Weber

Abstract

When mammals such as mice, cats, monkeys, or humans act in the world, they continually make behaviorally relevant decisions based on perceived sensory information and memorized experiences and they constantly adapt to outside challenges through learning. These cognitive capabilities largely arise from neural processing in the outermost thin sheet of the forebrain called the neocortex. Although the mammalian neocortex has been studied extensively, the astounding complexity of both its structure and dynamics has precluded a comprehensive understanding of its function so far. Higher cortical function emerges from the interplay of myriads of diverse neocortical cells, organized across multiple hierarchical levels from local neuronal networks ("microcircuits") to communicating brain regions ("macrocircuits"). It remains elusive how these neural circuits operate—assisted by glial networks and fuelled by the vascular system—to generate intelligent behavior and ensure adequate learning. Advances in experimental methodology are essential to further unravel cortical function and in this book we highlight the rapid recent progress in optical methods for measuring and controlling neocortical dynamics, complementing classic electrophysiological approaches. In this chapter we provide a brief overview of the functional organization of the neocortex, its tissue constituents, and current concepts of neocortical dynamics. In preparation of subsequent chapters, we summarize the manifold ways photons can be used to study neocortical function, utilizing specially designed molecular tools and various imaging technologies. We conclude with a brief future outlook. Putting neocortex literally "into the spotlight" may help uncover its intriguing mysteries.

Key words Neocortex, Neuron, Glia, Fluorescence, Optogenetics, Microscopy, Two-photon, Imaging

1 Introduction

The neocortex covers a large portion of the mammalian forebrain, exceedingly so in primates and humans. One of the outstanding features of the cerebral cortex is its remarkable ability and flexibility to fulfill a wide range of tasks, such as orchestrating motor actions, integrating different sensory inputs, and enabling higher cognitive functions such as decision-making and consciousness. The rodent neocortex sheet is about 1–2 mm thick—depending on species and cortex region—and organized in a generic, six-layered layout with

Bruno Weber and Fritjof Helmchen (eds.), *Optical Imaging of Neocortical Dynamics*, Neuromethods, vol. 85, DOI 10.1007/978-1-62703-785-3_1,

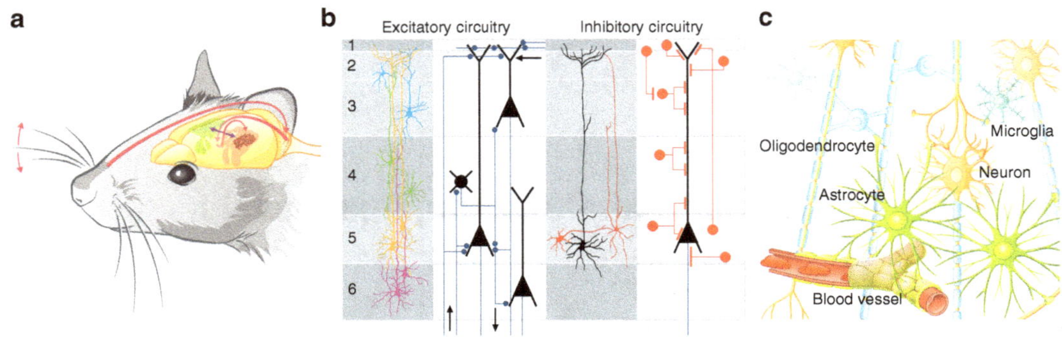

Fig. 1 Levels of neocortical circuit dynamics. Circuit dynamics and interactions occur at different spatial scales in neocortex. (**a**) On the large scale, information processing is based on the dynamic and parallel interaction between different areas (e.g., motor and sensory areas). Here, primary somatosensory cortex and motor cortex are indicated as red and green area, respectively. The barrel cortex is highlighted. (**b**). On the local network level, information processing is based on neocortical circuits that span throughout the six layers of cortex. The *left* two panels schematically depict the main excitatory neurons in the cortical layers (cellular morphologies adapted from [1]) and their principal connections [2]. On the *right panels*, major inhibitory inputs onto pyramidal neurons are depicted, comprising soma-, axon-, and various dendrite-targeting interneurons. Note the disinhibitory circuit motif. As exemplary microcircuitry the peri-somatic inhibition of a layer V pyramidal neurons and the feedback inhibition onto dendrites provided by deep Martinotti cells is shown (adapted from [3]). (**c**) On the cellular level, not only neurons but also glial cells are organized in networks and there is increasing evidence of manifold interactions between these neuronal and glial networks. Furthermore, there is a tight interaction between neurons, glia, and the vascular system to ensure metabolic homeostasis.

division into specialized regions, e.g., motor-related regions in frontal areas and sensory-related areas in the posterior part (Fig. 1a). The relative thickness and composition of layers is not uniform across all cortical regions and different areas have distinct communication lines with subcortical brain regions. Nonetheless, there are striking commonalities that have led to the concept of a basic local circuit architecture, a "canonical circuit" [4], presumed to instantiate fundamental computations and to repeat itself across the entire cortical sheet. Such basic circuitry likely spans throughout all layers, conceptualized in the idea of "cortical columns" as computational units [5]. On the other hand, local neural circuits (columns) never operate in isolation and are far from being "black boxes" with only a few inputs and outputs. Rather, information processing is distributed across wide areas of neocortex, with many operations happening in parallel through dynamic interactions of multiple regions. Overall, the excitable neuronal networks need to maintain their operation in adequate dynamic states, which not only requires appropriate excitation–inhibition balance and neuromodulatory tone but also depends on proper interaction with surrounding glial networks and the vascular system.

The functional architecture of a cortical column as well as the transversal organization of the entire cortical sheet—and thus the operational principles of both local "microcircuits" and larger-scale "macrocircuits" (Box 1)—are still poorly understood.

Box 1 Microcircuits and Macrocircuits.

Nervous systems are hierarchically organized cellular networks, spanning multiple spatial scales. In the neocortex this hierarchy is reflected by at least two levels: (1) the local (short-range) networks of interconnected neurons and (2) the larger (long-range) networks of communicating cortical and subcortical regions. Short-range connectivity patterns among neighboring neurons establish "microcircuits" that may serve specific information processing purposes, e.g., by implementing particular computations similar to filter functions. The output of microcircuit processing is then further transmitted via projection neurons, which send long axons to distant brain regions and thereby establish "macrocircuits," larger-scale networks of interconnected microcircuits. Macrocircuits involve many different brain areas and their orchestrated activity controls behavior. The entirety of connections among cells in microcircuits (or areas in macrocircuits) is nowadays often referred to as "connectome" [6, 7] and obviously is a key constraint of network dynamics. The functional distinction between local and larger-scale circuitry is, however, not as clear as it may seem. For example, each local neocortical region is heavily interconnected with other brain areas, questioning the concept of isolated, local processing units and rather emphasizing the distributed manner of cortical network operation.

Classic work during the last century, using mainly electrophysiological recordings, has established the fundament of our understanding of cortical function and its relationship to behavior; however, many questions could not be addressed due to a lack of suitable experimental methods. Fortunately, further development of neurophysiological techniques over the past two decades has greatly expanded our possibilities to measure various aspects of cortical dynamics and impose specific manipulations onto neocortical components. A central aim of such endeavors is to reveal principles of circuit dynamics that explain key aspects of behavior. Optical imaging techniques have emerged as a particularly suitable class of methods that has experienced enormous progress over the years and now enables a wide range of applications to investigate neocortical function. The aim of this book is to provide an overview of the state-of-the-art of optical methods for studying neocortical dynamics with a particular focus on in vivo studies in rodents, which have especially benefitted from these developments.

At present, optical methods continue to experience rapid further developments in technology (light sources, detectors, optics, microscopes, dyes, molecular probes, etc.). For several reasons, they take a special place in enabling conceptually novel approaches. First, light-based methods can provide information over a wide range of spatiotemporal scales, from synaptic structures to entire hemispheres, and from millisecond dynamics to slow processes occurring over a significant portion of an animal's lifespan. Second, they are applicable in a noninvasive (or minimally invasive) manner so that brain dynamics can be studied in the living animal (in vivo) and during natural behaviors. Third, they enable highly versatile approaches, for example by targeting specific cell types or employing special optical indicators that reveal various functional aspects, including those beyond electrical excitation of neuronal networks. Fourth, they are also unique in supporting simultaneous live observation of morphological changes and functional activation. Such co-observations will be essential to fully grasp cortical plasticity. Finally, but not least, imaging methods often permit a direct visualization of the dynamic processes in the brain, convincingly bringing forward an intuitive understanding in the sense of "seeing is believing."

In this chapter we provide basic background information on neocortical organization and then briefly introduce key aspects of optical methods that will be treated in detail in subsequent chapters. On the way we refer to individual chapters to allow the reader to directly jump to topics of interest. We end with a brief discussion of future prospects.

2 Functional Organization of the Neocortex

2.1 Parts and Compartments

A highly interconnected network of diverse types of neurons is responsible for information processing in the cerebral cortex [8]. Depending on region and species, a cubic millimeter of cerebral cortex consists of several tens to hundred thousand neurons and tens of millions to a billion synapses. The most prominent excitatory neuronal cell type is the so-called pyramidal neuron with characteristic dendrites as input structures and an axonal process as output structure [9] (Fig. 1b). Pyramidal neurons are located mainly in superficial layers 2 and 3 as well as in deep layers 5 and 6, while excitatory stellate cells are found in granular layer 4. Intermingled with the glutamatergic excitatory neurons are GABAergic interneurons of various types, which constitute about 15 % of the neuronal population. The systematic classification of GABAergic neurons has progressed significantly [10, 11] but many aspects of the functional role of inhibitory interneurons still remain to be resolved. Presumably, different types of GABAergic interneurons establish circuit motifs through specific connections with neighboring neurons to subserve special computational functions. For example, soma-targeting basket cells can control neuronal output while dendrite-targeting Martinotti cells impact on dendritic input sites [12] (Fig. 1b). In the adult neocortex, communication in neuronal networks mostly occurs via neurotransmission at chemical synapses, although some types of interneurons form sub-networks via electrical synapses [13]. Whereas synaptic signaling and dendritic integration operate in analogue regimes, the "digital output" of each neuron is the train of action potentials that is eventually generated (typically close to the soma) and then transmitted along myelinated or unmyelinated axons.

Besides these neuronal "parts," glia constitute the second major cell class of the cerebral cortex [14] (Fig. 1c). Myelin production and maintenance is provided by oligodendrocytes, one of the three major glia subtypes of the central nervous system (CNS). In addition, astrocytes and microglia densely populate the grey matter of neocortex. Astrocytes make contact with neurons at synapses and cover nearly the entire vasculature. They are involved in homeostatic functions, such as energy metabolism, transmitter recycling and blood flow regulation, and they also exert direct control over neuronal signaling, as they can release gliotransmitters that influence synaptic transmission [15, 16]. Finally, microglia serve as active immune cells of the CNS including the cerebral cortex.

Information processing in the cerebral cortex comes at a very high price in terms of energy use. From what would be expected on a weight basis, rates of glucose and oxygen used are at least one order of magnitude higher compared to the rest of the body. The maintenance of the electrochemical gradient across the neuronal membranes by activating the Na^+/K^+ pump is thought to be the major energy-consuming process. This remarkable energy demand in combination with the fact that brain tissues lack substantial capacity to store energy requires a tight spatiotemporal control of the energy supply via the blood stream. Indeed, the vascular linear density is about half a meter of vessels per cubic millimeter [17, 18]. Overall, the neuronal and glial networks as well as the vascular system form the three major interacting neocortical compartments.

2.2 Cerebral Cortex, a Highly Dynamic System

Modern neuroanatomy takes advantage of molecular, genetic, and high-resolution microscopy methods to examine cortical circuits [19, 20], in particular enabling identification of specific cell types and quantification of cellular distributions and connectivity in the mouse cortex. In addition to expanding our neuroanatomical knowledge, we also need to obtain a more profound view of the dynamic aspects of neocortical organization. One of the major tasks of the neocortex is to incessantly react to an ever-changing environment. In the evolution of nervous systems, structure and function always went hand in hand simply because the dynamic interactions with the environment determined the probability of the organism to produce offspring and survive. Therefore, by its very nature the neocortex is a highly dynamic system and a theory of neocortex inevitably will have to be a dynamic theory. Throughout this book it will be emphasized that light-based approaches offer a wide spectrum of methods ("How") to study neocortical dynamics through measurement of various dynamic signals ("What") that originate at well-defined levels of CNS organization

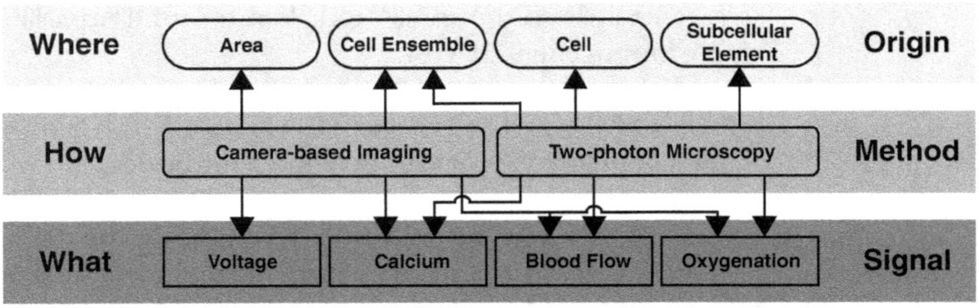

Fig. 2 Coarse overview of optical methods used for in vivo imaging of cortical dynamics. Most in vivo optical imaging methods are either based on sensitive cameras or involve two-photon excited fluorescence microscopy. The signals investigated span a broad range of spatial scales, from subcellular elements to entire cortical areas. The signals' sources range from voltage and calcium concentration to blood flow and oxygenation of hemoglobin and tissue.

("Where") (Fig. 2). Notably, optical methods also make it possible to study dynamic features of electrically silent glial cells and the vascular system. Optical imaging of neocortex function in living brains thus perfectly complements the expanding high-resolution neuroanatomical methods in order to bring structure and function together.

A hallmark of neocortical dynamics is the extensive range of spatiotemporal scales it covers. Relevant temporal scales range from fast electrical signaling (milliseconds), over short-term adaptation (seconds), to functional modulation (minutes to hours), to long-term plasticity and memory (days to years). In addition, maintenance of ionic concentration gradients as well as neurotransmitter synthesis, among other processes, requires energy. Hence, cortical information processing is tightly linked to energy metabolism, including substrate transport and oxygen delivery, which are regulated via cerebral blood flow on the seconds to minutes time scale. Similarly, cortical dynamics happen over an impressive range of spatial scales. Nanometer changes continually occur at the molecular level, involving for example expression and assembly of new proteins. On the cellular level, structural plasticity involves creation and destruction of dendritic spines and cortical synapses [21]. On even larger scale, signal flow through local and global neuronal networks is continually regulated and optimized for generating adequate behavior, coping with the particular challenges of the given environmental conditions. Optical approaches are particularly well suited to cover these wide spatiotemporal scales. Fig. 3 gives examples of various dynamic signals and spatial scales in the neocortex, which will be treated in more detail in subsequent chapters.

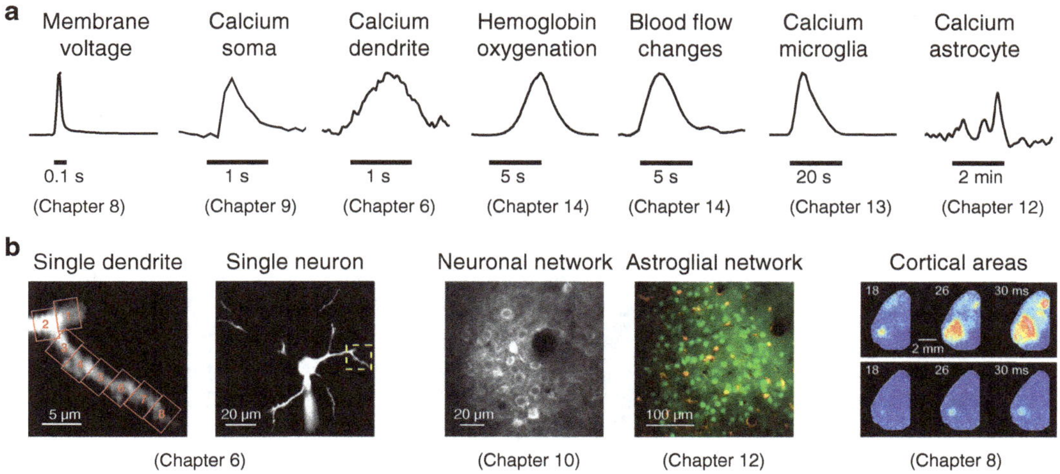

Fig. 3 Temporal and spatial range of cortical processes covered by optical imaging. The figure provides examples of cortical dynamics described in the indicated chapters of this volume, ranging from the sub-second to minute and from the micron to millimeter domain. (**a**) In vivo optical imaging methods can capture signals from a variety of underlying processes covering a large temporal dynamic range. (**b**) Likewise, a broad range of spatial scales is covered by current in vivo optical imaging methods.

3 Using Light to Study Neocortex Function

Optical techniques are among the most important innovations in the neurosciences in the last decades. Historically, most of what we know today about cortical dynamics has been obtained with the help of electrophysiology. While in vivo electrophysiological recordings provided a wealth of data and insights—through electroencephalographic (EEG) recordings, extracellular spike recordings, or by measuring membrane potential changes in individual neurons—optical methods have remarkable advantages on their own, enabling novel, complementary views on cortical dynamics [22]. First, optical approaches can be applied in a relatively noninvasive manner because photons, not electrodes, penetrate the tissue in order to interrogate neural circuits. Optical imaging therefore has become an indispensable method especially for in vivo studies of cortical function. Of course, aside from techniques based on intrinsic signals, dye molecules typically need to be first introduced into the tissue, for which various techniques—in particular genetic methods—now exist. Second, a large and diverse set of molecular tools has been developed, which make it possible to optically probe various aspects of cellular activities and cortical dynamics, ranging from intracellular ion concentrations, membrane voltage, enzyme activities, to oxygen and energy substrate levels. Development of optical probes has become a very active research field on its own, and the toolbox is rapidly growing, especially for genetically encoded probes that can be expressed in specific cell types. Third, a multitude of microscope technologies and optical methods—from simple to

Box 2 Light and (Grey) Matter

The interaction of light and matter is a fundamental aspect of the physical world. For experimental neuroscientists, the interaction of light and "grey matter" is of particular importance and two major aspects of this topic will recur throughout the book. First, light can interact with molecules that are present in grey matter. Photons are absorbed and cause molecular excitation and subsequent reactions, e.g., the generation of fluorescence or luminescence or conformational changes in proteins. The affected molecules either can be intrinsic to the tissue or they may have been artificially introduced into the system. The latter class of molecules comprises the large group of optical sensors and actuators, which are utilized to probe cortical function. A second important aspect of light interaction with grey matter concerns the way light beams are affected when they enter the tissue. Here, cortical tissue needs to be understood as a highly complex 3D structure, an amorphous wet block with variations of the refractive index down to the sub-micrometer scale. As a consequence, light scattering is abundant and therefore often hinders the experimenters in their goals to obtain precise spatial information. Recognizing the relevance and understanding the physics of light scattering in cortex, however, can help to exploit it for gaining specific insights. Moreover, nonlinear excitation—as used in two-photon microscopy—can trick the "scattering demon" and has therefore enabled crisp views in scattering tissue at imaging depth previously unthinkable.

highly sophisticated—has been developed, enabling functional recordings across a wide range of spatial scales. Some techniques provide superb spatial resolution and thus enable dynamic imaging of single cells and their dendritic and axonal structures (which are hardly directly accessible with electrodes). Other imaging methods provide large field-of-views (FOVs) spanning large cell populations or even several areas across the entire cortex. Although the temporal resolution of some imaging techniques still is amendable, modern camera technologies and new laser-scanning methods support high-speed data acquisition for capturing neural dynamics on sufficiently fast time scales. Thus, working with photons has complementary advantages compared to electrophysiology (and both often can be combined in beneficial ways) and optical methods offer a high degree of versatility and flexibility.

The application of optical imaging methods should be based on a solid understanding of the physics of the interactions of light with cortical tissue and an appreciation of its complexity (Box 2). These topics will thus be covered in Part I of the book in the next two chapters, focusing on measurement techniques (Chapter 2) and tissue properties (Chapter 3), respectively.

3.1 Molecular Tools: Lamps and Switches

Whereas other parts of the nervous system, like the retina, naturally interact with light, neocortical tissue is hardly light-sensitive. Nonetheless, several endogenous light-absorbing molecules can be used for read out of "intrinsic signals," enabling dynamic measurements without any extra tissue staining. Most often, however, artificially designed dye molecules are introduced for probing cortical function. These molecular tools comprise "lamps" that signal particular aspects of cell activity as well as "switches" that enable us to impose specific manipulations on the neural networks. Fortunately, the palette of available molecules has greatly expanded over the past four decades, first through the generation of suitable small organic indicator dyes that report cell activity in one or the other way, later through the development of fluorescent protein variants. The latter have the advantage that they can be specifically and stably expressed in genetically defined subsets of cells, allowing for repeated ("chronic") monitoring of specific cellular functions [23]. Chapter 4 provides an overview of the various types of fluorescent indicators that are currently in use. Calcium indicators are especially popular and examples of their application are given in several chapters in Parts II and III of this book. Functional imaging with fluorescent indicators may also be combined with other staining techniques to extract further neuroanatomical details, for example by application of anterograde or retrograde tracers [24] or by performing post hoc immunohistochemistry [25, 26].

Just as electrodes are not only used for electrical recordings but in addition for stimulation of neurons, optical tools can also be used to excite (or inhibit) neurons. Light-induced release of caged molecules (e.g., neurotransmitters) is an established method to locally perturb cortical activity [27–29]. In addition, the use of light-controllable proteins [30] and light-activated membrane channels and pumps [31–34] has greatly expanded over the last decade, giving rise to the field of "optogenetics." In particular, a plethora of light-sensitive ion channels and pumps has been designed based on naturally occurring opsins [35]. These molecular "switches" can be employed as light-controllable actuators and greatly complement the arsenal of functional indicators. Most significantly, they enable evaluation of hypotheses about the mechanisms and causal relationships underlying observed neuronal network dynamics and animal behavior. An overview of novel optogenetic tools is presented in Chapter 5 with example applications highlighted in several other chapters, mainly in Chapter 11.

3.2 High-Resolution Imaging of Cellular Dynamics

A major application of imaging techniques is the study of cellular and subcellular dynamics. Here, the complementary aspects of "optophysiology" and electrophysiology are particularly evident. Optical imaging enables recordings from subcellular structures that are inaccessible to electrodes (e.g., dendritic spines) as well as simultaneous measurements from tens to hundreds of neighboring cells, which at present is beyond what electrophysiological methods can achieve. High-resolution optical imaging with sub-micrometer resolution in the intact neocortex is, however, not trivial. The development of confocal laser-scanning microscopy more than 30 years ago enabled diffraction-limited imaging with 3D resolution. However, confocal microscopy suffers from high susceptibility to light scattering because it relies on unscattered ("ballistic") photons and scattered photons impair performance both on the fluorescence excitation and detection side. Confocal microscopy therefore is well suited for applications in thin tissue sections or cell cultures but it does not allow to "look" very deep inside tissue and rarely has been applied in vivo. The solution to the scattering problem came with the invention of two-photon laser-scanning microscopy by Denk and colleagues in 1990 [36]. Two-photon microscopy is based on localized fluorescence excitation and thus is less sensitive to scattering. As a result, it increased the attainable depth penetration several fold, now enabling high-resolution fluorescence imaging at depths of up to a millimeter and more in living, scattering tissue [37–39]. Not surprisingly, two-photon imaging application in the neocortex is a central recurring topic in many chapters of this volume.

Once the advantages of in vivo two-photon microscopy had become obvious, its application in neuroscience and other fields expanded enormously. For the study of neocortical dynamics, it opened many new avenues. First, due to its high-resolution capability, it enabled detailed investigations of synaptic structure such as dendritic spines and axonal boutons (for review, see [21]). Most recently, it became possible to study spine activation patterns on individual neurons upon sensory stimulation, which is covered in Chapter 6. Second, two-photon microscopy made it possible to comprehensively image local neuronal populations using various laser-scanning techniques (Chapter 9). Such population imaging studies are currently conducted in various regions of the neocortex (both sensory - and motor-related areas), enabling analysis of network dynamics under anesthesia or in awake, behaving animals and thus providing first glimpses of neuronal networks "at work." With stably expressed genetically encoded calcium indicators, the activity of the exact same neurons can be followed repeatedly over many weeks and months, promoting studies of cortical stability and plasticity (Chapter 10). Third, many other dynamic aspects of cortical function became amenable for high-resolution studies using in vivo two-photon microscopy, including activity of non-neuronal cells such as astrocytes (Chapter 12) and microglia (Chapter 13), hemodynamics (Chapters 14 and 16), and partial pressure of oxygen (Chapter 14).

Today, many of the new experimental possibilities offered by in vivo two-photon microscopy are utilized to investigate specific aspects of cortical function. At the same time, technologies continue to advance at a rapid pace, both in terms of molecular tools and imaging methods. Presumably, the possibilities to probe dynamic features of cortical micro- and macrocircuits will expand even further in upcoming years, on the one hand revealing finer and finer details (e.g., using super-resolution nanoscopy [40]) and on the other hand expanding the observation and analysis of neuronal population activity with cellular resolution to larger and larger FOVs.

3.3 Large-Scale Optical Imaging

Two-photon microscopy has complemented methods for large-scale optical imaging that lack single-cell resolution but have been already in use for a long time. The seminal work of Britten Chance initiated the advent of optical tools in biomedical research in the 1960s [41]. Later, the group of Amiram Grinvald made use of the properties of intrinsic chromophores and their relationship to dynamic changes of cerebral blood flow [42]. The rapid advance and availability of highly sensitive and fast electronic cameras, the so-called charge-coupled devices (CCDs), was key for the breakthrough of optical imaging for neuroscience applications. Optical imaging based on intrinsic signals is still widely used to obtain functional maps of cortical activity and to study cortical processing. Because it relies on the dynamic balance between oxygenated and deoxygenated hemoglobin contained in red blood cells, this method is based on the hemodynamic response to neural activation. Similarly, laser speckle imaging (LSI), another camera-based imaging technique, measures relative changes of cerebral blood flow [43, 44]. A moving object inside the tissue, such as a red blood cell, leads to a Doppler shift of the frequency of scattered light, which is directly proportional to the blood cell velocity. Rather than analyzing frequency shifts LSI looks at temporal fluctuations of the scattered light in order to map blood flow (discussed in Chapter 15).

Moreover, mitochondrial redox states and with it oxidative metabolism can be monitored in vivo using autofluorescence imaging [45, 46]. Both NADH and flavoproteins change their fluorescence properties from the oxidized to the deoxidized form so that changes in autofluorescence can report cortical dynamics in vivo [47]. The camera-based methods described so far all rely on a surrogate marker for electrical activity of neurons. In contrast, voltage-sensitive dye (VSD) imaging makes use of voltage indicators that change their fluorescence or absorption in direct dependence on the cells' membrane potential [48, 49]. Whereas the hemodynamic-based imaging tools can capture dynamics on the temporal scale of seconds, VSD imaging is able to map millisecond changes (see Chapter 8). The monitoring of these rapid changes relies on ultrafast cameras, such as CMOS cameras.

Calcium indicator loading can also be applied for reading out bulk population activity, either using wide-field camera imaging similar to VSD imaging [50, 51] or via optical fibers placed on areas of interest [3, 52–54]. Fiber-optic fluorescence recordings have in particular allowed the study of bulk dendritic activity in vivo (see Chapter 7). Finally, a number of other large-scale optical imaging tools exist, including for example opto-acoustic imaging, optical coherence tomography, or diffuse coherence tomography, which can also be used to probe certain aspects of cortical dynamics [55] but are beyond the scope of this book.

4 Bringing All Together

Understanding cortical function will require bringing all these dynamic aspects together in a coherent view of distributed signal flow and information processing. To tackle the complexity of neocortex, we need to extract principles of functional interactions among molecules and cells on the lower levels (e.g., within microcircuits) and comprehend how these determine emergent dynamic properties on next-higher levels. Eventually, this may lead to a dynamic theory of brain function, in which the computational principles across the hierarchical levels are worked out. We argue here that light-based methods, due to their special features summarized above, should be particularly helpful in this endeavor in the coming years.

First, better than ever before, we can identify and characterize particular neuronal cell types. Thus, we can also target them specifically with optical tools. Imaging methods can then be used to dissect the computational roles of these neuronal subsets within micro- and macrocircuits and their relevance for animal behavior. For example, with regard to the diverse types of inhibitory interneurons such a characterization is already on the way [56]. Likewise, subsets of neurons with specific long-range projection patterns can now be functionally investigated in a targeted fashion [24]. Optophysiology (alone or in combination with electrophysiology) will be the perfect counterpart to neuroanatomical projects that aim to resolve a complete wiring diagram ("connectome") for brain regions and ultimately entire brains. Whereas optical microscopy may contribute to the acquisition of "connectomics" data to a certain extent [57], highest-resolution connectivity maps will rely on electron-microscopic reconstruction with unambiguous resolution of individual synaptic contacts [58]. The real strength of optical methods will be, however, to fill the "snapshots" of full anatomical neural circuit reconstruction again "with life," by providing the functional data that uncover signaling dynamics implemented in these circuits. Along this line, several seminal studies recently obtained connectivity information from networks that previously had been investigated using in vivo functional imaging [59–61].

Hence, as always in neuroscience, going back and forth between structure and function should be the key approach to a better understanding of cortical circuitry [62]. Of course, further technological developments, such as those targeted by the new BRAIN initiative in the USA [55, 63, 64], will help meet these scientific challenges.

Second, as outlined above, light-based methods will be especially advantageous for revealing dynamic processes beyond pure neuronal activity, such as glial signaling, metabolic processes, and vascular dynamics. Recently, the potentially underestimated role of glia in information processing, and especially in cortical plasticity, has been increasingly appreciated [65, 66]. We expect that these non-neuronal aspects will shift even further into focus in coming years. A coherent view of cortical functioning definitely will have to include neuro-glial interactions as well as an improved understanding of neuro- and glio-vascular coupling, which will also help in interpreting fMRI data. With their ability to bridge microscopic and macroscopic views and their flexibility to reveal all the diverse dynamics, optical techniques will be perfectly suited to address these questions. Similar to the increasing use of electrical methods for measurements and interventions in human brains, one may also expect the expansion of applications of optical methods in humans (in addition to fMRI, etc.), especially for potential treatments of neurological or psychiatric disorders.

Finally, the power of optical methods also has huge potential for the study of pathological conditions. Detailed functional characterization of cortical dynamics not only will advance our understanding of normal cortex function but at the same time will be immediately useful to study dysfunctional signaling in animal models of brain trauma, neurodegeneration, neurological disorders, and psychiatric disorders [67, 68]. Whereas longitudinal functional studies have been limited when using electrophysiology, repeated two-photon calcium imaging of the exact same cells is now possible over months with GECIs [23, 69]. Hence, the ability of optics to bridge small and large scales (in time and in space) will be highly beneficial for pinpointing major defects in neural circuitry or in neuro-glia-vascular communication that underlie particular diseases. Direct observations with cellular resolution of the immediate tissue reactions, involving for example microglia [67, 70], or the slow progression of circuit dysfunction in parallel to cognitive decline, could prompt novel ideas on potential therapeutic approaches, which then also can be tested in the respective animal models. We should therefore take full advantage of the great potential of optical imaging to study neocortical dynamics, foster further technical developments (probes and imaging techniques) and press on with the applications in research and for diagnosis and therapy.

References

1. Oberlaender M, de Kock CP, Bruno RM, Ramirez A, Meyer HS, Dercksen VJ, Helmstaedter M, Sakmann B (2012) Cell type-specific three-dimensional structure of thalamocortical circuits in a column of rat vibrissal cortex. Cereb Cortex 22:2375–2391
2. Petersen CC (2009) Barrel cortex circuits. In: Larry RS (ed) Encyclopedia of neuroscience. Academic, Oxford, pp 41–45
3. Murayama M, Perez-Garci E, Nevian T, Bock T, Senn W, Larkum ME (2009) Dendritic encoding of sensory stimuli controlled by deep cortical interneurons. Nature 457:1137–1141
4. Douglas RJ, Martin KA (2004) Neuronal circuits of the neocortex. Annu Rev Neurosci 27:419–451
5. Mountcastle VB (1978) An organizing principle for cerebral function: the unit model and the distributed system. In: Edelman GM, Mountcastle VB (eds) The mindful brain. MIT, Cambridge
6. Bargmann CI, Marder E (2013) From the connectome to brain function. Nat Methods 10:483–490
7. Helmstaedter M (2013) Cellular-resolution connectomics: challenges of dense neural circuit reconstruction. Nat Methods 10:501–507
8. Shepherd GM, Grillner S (eds) (2010) Handbook of brain mircocircuits, 1st edn. Oxford University Press, New York
9. Spruston N (2008) Pyramidal neurons: dendritic structure and synaptic integration. Nat Rev Neurosci 9:206–221
10. Ascoli GA, Alonso-Nanclares L, Anderson SA, Barrionuevo G, Benavides-Piccione R, Burkhalter A, Buzsaki G, Cauli B, Defelipe J, Fairen A, Feldmeyer D, Fishell G, Fregnac Y, Freund TF, Gardner D, Gardner EP, Goldberg JH, Helmstaedter M, Hestrin S, Karube F, Kisvarday ZF, Lambolez B, Lewis DA, Marin O, Markram H, Munoz A, Packer A, Petersen CC, Rockland KS, Rossier J, Rudy B, Somogyi P, Staiger JF, Tamas G, Thomson AM, Toledo-Rodriguez M, Wang Y, West DC, Yuste R (2008) Petilla terminology: nomenclature of features of GABAergic interneurons of the cerebral cortex. Nat Rev Neurosci 9:557–568
11. Rudy B, Fishell G, Lee S, Hjerling-Leffler J (2011) Three groups of interneurons account for nearly 100% of neocortical GABAergic neurons. Dev Neurobiol 71:45–61
12. Silberberg G (2008) Polysynaptic subcircuits in the neocortex: spatial and temporal diversity. Curr Opin Neurobiol 18:332–337
13. Connors BW, Long MA (2004) Electrical synapses in the mammalian brain. Annu Rev Neurosci 27:393–418
14. Kettenmann H, Ransom BR (eds) (2012) Neuroglia, 3rd edn. Oxford University Press, New York
15. Parpura V, Heneka MT, Montana V, Oliet SH, Schousboe A, Haydon PG, Stout RF Jr, Spray DC, Reichenbach A, Pannicke T, Pekny M, Pekna M, Zorec R, Verkhratsky A (2012) Glial cells in (patho)physiology. J Neurochem 121:4–27
16. Parpura V, Zorec R (2010) Gliotransmission: Exocytotic release from astrocytes. Brain Res Rev 63:83–92
17. Tsai PS, Kaufhold JP, Blinder P, Friedman B, Drew PJ, Karten HJ, Lyden PD, Kleinfeld D (2009) Correlations of neuronal and microvascular densities in murine cortex revealed by direct counting and colocalization of nuclei and vessels. J Neurosci 29:14553–14570
18. Hirsch S, Reichold J, Schneider M, Szekely G, Weber B (2012) Topology and hemodynamics of the cortical cerebrovascular system. J Cereb Blood Flow Metab 32:952–967
19. Luo L, Callaway EM, Svoboda K (2008) Genetic dissection of neural circuits. Neuron 57:634–660
20. O'Connor DH, Huber D, Svoboda K (2009) Reverse engineering the mouse brain. Nature 461:923–929
21. Holtmaat A, Svoboda K (2009) Experience-dependent structural synaptic plasticity in the mammalian brain. Nat Rev Neurosci 10: 647–658
22. Scanziani M, Häusser M (2009) Electrophysiology in the age of light. Nature 461:930–939
23. Lütcke H, Margolis DJ, Helmchen F (2013) Steady or changing? Long-term monitoring of neuronal population activity. Trends Neurosci 36:375–384
24. Chen JL, Carta S, Soldado-Magraner J, Schneider BL, Helmchen F (2013) Behaviour-dependent recruitment of long-range projection neurons in somatosensory cortex. Nature 499:336–340
25. Kerlin AM, Andermann ML, Berezovskii VK, Reid RC (2010) Broadly tuned response properties of diverse inhibitory neuron subtypes in mouse visual cortex. Neuron 67:858–871
26. Langer D, Helmchen F (2012) Post hoc immunostaining of GABAergic neuronal subtypes following in vivo two-photon calcium imaging in mouse neocortex. Pflügers Arch 463:339–354

27. Adams SR, Tsien RY (1993) Controlling cell chemistry with caged compounds. Annu Rev Physiol 55:755–784
28. Callaway EM, Yuste R (2002) Stimulating neurons with light. Curr Opin Neurobiol 12:587–592
29. Sarkisov DV, Wang SS (2006) Alignment and calibration of a focal neurotransmitter uncaging system. Nat Protoc 1:828–832
30. Haupts U, Tittor J, Bamberg E, Oesterhelt D (1997) General concept for ion translocation by halobacterial retinal proteins: the isomerization/switch/transfer (IST) model. Biochemistry 36:2–7
31. Ernst OP, Sanchez Murcia PA, Daldrop P, Tsunoda SP, Kateriya S, Hegemann P (2008) Photoactivation of channelrhodopsin. J Biol Chem 283:1637–1643
32. Essen LO (2002) Halorhodopsin: light-driven ion pumping made simple? Curr Opin Struct Biol 12:516–522
33. Hegemann P, Ehlenbeck S, Gradmann D (2005) Multiple photocycles of channelrhodopsin. Biophys J 89:3911–3918
34. Schobert B, Lanyi JK (1982) Halorhodopsin is a light-driven chloride pump. J Biol Chem 257:10306–10313
35. Fenno L, Yizhar O, Deisseroth K (2011) The development and application of optogenetics. Annu Rev Neurosci 34:389–412
36. Denk W, Strickler JH, Webb WW (1990) Two-photon laser scanning fluorescence microscopy. Science 248:73–76
37. Helmchen F, Denk W (2005) Deep tissue two-photon microscopy. Nat Methods 2:932–940
38. Svoboda K, Yasuda R (2006) Principles of two-photon excitation microscopy and its applications to neuroscience. Neuron 50:823–839
39. Kerr JN, Denk W (2008) Imaging in vivo: watching the brain in action. Nat Rev Neurosci 9:195–205
40. Testa I, Urban NT, Jakobs S, Eggeling C, Willig KI, Hell SW (2012) Nanoscopy of living brain slices with low light levels. Neuron 75:992–1000
41. Chance B, Cohen P, Jobsis F, Schoener B (1962) Intracellular oxidation-reduction states in vivo. Science 137:499–508
42. Malonek D, Grinvald A (1996) Interactions between electrical activity and cortical microcirculation revealed by imaging spectroscopy: implications for functional brain mapping. Science 272:551–554
43. Dunn AK, Bolay H, Moskowitz MA, Boas DA (2001) Dynamic imaging of cerebral blood flow using laser speckle. J Cereb Blood Flow Metab 21:195–201
44. Zakharov P, Völker A, Buck A, Weber B, Scheffold F (2006) Quantitative modeling of laser speckle imaging. Opt Lett 31:3465–3467
45. Lothman E, Lamanna J, Cordingley G, Rosenthal M, Somjen G (1975) Responses of electrical potential, potassium levels, and oxidative metabolic activity of the cerebral neocortex of cats. Brain Res 88:15–36
46. Mayevsky A, Chance B (1975) Metabolic responses of the awake cerebral cortex to anoxia hypoxia spreading depression and epileptiform activity. Brain Res 98:149–165
47. Shibuki K, Hishida R, Murakami H, Kudoh M, Kawaguchi T, Watanabe M, Watanabe S, Kouuchi T, Tanaka R (2003) Dynamic imaging of somatosensory cortical activity in the rat visualized by flavoprotein autofluorescence. J Physiol 549:919–927
48. Ferezou I, Haiss F, Gentet LJ, Aronoff R, Weber B, Petersen CC (2007) Spatiotemporal dynamics of cortical sensorimotor integration in behaving mice. Neuron 56:907–923
49. Grinvald A, Hildesheim R (2004) VSDI: a new era in functional imaging of cortical dynamics. Nat Rev Neurosci 5:874–885
50. Berger T, Borgdorff A, Crochet S, Neubauer FB, Lefort S, Fauvet B, Ferezou I, Carleton A, Luscher HR, Petersen CC (2007) Combined voltage and calcium epifluorescence imaging in vitro and in vivo reveals subthreshold and suprathreshold dynamics of mouse barrel cortex. J Neurophysiol 97:3751–3762
51. Minderer M, Liu W, Sumanovski LT, Kügler S, Helmchen F, Margolis DJ (2012) Chronic imaging of cortical sensory map dynamics using a genetically encoded calcium indicator. J Physiol 590:99–107
52. Adelsberger H, Garaschuk O, Konnerth A (2005) Cortical calcium waves in resting newborn mice. Nat Neurosci 8:988–990
53. Schulz K, Sydekum E, Krueppel R, Engelbrecht CJ, Schlegel F, Schroter A, Rudin M, Helmchen F (2012) Simultaneous BOLD fMRI and fiber-optic calcium recording in rat neocortex. Nat Methods 9:597–602
54. Stroh A, Adelsberger H, Groh A, Ruhlmann C, Fischer S, Schierloh A, Deisseroth K, Konnerth A (2013) Making waves: initiation and propagation of corticothalamic Ca^{2+} waves in vivo. Neuron 77:1136–1150
55. Devor A, Boas D (2012) Neurovascular imaging. Front Neuroenergetics 4:1
56. Petersen CC, Crochet S (2013) Synaptic computation and sensory processing in neocortical layer 2/3. Neuron 78:28–48
57. Osten P, Margrie TW (2013) Mapping brain circuitry with a light microscope. Nat Methods 10:515–523

58. Denk W, Briggman KL, Helmstaedter M (2012) Structural neurobiology: missing link to a mechanistic understanding of neural computation. Nat Rev Neurosci 13:351–358
59. Briggman KL, Helmstaedter M, Denk W (2011) Wiring specificity in the direction-selectivity circuit of the retina. Nature 471:183–188
60. Bock DD, Lee WC, Kerlin AM, Andermann ML, Hood G, Wetzel AW, Yurgenson S, Soucy ER, Kim HS, Reid RC (2011) Network anatomy and in vivo physiology of visual cortical neurons. Nature 471:177–182
61. Ko H, Hofer SB, Pichler B, Buchanan KA, Sjostrom PJ, Mrsic-Flogel TD (2011) Functional specificity of local synaptic connections in neocortical networks. Nature 473:87–91
62. Lichtman JW, Denk W (2011) The big and the small: challenges of imaging the brain's circuits. Science 334:618–623
63. Alivisatos AP, Chun M, Church GM, Greenspan RJ, Roukes ML, Yuste R (2012) The brain activity map project and the challenge of functional connectomics. Neuron 74:970–974
64. Devor A, Bandettini PA, Boas DA, Bower JM, Buxton RB, Cohen LB, Dale AM, Einevoll GT, Fox PT, Franceschini MA, Friston KJ, Fujimoto JG, Geyer MA, Greenberg JH, Halgren E, Hämäläinen MS, Helmchen F, Hyman BT, Jasanoff A, Jernigan TL, Judd LL, Kim SG, Kleinfeld D, Kopell NJ, Kutas M, Kwong KK, Larkum ME, Lo EH, Magistretti PJ, Mandeville JB, Masliah E, Mitra PP, Mobley WC, Moskowitz MA, Nimmerjahn A, Reynolds JH, Rosen BR, Salzberg BM, Schaffer CB, Silva GA, So PTC, Spitzer NC, Tootell RB, Van Essen DC, Vanduffel W, Vinogradov SA, Wald LL, Wang LV, Weber B, Yodh AG (2013). The Challenge of Connecting the Dots in the B.R.A.I.N. Neuron 80:270–274
65. Volterra A, Meldolesi J (2005) Astrocytes, from brain glue to communication elements: the revolution continues. Nat Rev Neurosci 6:626–640
66. Clarke LE, Barres BA (2013) Emerging roles of astrocytes in neural circuit development. Nat Rev Neurosci 14:311–321
67. Garaschuk O (2013) Imaging microcircuit function in healthy and diseased brain. Exp Neurol 242:41–49
68. Tye KM, Deisseroth K (2012) Optogenetic investigation of neural circuits underlying brain disease in animal models. Nat Rev Neurosci 13:251–266
69. Aramuni G, Griesbeck O (2013) Chronic calcium imaging in neuronal development and disease. Exp Neurol 242:50–56
70. Aguzzi A, Barres BA, Bennett ML (2013) Microglia: scapegoat, saboteur, or something else? Science 339:156–161

Chapter 2

Principles and Fundamentals of Optical Imaging

Frank Scheffold

Abstract

In this chapter I will give a brief general introduction to optical imaging and then discuss in more detail some of the methods specifically used for imaging cortical dynamics today. Absorption and fluorescence microscopy can be used to form direct, diffraction-limited images but standard methods are often only applicable to superficial layers of cortical tissue. Two-photon microscopy takes an intermediate role since the illumination pathway is diffraction-limited but the detection pathway is not. Losses in the illumination path can be compensated using higher laser power. Since the detection pathway does not require image formation, the method can substantially increase the imaging depth. Understanding the role of scattering is important in this case since non-descanned detection can substantially enhance the imaging performance. Finally, I will discuss some of the most widely used imaging methods that all rely on diffuse scattering such as diffuse optical tomography, laser speckle imaging, and intrinsic optical imaging. These purely scattering-based methods offer a much higher imaging depth, although at a substantially reduced spatial resolution.

Key words Optical imaging, Fluorescence microscopy, Scattering, Diffuse optical tomography, Laser Speckle Imaging

1 Introduction

Modern optical imaging relies on a number of optical phenomena that provide access to spatial and temporal information about the object under study. Sample-specific information can be extracted not only from the intensity or phase but also from the analysis of photon statistics, fluorescence lifetime, polarization, and so on [1]. Here I will give a brief general introduction to optical imaging and then discuss in more detail some of the methods specifically used for imaging dynamics of cerebral cortex today. Imaging cortical dynamics often mandates the use of specific indicators, requires deep tissue penetration or high temporal resolution. Methods prominently discussed in this volume are two-photon microscopy, absorption spectroscopy, and laser speckle imaging. Although these methods are based on different principles, they all are intimately related to (multiple) light scattering and absorption, topics

Bruno Weber and Fritjof Helmchen (eds.), *Optical Imaging of Neocortical Dynamics*, Neuromethods, vol. 85,
DOI 10.1007/978-1-62703-785-3_2,

that often are not treated in classical optical imaging books and reviews and therefore are highlighted here.

I will first discuss the main mechanism of light–matter interactions in biological tissue, namely scattering, absorption, and fluorescent light emission. In the next step, I address the fundamentals of image formation for some selected methods and discuss their advantages and limitations.

2 Imaging and Microscopy Principles

An image in traditional light microscopy is formed by absorption or scattering of light waves. In essence the microscope records the transmission shadow of a section of the sample that is in focus. The depth of field Z denotes the thickness of a z-section around the focal plane for which object points are imaged with acceptable sharpness. Light beams scattered or absorbed at a given lateral (x–y) position, within this section Z, are removed and therefore this x–y point appears dark in the image plane. Light beams unaffected by the sample lead to a bright background. The difference between dark and bright areas, or low and high light amplitudes, is called "contrast."

Absorption is a molecular process where the photon energy is converted via radiative and non-radiative pathways in the molecule (see below). Refractive index variations $n(x,y,z)$ in the sample can lead to scattering or phase changes or both. Phase contrast originates from differences in the optical light path, defined as $s = xc$. Here x is the geometrical path, $c = c_0/n$ the speed of light and c_0 denotes the vacuum speed of light. Roughly speaking one can say that refractive index variations on length scales comparable to the wavelength λ, here approximately 500 nm, lead to scattering while variations on much larger length scales mainly lead to phase shifts. Phase shifts can be converted to amplitude changes using the so-called phase plates in phase contrast microscopy. Applications are mainly found for unstained biological sections that are weakly scattering objects. For details, please refer to the standard textbooks.

For all these cases, image formation can be treated with Abbe's classical coherence theory for image formation: The lateral resolution is set by the effective numerical aperture (NA) of the objective and the condenser to $\rho_{min} = 0.61\ \lambda/\text{NA}$ and the depth of field is $Z = n\ \lambda/\text{NA}^2$ [2]. In addition to the lateral resolution limit set by the diffraction of light the utility of classical microscopy for biomedical imaging is severely limited by other factors. In particular the lack of specificity, the limited axial resolution and the poor penetration depth in optically dense media have led to the development of a large number of new methods based on other principles of contrast formation [1].

Fluorescence microscopy is nowadays the method of choice in many imaging applications. The interaction of light with fluorescent molecules offers several advantages. Firstly the molecules can be attached to specific sites and act as specific indicators for biological function. Since each molecule emits light at wavelengths within a certain spectral window, it is possible to separate several different fluorescent markers in a single experiment. As a matter of fact, biological specimen often display limited intrinsic amplitude or phase contrast, thus fluorescence microscopy often provide images of unrivalled quality.

Although a source of contrast for thin samples, absorption and scattering usually impose severe limitations on the penetration depth when imaging thick specimen. If the absorbed light is not reemitted (e.g., via fluorescence), the energy relaxes via vibrational pathways and the photons are lost. In the case of scattering however, the light is merely redirected. Although this precludes classic image formation in a microscope, it still allows to reconstruct lower resolution images based on diffuse propagation of scattered photons [3]. Collection of scattered photons is also an important issue in the detection pathway of two-photon microscopy [4]. In cases where the scattering has an inelastic or Doppler-contribution due to microscopic internal motion, additional information is encoded in the propagating light fields that carry information, for example, about blood flow in tissue [5]. In the following I describe in more detail the main principles of image formation for a number of cases that are of importance for cortical imaging.

2.1 Absorption and Fluorescence

Light energy can be "absorbed" by matter and subsequently released via radiative and non-radiative pathways. The latter process transforms the entire energy of an incident photon to thermal energy. Absorption without emission can provide amplitude contrast in image formation but it can also severely limit the penetration depth of optical imaging methods. Radiative decays result in the emission of photons via fluorescence or phosphorescence. In complex molecules both radiative and non-radiative processes occur sequentially or concurrently. As a consequence, for single photon absorption, the emitted photon possesses a lower energy (or has a red-shifted wavelength) compared to the absorbed photon. The situation can be described using the so-called Jablonski diagrams depicting the energy states of a molecule and its relaxation pathways (Fig. 1) [6].

In contrast to a single atom, the electronic states of a complex molecule have many associated vibrational energy levels. In practice a molecule is excited from one of these thermally excited vibrational levels of a singlet ground state S_0 to an exited single state S_1 or S_2 by absorbing a photon. The molecular absorption spectrum is rather broad, typically tens of nanometers. The excited molecule can relax to the ground state via different pathways. Non-radiative transitions are described via dashed lines whereas solid line arrows describe radiative pathways (Fig. 1). "Fluorescence" is the process

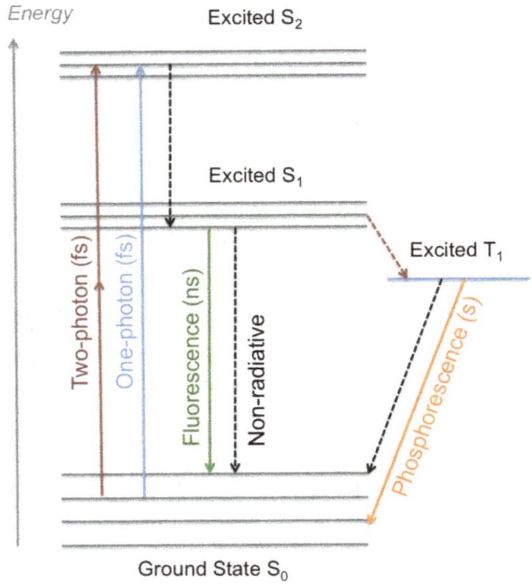

Fig. 1 Jablonski diagram. Schematic representation of the energy levels of an electron in a fluorescent molecule (typical relaxation time scales in *brackets*). The molecule can be excited by 1p or 2p absorption to a higher electronic state S_2 or S_1. Several radiative and non-radiative pathways eventually lead to a collapse to the ground state S_0. The term "fluorescence emission" denotes the fast radiative decay from the lowest vibrational level of S_1 to an excited ground state S_0 level (*green arrow*).

of fast emission of a photon via a direct relaxation from a low energy vibrational state of S_1 to S_0, which typically takes place on the nanosecond scale. The exact lifetime of the excited state of a fluorescent molecule depends on its immediate environment and thus carries relevant morphological and often also functional information. An image of these properties is constructed by point-scanning in the frame the fluorescence lifetime imaging microscopy (FLIM) approach [7]. If the molecule first undergoes a spin conversion, it can end up in a "forbidden" triplet state T_1. A subsequent radiative decay is slow and the photon energy is emitted over time scales of a millisecond or more, a process known as "phosphorescence." For imaging purposes, mostly fluorescence is employed whereas the other processes merely lead to a reduced quantum efficiency and increased background. In some cases, however, these other processes can carry specific information. Fluorescence can be suppressed by resonant energy transfer (FRET), provided the molecules are sufficiently close to each other [8]. The efficiency of this process scales with the sixth power of the distance and the signal (or the absence of the signal) thus carries information about the location of the molecules at the nanoscale. Equally phosphorescence can be quenched, for example, by the presence of oxygen [6, 9, 10].

Standard fluorescence microscopy is based on the absorption of one high energy photon, some non-radiative losses and the subsequent emission of a lower energy photon: for example, blue excitation light is turned into green emission light (Fig. 1). In contrast, in two-photon microscopy a pair of photons is absorbed simultaneously and the fluorescent molecule emits a single higher energy photon: for example, near infrared (NIR) light is turned into green light (Fig. 1). In the latter case the fluorophores normally do not absorb single low energy NIR-photons and thus two-photon excited fluorescence is the only allowed process.

2.2 Scattering and Multiple Scattering

The scattering of light refers to the elastic or quasi-elastic interaction of light with matter where the exchange of energy is minimal. In the latter case, for all cases of practical interest, the frequency shift $\Delta\nu$ due to scattering from moving objects is very small (much smaller than the line-width of a single frequency laser of typically 5–10 MHz), which means that the scattering process itself does not alter the properties of the scattered light. Scattering occurs due to variations in the refractive index of the medium on length scales of the wavelength of light. The incident wave induces oscillating dipoles in the medium that can reemit part of the light energy in some direction. Both scattering and absorption perturb the light propagation such that for a given path of light z the stream of photons is attenuated according to Lambert–Beers law $I(z) = I_0 \exp(-A) = I_0 \exp(-\mu z)$. The scattering and absorption strengths are characterized by the parameters μ_s and μ_a, respectively (see Box 1).

Box 1 Scattering Efficiency

The efficiency of scattering is directly related to the polarizability α of the scattering object [11]. In the weak scattering limit $[4\pi R\,(n/n_0-1) \ll \lambda]$, α is proportional to the refractive index contrast (n/n_0-1). Here n denotes the refractive index of the object and n_0 of the background medium and λ is the wavelength of light. For small objects of size R, one finds $\alpha \propto R^3$ $(R<\lambda/2\pi)$ and for larger particles $\alpha \propto R^2$ $(R>\lambda/2\pi)$. The total scattering cross section is proportional to α^2. Both the influence of absorption and scattering are usually measured in terms of the parameters μ_a and μ_s with the latter being directly proportional to α^2 and the number density of scatterers. The parameter $\mu_s = 1/l_s$ is equal to the inverse of the mean free path l_s defined as the distance a wave can travel on average before being scattered. In other words the unscattered or "ballistic" light intensity decreases exponentially $I = I_0 \exp[-z/l_s]$ as a function of the penetration depth z (Lambert–Beer law). Objects of size comparable or larger than the wavelength of light, such as cells and other tissue structures, scatter preferentially in forward direction. A measure of this effect is the scattering anisotropy parameter defined as the average of the cosine of the scattering angle $g = \langle\cos\Theta\rangle$. For objects much smaller than the wavelength $g \simeq 0$ whereas for biological tissue and blood, it is typically of the order $g = 0.9$. Diffuse propagation of light is characterized by the parameter $\mu_t = 1/l_t$ that takes account of the scattering anisotropy via a renormalized path length $l_t = l_s\, 1/(1-g)$.

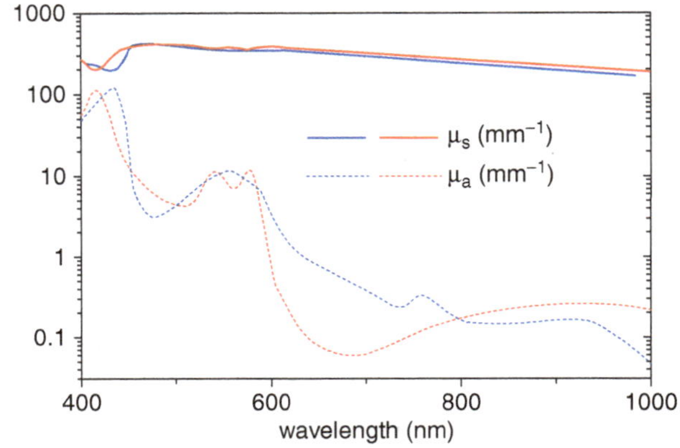

Fig. 2 Optical properties of blood. Scattering coefficient μ_s and absorption coefficient μ_a of oxygenated (*red*) and deoxygenated (*blue*) whole blood, both adapted from Faber et al. [13].

Scattering in biological tissue is either due to refractive index variations in the tissue or due to scattering from blood cells. Scattering from blood can play an important role, for example, for in vivo two-photon imaging, given the high vascularization of the brain cortex [12]. Figure 2 shows the scattering properties of whole blood for light in the visible (VIS) and NIR region. For blue and green light, absorption and scattering are of similar strength while in the red and NIR region absorption decreases to a level where it becomes negligible for most applications. Scattering, however, remains strong in this wavelength range and limits the penetration depth of light in many optical imaging methods. As can be seen from Fig. 2 the scattering coefficient μ_s for whole blood never drops below 200 mm^{-1} and thus mean free path l_s never exceeds a value of 5 μm. Therefore whole blood is almost impenetrable for microscopy methods. The blood content in tissue varies substantially but is typically of the order of 2–5 %. At the same time refractive index variations in (cortical) tissue also provide scattering contrast. As a consequence, regular tissue, such as brain grey matter, for example, has a scattering mean free path of about 100–200 μm in vivo and negligible absorption in the NIR [4, 14] (see also Chapter 3 in this volume).

3 Single-Photon Microscopy

There are two distinct ways of obtaining an image from a fluorescently labeled specimen upon single-photon excitation: (a) wide field detection and (b) laser scanning combined with confocal detection. Wide-field fluorescence microscopy essentially works the same way as standard microscopy. The only difference is the wavelength-selective illumination and detection pathways. The main advantages are reduced background, increased contrast, and the possibility to specifically label (multiple) sites of interest in the specimen. The sectioning capabilities are, however, severely compromised by the contribution of out-of-focus fluorescence excitation if the entire sample is illuminated. The latter also leads to blurring of the image and increased background. Numerical reallocation of out-of-focus light by recording a series of focal planes is known as "deconvolution." It allows 3D reconstruction and can improve the focal plane image sharpness [1]. Selective Plane Illumination Microscopy (SPIM) is an interesting new wide-field imaging approach with improved sectioning capabilities. The method is based on creating a sheet of light, which coincides with the focal plane of the objective. With this approach, fluorescence is excited only on the plane of measurement, and by moving the object through this excitation plane it is possible to recover 3D information with microscopic resolution [15, 16].

The most popular approach to achieve 3D high-resolution microscopy is confocal laser scanning microscopy. In CLSM out-of-focus light is reduced by physical mechanism: First, the sample is illuminated with a focused laser beam that is scanned across the specimen. Second, out-of-focus emitted light is rejected by placing a small aperture (the "pinhole") in the image plane of the focal spot. The light that passes the pinhole is then recorded using an electronic detector. A 2D or 3D image is obtained by scanning the sample or the beam [2]. Confocal microscopy allows 3D microscopy at the Abbe resolution limit for multicolor-labeled samples and is thus widely used in biomedical imaging. The penetration depth in tissue is however limited in scattering samples since a sharp image of the focal plane must be formed on the detection aperture (Fig. 3).

4 Two-Photon Microscopy

Two-photon laser scanning microscopy also uses a raster-scanning approach and an objective lens to focus the incident light [4, 12, 17, 18]. Firstly it is necessary to provide a high enough light intensity to achieve a decent transition probability for two-photon absorption. In practice this can only be achieved using pulsed lasers with pulse durations in the sub-picosecond range and an average power ranging from tens of mW to several hundreds of mW for in vivo applications. Provided two-photon excitation can be achieved, the method offers some particular advantages. The two-photon absorption probability scales with the light-intensity square and thus the region where absorption takes place is strongly confined to the focal spot of the focused laser beam. Both axially and laterally the probability decays sharply. All fluorescent light emitted can

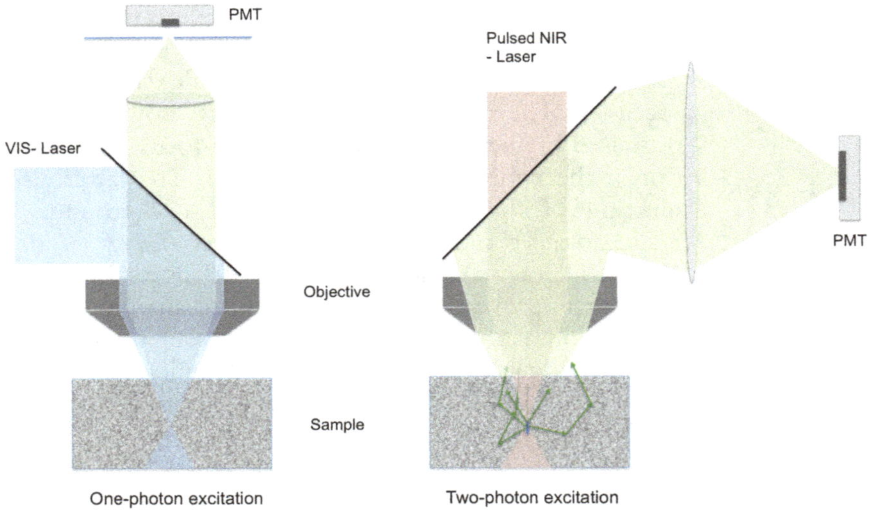

Fig. 3 Laser scanning optical microscopy based on one-photon (*left*) and two-photon (*right*) excitation. In both cases scattering reduces the incident power reaching the focal spot. While in one-photon microscopy this leads to background fluorescence, in two-photon microscopy scattered incident light is invisible. In one-photon microscopy (CLSM) illumination and detection pathways are equivalent. Scattered light cannot be imaged sharply on the confocal aperture and is thus lost when forming an image. In two-photon microscopy all emitted photons can be used to construct the image.

therefore be attributed to absorption at the position of the illumination focus. Rejection of out-of-focus light is not required and even scattered light can be collected when constructing an image (Fig. 3). Overall the setup is largely simplified since for the detection pathway a maximum amount of the emitted light is collected. This approach is often called non-descanned detection since, in contrast to confocal microscopy, the reverse pathway does not require the formation of an image of the focal spot. Another important advantage, in particular for biomedical imaging, is the fact that the incident light can be from the NIR region of the spectrum while the emitted light is typically in the VIS range. Both NIR illumination and non-descanned detection strongly increase imaging depth of two-photon microscopy compared to confocal microscopy [4]. Figure 3 summarizes the detection pathways of single- and two-photon excited fluorescence microscopy.

5 Diffuse Light Imaging

Imaging using microscopy requires a well-defined optical path from illumination to detection. If scattering or absorption is strong, this imposes severe limitations. For in vivo microscopy on small animals or humans, access is commonly restricted to the reflection measurements, limiting the optical penetration depth to 0.5–1 mm even for best case of two-photon microscopy.

It is however still possible to obtain low-resolution images by monitoring scattered light only. The propagation of light in the multiple scattering regime can be described as a random walk of photons with a step length of $l_t = l_s\ 1/(1-g)$. For sufficiently large number of steps, light propagation is diffusive with a diffusion coefficient $D = cl_t/3$ [19]. Diffusely reflected light penetrates the sample up to a depth of a few l_t and when leaving the sample spreads out over an area of roughly $10(l_t)^2$ [20, 21]. In turn, light detected at a given point (x,y) on the surface might originate from any position within the same surface area around this point. The length l_t therefore sets a lower bound for the spatial resolution of diffuse reflectance imaging techniques such as optical intrinsic or laser speckle imaging. Imaging contrast can be generated by absorption, fluorescence, and local motion in the sample (Box 2). In the presence of heterogeneities on length scales comparable or larger than l_t, knowledge about the effective path length is needed as discussed in more detail below. Such knowledge is however often difficult to obtain experimentally. For small animals, this limitation can be overcome using tomography techniques (both for absorption [22] and fluorescence [23]). The situation is much more involved for larger animals and humans when only reflection measurements can be performed. The mathematical description of diffuse light propagation is fairly complicated. A brief primer is given in Box 2.

Box 2 Diffuse Light Propagation: Scattering, Absorption of Light, and Correlation

Contrast in diffuse light imaging can be generated by spatial heterogeneities with respect to scattering (μ_t), absorption ($\mu_a > 0$), and internal motion $f(\tau)$. All four phenomena can be treated within the same light diffusion formalism [3, 24, 25] and the position-dependent sample properties are characterized by the expression $3\mu_t\mu_a + f(\tau)$. Stimulated emission can be treated as negative absorption ($\mu_a < 0$) while fluorescence can also be included in a slightly expanded approach by taking account for the difference in the absorption and emission wavelengths. In the presence of locally increased absorption or fluorescence the mean intensity of light transmitted or reflected from the sample displays spatial variations that can be detected. The function $f(\tau)$ is a measure of the time-dependent microscopic motion, with $f(0) = 0$ [26]. Formally it can be considered as an absorption coefficient for the "correlation" of optical fields. Motion gradually destroys the perfect phase correlation of multiply scattered optical fields. For coherent laser light, in the presence of internal motion, light fields fluctuate and $f(\tau)$ increases as a function of time τ. The associated loss of correlation can be probed by studying the optical speckle pattern [27, 28].

Diffuse light imaging can be implemented in a number of different ways as will be described in the following sections [3, 29].

6 Optical Imaging of Intrinsic Signals

Optical imaging of intrinsic signals (OIS) determines the absorption of reflected light, such as the exposed brain cortex, for one or more often several characteristic wavelengths ([30, 31]). Due to its simplicity, it has been widely applied in an attempt to determine the cortical concentration of oxy- and deoxy-hemoglobin (oxy-Hb, deoxy-Hb) during cortical activation [32, 33]. Using light in the NIR range, optical imaging can also be applied in humans to noninvasively monitor hemoglobin oxygenation [33, 34].

A major complication of the OIS technique is the fact that the interpretation of recorded data is not straightforward at all. Since the light propagation in VIS and NIR spectral regions is dominated by the process of multiple scattering, the actual path length within the absorbing tissue, required for quantification, cannot be easily found (see also Box 3). The situation is illustrated in Figs. 2 and 4. Figure 2 displays the blood absorption and scattering properties for the relevant spectral region. Figure 4 illustrates that the light path can be substantially longer than the penetration depth z, which means that the meandering photons have a higher probability of getting absorbed. The situation is further complicated by the fact that the relevant distance travelled within a region of activity (absorption changes or motion) might be much smaller than the total path length [37].

The main problem with the differential path length approach is that experimental access [38] and modeling of $\mathrm{DP}(\lambda)$ in reflection from a heterogeneous medium is very complicated and to date the interpretation of intrinsic optical imaging data remains controversial.

Box 3 The Differential Path Length

Quantification in optical intrinsic imaging (OIS) relies on the concept of the so-called wavelength-dependent *differential path length* $\mathrm{DP}(\lambda)$. It is defined such that for the case of absorption the attenuation coefficient is $A(\lambda) = \mathrm{DP}(\lambda)\ \mu_a$ with $A(\lambda) = -\ln[I/I_0]$. In other words, $\mathrm{DP}(\lambda)$ is the mean path length travelled in a corresponding homogeneous medium with an absorption coefficient μ_a (Lambert–Beer law). In an actual experiment, it is almost impossible to measure absolute values of $A(\lambda)$ and therefore the analysis is usually restricted to relative changes of the attenuation coefficient: $\partial A(\lambda)/\partial\mu_a = \mathrm{DP}(\lambda)$ or $\Delta\mu_a = \Delta A(\lambda)\mathrm{DP}(\lambda)^{-1}$, where only small changes of μ_a are considered such that $\mathrm{DP}(\lambda)$ remains unchanged. The knowledge of $\mathrm{DP}(\lambda)$ therefore would allow to relate absorption changes to changes in concentration of the absorbing substance since $\Delta\mu_a \propto$ concentration.

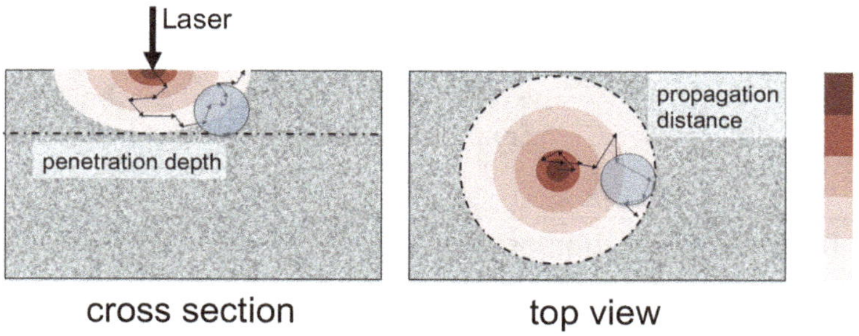

Fig. 4 Diffuse scattering of light in reflection. The incident wave is dispersed and scattered along a path of a certain contour length. The scattered wave leaves the sample at a distance $\rho_{x,y} \leq 3l_t$. The propagating waves reach a typical depth z of about half this distance [35, 36]. If only a section of the light path crosses a region of interest (*blue shaded circle*), the effective path length inside this region needs to be considered only. Modeling the latter is difficult if the sample is accessed only from a single perspective. The problem can however be overcoming using optical tomography based on a numerical analysis of the signal obtained from a number of different source/detector pairs placed around the object [24].

7 Laser Speckle Imaging

Laser speckle imaging is a related approach that uses coherent laser light and measures the "absorption of correlation" [5, 27, 39–42] in the process of multiple scattering. It provides a measure for the local dynamic properties in a scattering medium. The underlying principle of LSI is equivalent to the one exploited in laser Doppler flowmetry, a method widely used in biomedical functional imaging of blood flow [5]. A moving object inside the tissue, such as a red blood cell, leads to a Doppler shift of the frequency of scattered light. Rather than analyzing Doppler frequency shifts LSI looks at temporal fluctuations of the scattered light in order to map dynamics. In LSI a digital camera records the surface intensity pattern of light reflected from the tissue. In the absence of movement (or for very short exposure time), there is no detectable fluctuation and the typical granular interference pattern, called speckle, has the same properties in the entire field of view. The spatially resolved speckle properties evolve in the presence of movement and thus can be a sensitive probe of blood flow dynamics. The strength and limitations of LSI are similar to the ones found in OIS. This is not surprising since both methods are founded on exactly the same physical principles (absorption and scattering of diffusely reflected light). The LSI measurement is rather easy to implement and the data acquisition is straightforward. Quantification, however, faces the same problems as OIS since knowledge about the exact path lengths is difficult to obtain from a single wide-field experiment. Moreover, the penetration depth is limited to the mm-range due to the wide-field illumination usually applied. Additional complications, specific to LSI and not present in OIS, are the limited signal to noise ratio and the sensitivity of the method to vibrations and body movements, for example, due to breathing and heartbeat [43].

8 Diffuse Optical Tomography

In contrast to the planar imaging methods discussed in the previous sections, diffuse optical tomography (DOT) attempts to reconstruct 3D images by acquiring multiple images from different perspectives or by using a set of point sources and detectors. The tomographic scheme consists in solving a number of equations for the unknown spatial distribution of scattering properties. It has been implemented for absorption, fluorescence, and speckle-decorrelation [3, 5, 24]. Elaborate theoretical frameworks have been developed since the early 1990 [44]. Due to the additional information available, a numerical reconstruction is possible within some approximation (such as the diffusion equation for light propagation) without any prior knowledge about the medium. For small animals, NIR light is able to penetrate the whole body and thus diffuse transmission measurements can be performed for most possible combinations of source and detector position. Using fluorescence markers, the signal to noise ratio can be substantially improved. The existence of parameter-free theoretical model moreover allows quantification of the signal, for example, to extract the fluorophore density [24].

DOT type approaches have been applied to humans and neonates since the early 1990s [45]. Due to the limited penetration depth of NIR light transmission measurements are precluded and thus an optical topography approach is usually applied. It consists of placing an array of sources and detectors around the skull. With a spacing of typically 3 cm this provides a penetration depth of roughly 1.5 cm. Based on tomographic reconstruction methods, a map of the absorption parameter can be obtained. Since blood is a strong absorber (Fig. 2), this approach allows mapping of the hemodynamic response in the brain cortex with mm resolution [36, 46].

9 Summary and Conclusions

The aim of this introductory chapter was to highlight some of the main physical and optical mechanism that make up the foundation of modern neuroimaging methods. Naturally, the limited space available only allows touching the very surface of this broad field. My modest goal was to remind the reader of some basic principle of image formation. I then focused on the important role of scattering in many modern imaging applications. As I have shown for the case of diffuse light general concepts exist that can be applied to a variety of imaging techniques. The complexity of applying these concepts to light propagation in a complicated heterogeneous medium such as living tissue, however, remains a challenging task until today.

Acknowledgments

Financial support from the Swiss National Science Foundation (Grants No. 126772 and 132736) and the Adolphe Merkle Foundation is gratefully acknowledged. I am grateful to Bruno Weber, Fritjof Helmchen, and Jorge Ripoll for valuable comments and discussions.

References

1. Pawley JB (2006) Handbook of biological confocal microscopy, 3rd edn. Springer, New York, NY
2. Murphy DB (2001) Fundamentals of light microscopy and electronic imaging. Wiley-Liss, New York
3. Yodh A, Chance B (1995) Spectroscopy and imaging with diffusing light. Phys Today 48(3): 34–40
4. Helmchen F, Denk W (2005) Deep tissue two-photon microscopy. Nat Methods 2(12): 932–940
5. Briers JD (2001) Laser Doppler, speckle and related techniques for blood perfusion mapping and imaging. Physiol Meas 22(4):R35–R66
6. Atkins PW, De Paula J (2002) Physical chemistry, 7th edn. W.H. Freeman, New York
7. Bastiaens PIH, Squire A (1999) Fluorescence lifetime imaging microscopy: spatial resolution of biochemical processes in the cell. Trends Cell Biol 9(2):48–52
8. Miyawaki A (2005) Innovations in the imaging of brain functions using fluorescent proteins. Neuron 48(2):189–199
9. Lakowicz JR (1999) Principles of fluorescence spectroscopy, 2nd edn. Kluwer, New York
10. Sakadzic S, Roussakis E, Yaseen MA, Mandeville ET, Srinivasan VJ, Arai K, Ruvinskaya S, Devor A, Lo EH, Vinogradov SA et al (2010) Two-photon high-resolution measurement of partial pressure of oxygen in cerebral vasculature and tissue. Nat Methods 7(9):755–759
11. Bohren CF, Huffman DR (1983) Absorption and scattering of light by small particles. Wiley, New York
12. Haiss F, Jolivet R, Wyss MT, Reichold J, Ben-Braham NB, Scheffold F, Krafft MP, Weber B (2009) Improved In vivo two-photon imaging after blood replacement by perfluorocarbon. J Physiol 587:3153–3158
13. Faber DJ, Aalders MCG, Mik EG, Hooper BA, van Gemert MJC, van Leeuwen TG (2004) Oxygen saturation-dependent absorption and scattering of blood. Phys Rev Lett 93(2):028102
14. Oheim M, Beaurepaire E, Chaigneau E, Mertz J, Charpak S (2001) Two-photon microscopy in brain tissue: parameters influencing the imaging depth. J Neurosci Methods 111(1):29–37
15. Huisken J, Swoger J, Del Bene F, Wittbrodt J, Stelzer EHK (2004) Optical sectioning deep inside live embryos by selective plane illumination microscopy. Science 305(5686):1007–1009
16. Mertz J (2011) Optical sectioning microscopy with planar or structured illumination. Nat Methods 8(10):811–819
17. Denk W, Strickler JH, Webb WW (1990) 2-Photon laser scanning fluorescence microscopy. Science 248(4951):73–76
18. Trachtenberg JT, Chen BE, Knott GW, Feng GP, Sanes JR, Welker E, Svoboda K (2002) Long-term in vivo imaging of experience-dependent synaptic plasticity in adult cortex. Nature 420(6917):788–794
19. Ripoll J (2012) Principles of diffuse light propagation: light propagation in tissues with applications in biology and medicine. World Scientific Publishing Company, Singapore
20. Crassous J, Erpelding M, Amon A (2008) Diffusive wave spectroscopy applied to the spatially resolved deformation of a solid. Phys Rev E Stat Nonlin Soft Matter Phys 78(4):046104

21. Boffety M, Allain M, Sentenac A, Massonneau M, Carminati R (2008) Analysis of the depth resolution limit of luminescence diffuse optical imaging. Opt Lett 33(20):2290–2292
22. Azizi L, Zarychta K, Ettori D, Tinet E, Tualle JM (2009) Ultimate spatial resolution with diffuse optical tomography. Opt Express 17(14): 12132–12144
23. Deliolanis N, Lasser T, Hyde D, Soubret A, Ripoll J, Ntziachristos V (2007) Free-space fluorescence molecular tomography utilizing 360 degrees geometry projections. Opt Lett 32(4):382–384
24. Ntziachristos V, Ripoll J, Wang LHV, Weissleder R (2005) Looking and listening to light: the evolution of whole-body photonic imaging. Nat Biotechnol 23(3):313–320
25. Boas DA, Yodh AG (1997) Spatially varying dynamical properties of turbid media probed with diffusing temporal light correlation. J Opt Soc Am A 14(1):192–215
26. Boas DA, Campbell LE, Yodh AG (1995) Scattering and imaging with diffusing temporal field correlations. Phys Rev Lett 75(9): 1855–1858
27. Briers JD (1996) Laser Doppler and time-varying speckle: a reconciliation. J Opt Soc Am A 13(2):345–350
28. Volker AC, Zakharov P, Weber B, Buck F, Scheffold F (2005) Laser speckle imaging with an active noise reduction scheme. Opt Express 13(24):9782–9787
29. Durduran T, Choe R, Baker WB, Yodh AG (2010) Diffuse optics for tissue monitoring and tomography. Rep Prog Phys 73(7):076701
30. Grinvald A, Lieke E, Frostig RD, Gilbert CD, Wiesel TN (1986) Functional architecture of cortex revealed by optical imaging of intrinsic signals. Nature 324(6095):361–364
31. Frostig RD, Lieke EE, Tso DY, Grinvald A (1990) Cortical functional architecture and local coupling between neuronal-activity and the microcirculation revealed by invivo high-resolution optical imaging of intrinsic signals. Proc Natl Acad Sci U S A 87(16): 6082–6086
32. Kohl M, Lindauer U, Royl G, Kuhl M, Gold L, Villringer A, Dirnagl U (2000) Physical model for the spectroscopic analysis of cortical intrinsic optical signals. Phys Med Biol 45(12): 3749–3764
33. Villringer A, Chance B (1997) Non-invasive optical spectroscopy and imaging of human brain function. Trends Neurosci 20(10):435–442
34. Joebsis FF (1977) Science 198 (4323): 1264–1267 DOI: 10.1126/science. 929199
35. Gibson AP, Hebden JC, Arridge SR (2005) Recent advances in diffuse optical imaging. Phys Med Biol 50(4):R1–R43
36. Strangman G, Boas DA, Sutton JP (2002) Non-invasive neuroimaging using near-infrared light. Biol Psychiatry 52(7):679–693
37. Tian PF, Devor A, Sakadzic S, Dale AM, Boas DA (2011) Monte Carlo simulation of the spatial resolution and depth sensitivity of two-dimensional optical imaging of the brain. J Biomed Opt 16(1):016006
38. Patterson MS, Chance B, Wilson BC (1989) Time resolved reflectance and transmittance for the noninvasive measurement of tissue optical-properties. Appl Optics 28(12):2331–2336
39. Dunn AK, Bolay T, Moskowitz MA, Boas DA (2001) Dynamic imaging of cerebral blood flow using laser speckle. J Cereb Blood Flow Metab 21(3):195–201
40. Ayata C, Ozdemir Y, Dunn A, Atochin DN, Huang PL, Muzykantov VR, Murciano JC, Boas DA, Moskowitz MA (2003) Laser speckle-flowmetry: a novel two-dimensional technique for the study of cerebral blood flow in normal and ischemic mouse brain, in vivo. Stroke 34(1):251
41. Cheng HY, Luo QM, Zeng SQ, Chen SB, Cen J, Gong H (2003) Modified laser speckle imaging method with improved spatial resolution. J Biomed Opt 8(3):559–564
42. Weber B, Burger C, Wyss MT, von Schulthess GK, Scheffold F, Buck A (2004) Optical imaging of the spatiotemporal dynamics of cerebral blood flow and oxidative metabolism in the rat barrel cortex. Eur J Neurosci 20(10):2664–2670
43. Peuser J, Belhaj-Saif A, Hamadjida A, Schmidlin E, Gindrat AD, Volker AC, Zakharov P, Hoogewoud HM, Rouiller EM, Scheffold F (2011) Follow-up of cortical activity and structure after lesion with laser speckle imaging and magnetic resonance imaging in nonhuman primates. J Biomed Opt 16(9):096011
44. Arridge SR, Schweiger M, Hiraoka M, Delpy DT (1993) A finite-element approach for modeling photon transport in tissue. Med Phys 20(2):299–309
45. Gratton G, Corballis PM, Cho EH, Fabiani M, Hood DC (1995) Shades of gray-matter—noninvasive optical-images of human brain responses during visual-stimulation. Psychophysiology 32(5):505–509
46. Koizumi H, Yamamoto T, Maki A, Yamashita Y, Sato H, Kawaguchi H, Ichikawa N (2003) Optical topography: practical problems and new applications. Appl Optics 42(16): 3054–3062

Chapter 3

Optical Properties of Neural Tissue

Andrew K. Dunn

Abstract

The optical properties of neural tissues play critical roles in all types of optical imaging methods. The wavelength-dependent absorption and scattering properties of tissue influence imaging resolution, penetration depth, and often provide sources of contrast. Therefore, quantitative interpretation of imaging data requires knowledge of the optical properties of neural tissues. Light scattering in tissue arises from nanometer-scale spatial variations in refractive index and requires a thorough electromagnetic description of light propagation through this complex medium. Unfortunately, the complexity of neural tissues and the difficulty in measuring refractive index values make such a complete description unrealistic. Therefore, approximations must be made in order to characterize the light scattering properties of neural tissue. This chapter summarizes the various approaches to assess and describe the optical properties of neural tissue and discusses their role for cortical imaging.

Key words Light scattering, Light absorption, Optical properties, Monte Carlo, Finite-difference time-domain

1 Introduction

As described throughout this book, many optical techniques have been developed for imaging cortical structure and function. Each technique utilizes different sources of contrast and has different limitations in terms of spatial and temporal resolution, imaging depth, and sensitivity (see Chapter 2 in this volume). However, these techniques share a common feature: all are strongly dependent on the optical properties of the cortex. In some cases the optical properties provide the source of contrast and therefore these imaging methods capitalize on spatial and/or temporal variations in cortical optical properties to provide functional information. For other imaging methods, the optical properties impose limitations on the penetration depth or resolution of the technique. The most well-known example is the limited penetration depth of high-resolution imaging methods that arises from scattering in the cortex, which fortunately has been partially overcome by in vivo two-photon excited fluorescence microscopy.

Bruno Weber and Fritjof Helmchen (eds.), *Optical Imaging of Neocortical Dynamics*, Neuromethods, vol. 85,
DOI 10.1007/978-1-62703-785-3_3,

Developing an improved understanding of the optical properties of neural tissue and their role in imaging of cortical function is necessary for proper interpretation of imaging data. Such an understanding is particularly important in applications such as multi-wavelength optical imaging of intrinsic signals (OIS) and diffuse optical tomography where quantitative values for hemoglobin oxygenation and volume changes are sought. These measurements are combined with a solution to an inverse problem and this solution is a strong function of the tissue optical properties [1]. In addition, development of new microscopy methods or refinement of existing techniques can be aided by a deeper understanding of optical properties.

This chapter reviews the physical origins of the optical properties of the cortex with a particular emphasis on light scattering. Some of the methods for modeling light interaction with tissue are described in the context of cortical imaging. Finally, several practical examples of the role of optical properties in cortical imaging are described.

2 Origins of Optical Properties of Neural Tissues

The optical properties of tissue are usually characterized by the degree of light scattering and absorption, both of which arise from the complex physical and biochemical structure of tissue. Structurally, cortical tissue comprises different neuronal and glial cell types, vasculature, blood cells, and a complex distribution of intracellular and extracellular proteins. The densities of neuronal cells and vasculature vary with cortical layers, and as a result, the absorption and scattering properties of cortical tissue vary spatially as well as spectrally. This structural complexity makes it very challenging to define specific scattering and absorption values for cortical tissue.

Absorption of light in neural tissues is generally the result of molecular absorption. Although a vast number of molecules and proteins contribute to the overall absorption, a few are the dominant absorbers in the visible and near-infrared spectrum. The strength of an individual chromophore is most commonly described by its molar extinction coefficient, ε, which has units of M^{-1} cm^{-1}. The dominant chromophores in neural tissue are oxy- and deoxy-hemoglobin, cytochrome-c-oxidase, water, NADH, and flavoproteins. Each of these chromophores has a unique molar extinction spectrum (Fig. 1). The total tissue absorption is described by the absorption coefficient, $\mu_a(\lambda)$ which has units of cm^{-1}. The quantity $\mu_a(\lambda)dl$ represents the probability that a photon of a particular wavelength will be absorbed along a distance dl. The absorption coefficient varies with wavelength and can be treated as a sum of the molar extinction coefficients of all individual chromophores weighted by their concentrations,

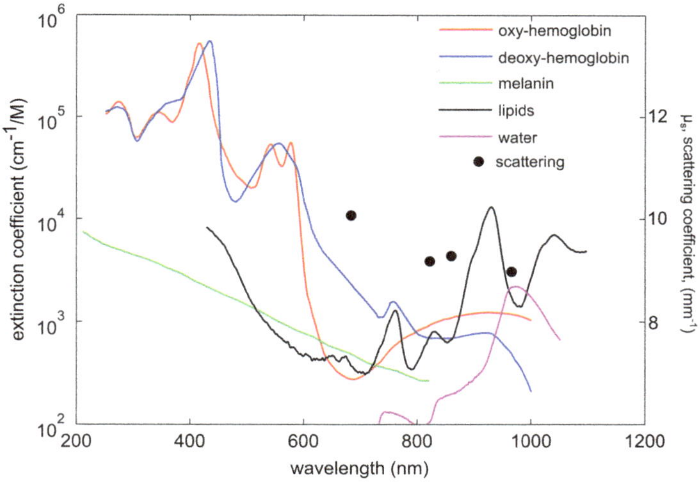

Fig. 1 Optical properties of common tissue components. The wavelength-dependent extinction coefficient ε is plotted for common tissue chromophores (*left axis*) and the scattering coefficient μ_s of neural tissue (*right axis*) is shown for four values measured from cortical tissue [2]. Note the different scales of the two axes.

$$\mu_a(\lambda) = 2.303 \sum_i \varepsilon_i(\lambda) C_i \quad (1)$$

where C_i is the concentration of a particular absorber. For wavelengths less than ~650 nm, oxy- and deoxy-hemoglobin are the dominant absorbers and this sum is usually truncated at two terms.

Although the absorption properties of neural tissues can be described as a linear combination of independent chromophores, the scattering properties of tissues are much more challenging to describe quantitatively. Scattering arises whenever light encounters a refractive index change. Therefore, the scattering properties of any tissue are determined by the wavelength of light and the spatial variations in refractive index. The spatial complexity of the refractive index of cortical tissue makes detailed descriptions of scattering extremely challenging. Neural tissue comprises many different cellular and noncellular structures each of which has a slightly different refractive index and size. Furthermore, most of these structures contain internal variations in refractive index that occur on the nanometer spatial scale. An exact description of the scattering properties of any tissue requires detailed knowledge of the spatial distribution of refractive index at the nanometer level, which is not feasible.

Although many studies have attempted to characterize the index of cell components, there are no definitive values for each component due to the difficulties inherent in these measurements, as well as the natural variations across cells and tissues. Table 1 summarizes some of the previously published refractive index

Table 1
Selected values of refractive index of tissue components taken from the literature.

Cell component	Refractive index	Reference
Water	1.33	[3]
Cytoplasm, rat liver cells	1.38	[4]
Mitochondria, rat liver cells	1.40	[4]
Lipid	1.48	[4]
Cytoplasm	1.35	[5]
Protein	1.50	[5]
Cytoplasm, hamster ovary cells	1.37	[6]
Mitochondria, rat liver	1.42	[7]
Melanin	1.7	[8]
Cytoplasm	1.358–1.374	[9]
Cortical cytoplasm	1.353–1.368	[10]
Dried protein	1.58	[11]

values for cellular components in general. Although many of these values are for non-neural tissues and cells, these values reveal that most organelles have refractive indices slightly above that of water.

The refractive index of cellular and noncellular components is largely determined by the local protein concentration within the component. Any component can be considered as a protein solution and its index can be written as [12]

$$n = n_o + \alpha C \tag{2}$$

where n_o is the refractive index of the solvent, which is approximately equal to water for cells, α is the specific refraction increment, and C is the concentration of the solute (g/100 ml). For protein, $\alpha = 0.0018$, and for other solutes found in cells such as sodium, $\alpha = 0.0016$ [11]. While the specific refraction increments are similar for proteins and other solutes, proteins play the largest role in determining the index of refraction because their concentrations in terms of weight per volume are considerably greater than other solutes [12].

The net scattering properties that arise from these spatial variations in refractive index are generally characterized in two ways. The first is the scattering cross section, which is a measure of the amount of light scattered by a single isolated tissue component such as a cell or organelle. The scattering cross section is related to the total amount of light scattered in all directions by a single particle and is denoted by σ_s. Although the units of σ_s are area (m^2), the scattering cross section is not the same as the geometrical cross section of the particle and, in fact no simple relationship exists between the cross sectional area and σ_s.

While the scattering cross section is used to define the degree of light scattering from a single particle, it has limited applicability in complex tissues such as neural tissue since it is very difficult to define what comprises a single particle. The scattering cross section of objects such as microspheres and metallic nanoparticles are very useful since these particles have well-defined geometries. However, tissues contain a continuously varying refractive index on the nanometer length scale. Previous studies have characterized the scattering cross sections of biological cells and subcellular organelles [13], but it is very challenging to relate these isolated scattering cross sections to tissue level descriptions of light scattering [14].

The scattering coefficient, μ_s, is the most common measure of macroscopic, or tissue level scattering. The quantity $\mu_s dl$ represents the probability of light scattering along a distance dl, and $1/\mu_s$ represents the scattering mean free path, l_s, i.e., the average distance a photon travels between scattering events. The scattering coefficient, μ_s, is a macroscopic measure of light scattering whereas the scattering cross section, σ_s, is a microscopic or single particle measure. For discrete particles, the scattering coefficient and cross section are related by $\mu_s = \sigma_s N_v$, where N_v is the volumetric particle density (m^{-3}). For neural tissues, the scattering cross section is difficult to define due to the complex spatial composition of the tissues, and as a result, the volumetric particle density is also difficult to define. Nevertheless, the scattering coefficient, μ_s, is still a useful quantity since it is an aggregate measure of the scattering strength over a region of neural tissue.

Because in vivo measurement of absolute scattering coefficients involves complex experimental measurements that must be carefully calibrated and coupled with robust models of light propagation in tissues, significant variability exists in scattering coefficients reported in the literature. In vivo measurement of optical properties typically involves measurement of reflected intensities at multiple points along the tissue surface. The scattering and absorption properties of the tissue that give rise to these signals must then be determined by solving an appropriate inverse problem. Decoupling the contributions of skull, scalp, gray, and white matter are particularly challenging and can be aided by time- and frequency-domain measurements as well as a priori knowledge of anatomical structure [15, 16]. Therefore, the accuracy of in vivo measurement of optical properties is often determined by the accuracy of the approach used to solve the inverse problem. Ex vivo measurement of optical properties is often performed using integrating spheres that measure the diffuse reflectance and transmission of thin tissue sections as a function of wavelength. Like in vivo measurements, the integrating sphere measurements must be combined with an appropriate inverse model to extract optical properties. These models include Monte Carlo models, analytical expressions based on the diffusion approximation, and the inverse adding doubling technique [17].

Experimental measurements of the scattering coefficient of cortex are in the range of $\mu_s = 8–12\ \text{mm}^{-1}$ at wavelengths between 650 and 950 nm and show a slight decrease with increasing wavelength [2, 18]. These values correspond to a mean free path of 80–125 μm, which represents the average distance between scattering events when scattering is treated as a macroscopic process. In other words μ_s attempts to encapsulate the microscopic scattering from a volume of tissue. However, when considering very small length scales, the macroscopic interpretation of light scattering can be somewhat misleading since light cannot really travel 100 μm without any deviation. Instead, the 100 μm mean free path attempts to combine all of the very small deviations due to the microscopic variations in refractive index.

3 Optical Tissue Properties and Their Relation to Optical Imaging of Cortical Function

Both scattering and absorption can provide in vivo functional information about the cortex. One of the earliest examples of such an application was the detection of membrane potential changes by changes in light scattering in cultured neurons [19]. Dark field microscopy was used to detect the very small changes in light scattering that resulted from induced changes in transmembrane potential and a linear relationship between the scattered light intensity and membrane potential was found. The underlying mechanism was found to be a change in the radial component of the refractive index that corresponded to membrane potential. Although this study and other similar studies in slices [20, 21] established experimentally a direct change in light scattering intensity with membrane potential, the magnitude of the intensity changes is very small ($<10^{-5}$/mV). These studies have led to a renewed interest in monitoring neural activity in intact cortex through light scattering changes. One of the first reports of noninvasive detection of this signal was by Gratton et al. [22] who showed a phase shift in frequency-domain diffuse optical tomography measurements. A number of more recent studies have demonstrated that measuring such a small signal noninvasively is extremely challenging due to the very low signal to noise ratio (see [23] for a review of recent literature).

Perhaps the most widely used optical method for imaging cortical function is OIS, which indirectly measures cortical activity by detecting the hemodynamic and metabolic changes that arise from neural activity [24]. Both scattering and absorption play important roles in these measurements. Light scattering is required to reflect light back to the camera, but it also limits the penetration depth of the measurements to the very superficial cortex. Absorption of light by oxygenated and deoxygenated hemoglobin provides the source of contrast in OIS. Therefore, the spatial and temporal dynamics of the hemodynamic response to activation can be determined qualitatively simply by examining the resulting changes in reflected light intensity. However, quantitative determination of the changes in oxy- and deoxy-hemoglobin requires measurements at multiple wavelengths and spectroscopic analysis of such measurements [25–28]. The optical properties of neural tissue play a very important role in such an analysis. In particular, both absorption and scattering effects must be accounted for. Assumptions must be made about baseline values of absorption and scattering coefficients. The absorption coefficient values are determined from the assumed baseline concentrations of oxy- and deoxy-hemoglobin. A value of the scattering coefficient and its wavelength dependence must also be assumed for calculation of the differential path length factor [27]. The controversy over the presence or absence of an initial increase in deoxy-hemoglobin during functional activation arose in part due to the role of the wavelength dependent scattering of neural tissue [25, 27].

The optical properties of neural tissues and tissue in general, particularly scattering, place constraints on the surgical preparation for most cortical imaging studies. Scattering by the skull and dura necessitates a thinned skull or full craniotomy preparation. Even when thinned skull preparations are used, mineral oil or glycerol is usually applied to the surface prior to imaging. These agents serve to improve image quality by reducing the scattering of the remaining skull. The applied oils have a relatively high refractive index that serves to reduce the spatial variations in refractive index, which in turn reduces the total scattering and improves image quality. Because scattering from the skull and cortical tissue strongly attenuates excitation light, microscopic methods such as in vivo two-photon imaging and confocal imaging usually require removal of the skull, although thinned-skull preparations can also be applied (see Chapter 16 in this volume). Maximum imaging depth however, is achieved by full removal of the skull or polishing the skull to transparency [29]. However, thinning the skull can reduce scattering sufficiently to allow imaging to depths of a few hundred microns while making the preparation slightly less invasive.

4 Methods for Modeling Light Propagation in Neural Tissues

Knowledge of the scattering and absorption properties of neural tissues alone is not sufficient to perform a quantitative analysis of imaging data or to predict maximum imaging depths.

For such analysis, a model of light propagation in tissue is required. Many different models have been developed and each is useful for modeling different spatial scales and measurement geometries. However, all approaches require knowledge of the scattering and absorption properties of tissue as inputs. In this section we review some of the modeling approaches that have been used and highlight how the role of optical properties of neural tissue in these models.

4.1 Monte Carlo Simulations

The Monte Carlo method for simulating light propagation in biological tissues is the most widely used modeling approach. In this approach photons are treated as particles that traverse the tissue in a stochastic process based on the radiative transport equation [30]. After tracing a large number of photons, information about reflected intensities and sampling distributions can be determined. Like all of the modeling approaches described here, results obtained from Monte Carlo simulations are only accurate when the spatial distribution and quantitative values of scattering and absorption coefficients are accurate.

Monte Carlo simulations require knowledge of the scattering coefficient, μ_s, the absorption coefficient, μ_a, and the scattering anisotropy, g, which characterizes the angular distribution of scattered light. The total attenuation coefficient, μ_t, is defined as the sum of the absorption and scattering coefficients, $\mu_t = \mu_s + \mu_a$. Anisotropy is the average cosine of the scattering angle and its values range from −1 to 1, where a value of 1 indicates completely forward scattering, 0 indicates isotropic scattering, and −1 indicates pure backscattering. Most soft tissues have anisotropy values around 0.9. In Monte Carlo simulations, a series of random numbers are generated to sample probability distributions that characterize photon propagation events such as scattering angle and distance between scattering events. Further details on Monte Carlo simulations of light propagation in tissues can be found in a number of reviews [31, 32].

Light propagation through complex geometries such as a human head can be modeled by assigning spatially varying scattering and absorption properties to anatomical geometries obtained with other imaging modalities such as MRI [33, 34], two-photon excited fluorescence microscopy [35], and optical coherence tomography [36, 37]. Figure 2 illustrates two examples of heterogeneous geometries used in Monte Carlo models. In one case, an anatomical MRI was segmented into different tissue types and optical properties of each tissue type were assumed [33]. In the other case, two-photon fluorescence images of cortical microvasculature were segmented to model light propagation in vascular fluorescence imaging (described in more detail below), and scattering and absorption coefficients were assigned to vascular and nonvascular tissues.

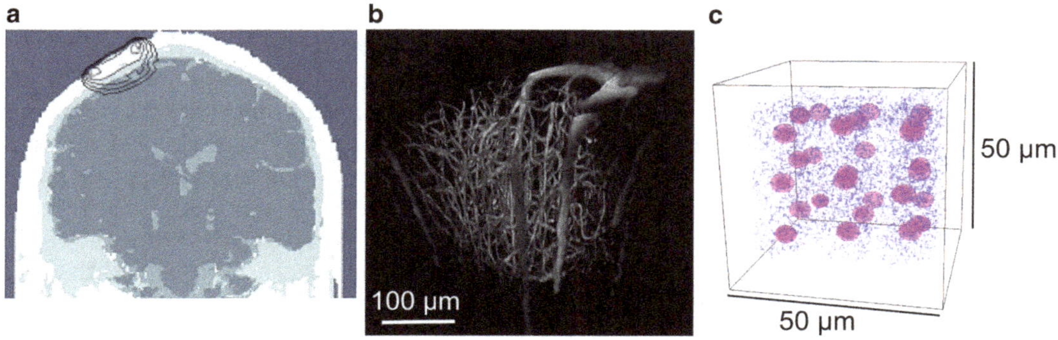

Fig. 2 Examples of geometries used in modeling light propagation through tissue. (**a**) Light transport through human head modeled using Monte Carlo simulations coupled with anatomical MRI data that was segmented into discrete tissue types (indicated by different *grey* levels). Scattering and absorption properties can then be assigned to each tissue type [33]. (**b**) Monte Carlo models can also be used to simulate light transport in invasive optical imaging geometries using segmented image stacks from two-photon microscopy as shown here. Separate optical properties can then be used for vascular and nonvascular tissues. (**c**) Electromagnetic modeling of light propagation enables subcellular structures to be considered assigning different refractive indices to each structure. The colors represent refractive indices ranging from 1.37–1.40 [38]. Each of the spherical structures (*magenta color*) represents cellular nuclei and the blue structures represent smaller organelles.

4.2 Finite-Difference Time-Domain Method

Monte Carlo simulations work well when simulating multiple scattering effects in tissues. However, when the tissue volume of interest is small, or when the effects of small-scale features such as subcellular components on light scattering are desired, Monte Carlo methods cannot usually be used. In order to properly model such small-scale features, an electromagnetic approach that takes into account the wave nature of light must be employed. Although numerous electromagnetic modeling methods have been developed, all require extensive computational resources. The most common method for modeling light scattering from complex biological tissues [13] is the finite-difference time-domain (FDTD) method [39] due to its relative simplicity and its ability to incorporate arbitrary three-dimensional spatial variations in refractive index. The FDTD method has been used extensively to solve a wide variety of electromagnetics and scattering problems [40]. This method is a full vector solution of the electric and magnetic fields in a small region surrounding an object. The electric and magnetic fields are discretized on a spatial grid and are updated in a time stepping manner. The full details of FDTD simulations can be found in a number of reviews [40].

Unlike Monte Carlo models where the macroscopic scattering and absorption coefficients are used to define the tissue, in FDTD models, the spatially varying dielectric properties (or refractive index) of tissue at the scale of a few nanometers are used to create the geometry. Absorption can be incorporated into such models by including a complex refractive index. Figure 2c illustrates an example of a three-dimensional multi-cell geometry used in an FDTD simulation of focused beam propagation [38]. In this example, the cells are constructed from cubic voxels and the refractive index at each voxel is used to specify individual cellular components. In the example in Figure 2c nuclei of cells have a different refractive index than smaller organelles and cytoplasm. The major drawback of FDTD models, however, is the large computational demand that is required to simulate realistic tissues. The computational requirements result from the stability criteria of the finite difference approach which limits the maximum voxel size to one tenth of the wavelength of light [39]. In practice, most simulations of light scattering by biological cells use cubic voxels that are $\lambda/20$ to $\lambda/15$. Due to the need to discretize geometries with such fine resolution, two problems arise in FDTD models. First, the refractive index distribution should be known on length scales of approximately 10 nm. As discussed earlier, refractive index values are very difficult to measure and therefore, they are assumed in most models. Despite this limitation, FDTD simulations can provide detailed information about the effects of organelles on light scattering from cells [13, 14, 41] as well as the effects of cellular structures on the beam profile of focused beams [38]. The second problem is that the computational requirements for simulation of volumes of tissue that are only ~50 μm on a side are extremely high. In particular, the memory requirements for such simulations require the use of distributed computing systems that can provide hundreds to thousands of gigabytes of memory [14].

5 Influence of Optical Properties on Cortical Imaging

In this section we describe several examples of how optical properties of neural tissue influence some common cortical imaging methods. These studies illustrate the strong influence of scattering and absorption on the depth sensitivity of camera-based reflectance and fluorescence methods and on the depth penetration limits of techniques such as two-photon excited fluorescence microscopy. These studies also highlight the importance of the modeling approaches described above.

5.1 Depth Sensitivity of Camera-Based Reflectance and Fluorescence Imaging of Cortex

Camera-based imaging of the cortex such as OIS, multi-wavelength OIS, laser speckle contrast imaging (see Chapters 14 and 15 in this volume), and voltage sensitive dye imaging (see Chapter 8 in this volume), provide valuable information about cortical activity on the macroscopic spatial scale (without cellular resolution). Rather than providing depth-resolved activation images, these techniques provide depth-integrated measures of cortical activation. However, the depth weighting is not a simple uniform weighting function. Instead, it is dependent on tissue optical properties, measurement geometry, and tissue structure. In vivo experimental determination of this depth sensitivity function and the role of optical properties are exceedingly challenging. As a result, modeling studies must be performed to systematically address these issues.

One example of such a modeling study is illustrated in Box 1 [42]. In this study the role of several experimental parameters on the depth sensitivity of camera-based reflectance and fluorescence imaging was investigated using Monte Carlo simulations.

Box 1 Spatial Resolution and Depth Sensitivity of Camera-Based Intrinsic Optical Imaging.

This box describes the main findings from the Monte Carlo simulation study by [42] focused on estimation of the spatial resolution and the depth sensitivity of the camera-based 2D optical imaging methods [42].

In OIS, the measured signal (the intensity of each pixel on the CCD) is a weighted sum of the contribution from tissues at different cortical depths. The term "depth sensitivity" is used to describe this weighting function. Due to both light scattering and defocus, an array of point sources along the cortical depth results in a blurred image on the CCD. The term "spatial resolution" is used to characterize the lateral size of this image.

Both properties, i.e., spatial resolution and depth sensitivity, are determined by configurations of the imaging system (numerical aperture, NA, and focal plane depth, FPD) and optical properties of the cortical tissue (refractive index n_0, the anisotropic factor g_0, the scattering and absorption coefficients, μ_{s0} and μ_{a0}). Monte Carlo simulations were used to characterize the spatial resolution and depth sensitivity with optical configurations typically encountered in functional brain imaging, i.e., NA of 0.1, 0.2, and 0.4; FPD from 20 to 600 μm and tissue parameters that represent cortical gray matter in visible wavelengths up to 670 nm: n_0, g_0, μ_{s0}, and μ_{a0}, of 1.4, 0.9, 35, and 0.27 mm, respectively.

If the image is in focus on the cortical surface (FPD = 0), then a point at the focal plane will be imaged onto a point on the detector (the CCD chip) with minimal blurring. However, if FPD>0, then an area with a certain radius around the point at the focal plane will contribute to a point on the detector; therefore, the profile on the detector will be "blurred." The larger the focal plane depth and NA, the more blurring this will induce, thus leading to a wider profile and a lower spatial resolution, as can be seen in Fig. B1a. This analysis demonstrates that *the spatial resolution of OIS is less than or comparable to the size of cortical columns.* For example, it is less than 200 μm for NA less than 0.2 or FPD less than 300 μm, and can be as high as ~20–30 μm for FPD less than 100 μm.

For functional imaging applications with hemodynamic activity spread over hundreds of microns (with lateral size comparable to or larger than spatial resolution), *increasing NA and focal plane depth ("focusing deeper in tissue") does not increase the contribution from deeper tissues, but rather reduces the spatial resolution* (Fig. B1b). Increasing NA and focal plane depth may improve the contribution from deeper tissues when the activity is localized

(continued)

Box 1 (continued)

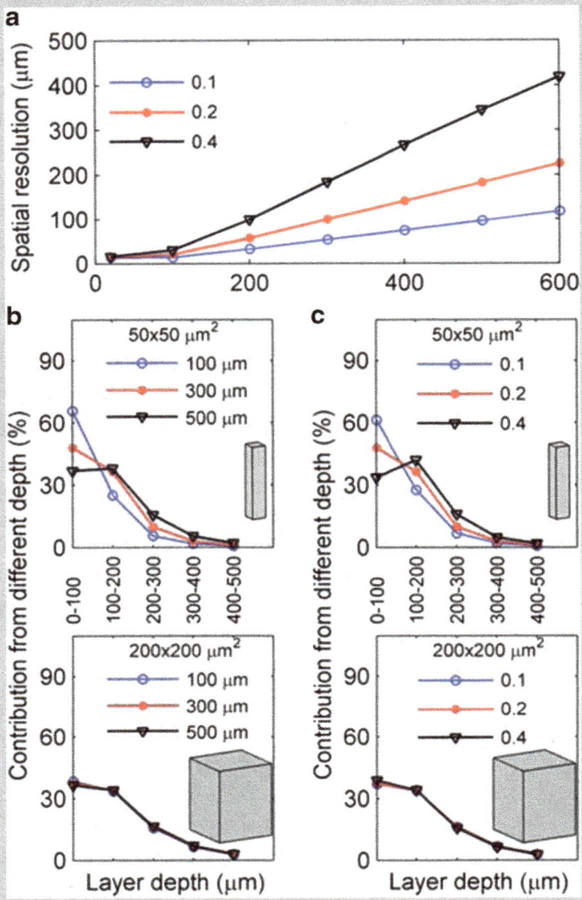

Fig. B1 Estimation of spatial resolution and depth sensitivity of OIS. (**a**) Spatial resolution versus focal plane depth for NA of 0.1 (*circle*), 0.2 (*star*), and 0.4 (*triangle*), respectively. (**b**) Depth sensitivity for functionally activated "columns" of two different cross-sectional areas: 50 × 50 and 200 × 200 μm². NA is fixed at 0.2 and the focal plane depth is varied between 100 (*blue*), 300 (*circle*), and 500 (*triangle*) μm. (**c**) The same as (**b**) but the focal plane depth is fixed at 300 μm and NA is varied among 0.1 (*circle*), 0.2 (*star*), and 0.4 (*triangle*).

within tens of microns spread in the *XY* plane (Fig. B1b). However, such limited spread does not naturally occur in the cerebral cortex.

More than 97 % of the OIS signal comes from the top 500 μm of the tissue, cortical layers 1 to 3. More specifically, about 38, 34, 16, 7, and 3 % of the signal are from tissues from 0–100, 100–200, 200–300, 300–400, and 400–500 μm of the cortical depths, respectively (Fig. B1b). In other words, *OIS has virtually no sensitivity to hemodynamic events deeper than 400 μm*, unless the activity propagates upstream within the arteriolar vessel wall [43].

The optical properties of the cortex also play an important role in fluorescence and phosphorescence-based measurements. A common assumption for Monte Carlo-based modeling studies is that the tissue can be approximated as a homogeneous medium. However the validity of this assumption has not been thoroughly investigated. Figure B1 shows an example of Monte Carlo simulations of camera-based fluorescence measurements in mouse cortex [35]. In these simulations the fluorescence signal was assumed to originate in the vasculature to mimic vascular fluorescence measurements. To accurately model this type of measurement, the detailed microvascular anatomy was incorporated into the simulation. The geometry was created from a stack of two photon fluorescence images of mouse cortical vasculature (blood plasma was labeled with fluorescein-dextran) that were segmented into two tissue types (vasculature and non-vasculature). Scattering and absorption properties were assigned to intra- and extravascular tissues, and a Monte Carlo model was used to propagate excitation and emission photons resulting from one photon absorption [34]. The simulated fluorescence intensity was used to calculate the spatial distribution of the origins of fluorescent signals. Vascular areas were assumed to contain a fluorophore to mimic measurements such as oxygen-dependent phosphorescence quenching [44–46] and indocyanine green fluorescence of vasculature [47]. Figure B1a and B1b shows the depth profile of the detected fluorescence signal at three different excitation wavelengths (415, 524, and 800 nm). The optical properties at each of these excitation wavelengths varied and the results indicate that for all wavelengths the majority of the detected signal originates in the first 400 μm of tissue, consistent with the results from a homogeneous geometry. However, the distribution as a function of depth is not smooth due to the heterogeneity of the cortical vasculature. Each of the peaks in the curves in Fig. 3a is the result of fluorescence contributions of single subsurface vessels. While this result may be specific to fluorescence originating only in the vasculature, it illustrates the importance of modeling with realistic geometries. Finally, Figure 3b illustrates the spatial distribution of the detected fluorescence or phosphorescence for a single pixel in a camera. Therefore, these results represent a measure of the spatial point spread function of camera-based fluorescence imaging of fluorescently labeled cortical vasculature.

5.2 Effects of Scattering on Resolution and Signal Strength in Two-Photon Imaging

As described in several chapters of this book, two-photon excited fluorescence microscopy is now a standard tool for in vivo investigation of many aspects of cortical function. Two-photon imaging enables a high spatial resolution and three dimensional imaging throughout much of the cortex, but the maximum penetration

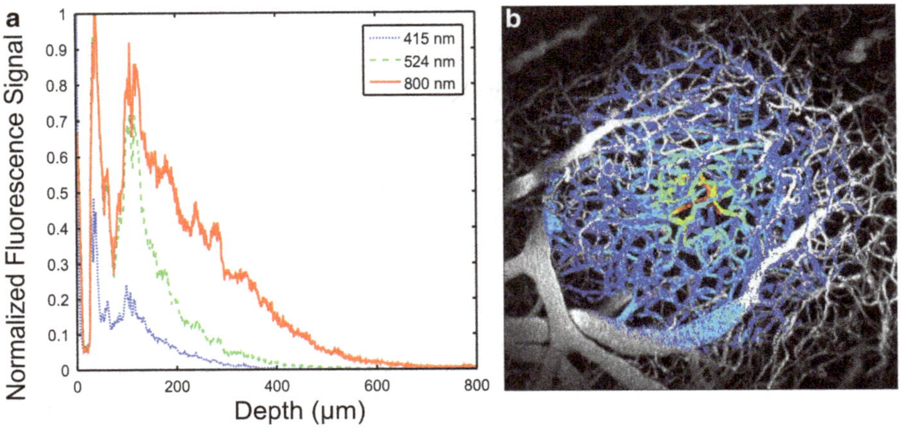

Fig. 3 Monte Carlo modeling of cortical fluorescence. (**a**) The depth dependence of fluorescence signals for camera-based imaging of fluorescently labeled cortical vasculature was modeled for visible and near-infrared fluorescent dyes with three different excitation wavelengths. (**b**) The three-dimensional geometry derived from the stack of two-photon fluorescence images shown in Fig. 2b was used. The majority of the measured fluorescence signal was found to originate in the superficial 200 μm of cortical vasculature. The three-dimensional rendering demonstrates the spatial distribution of fluorescence light for the visible and near-infrared fluorescent dyes measured by 1 camera pixel.

depth is ultimately limited by light scattering. Excitation wavelengths are in the near-infrared where scattering is lower than for single photon excitation. In addition, non-descanned detection enables detection of emission photons that have been multiply scattered. These two features enable significantly deeper imaging than with single-photon excited fluorescence confocal microscopy. Nonetheless, scattering can affect two-photon fluorescence signals in two main ways. First, scattering attenuates excitation and emission light resulting in a strong decay of signal strength with depth. Second, scattering could potentially lead to a spatial spreading of the focused beam spot leading to lower two photon absorption and degradation in spatial resolution. Experimentally, the first of these two effects has been shown to be the dominant factor. However, systematic modeling of these two effects and the role of scattering properties in signal attenuation and resolution has only been reported in a few studies [38, 48, 49].

Several Monte Carlo-based studies have investigated the role of optical properties on two-photon excited fluorescence signal levels and resolution [48–51]. These models simulate the focused beam geometry and propagate photons in a ray tracing approach that accounts for multiple scattering. Results from these studies have predicted models for signal attenuation. In particular, these simulations revealed that a single scattering event by an excitation photon is sufficient for that photon to "miss" the focal volume and therefore, not contribute to two-photon absorption even if the angular deviation of the scattering is very small [48]. Perhaps the most significant result of these Monte Carlo studies is the result demonstrating the significant advantage in detected signal levels using high NA, low magnification objectives [49]. Such an objective maintains spatial resolution, which is only dependent on NA and wavelength. However, the combination of high NA and low magnification greatly increases the collection efficiency of fluorescence emission in scattering samples by increasing the amount of multiply scattered fluorescent photons that would be lost with high magnification or lower NA objectives. For non-scattering samples, the high NA, low magnification objectives do not provide such an advantage since the collection efficiency is only dependent on NA in the absence of scattering (Fig. B1).

The effects of scattering on the beam profile in two-photon fluorescence microscopy are more challenging to systematically investigate than the effects of scattering on signal strengths due to the need to account for diffraction. Recently, the FDTD modeling approach has been used to simulate the propagation of focused beams in tissues [38]. Although computationally costly, FDTD modeling enabled full vector solutions of Maxwell's equations for focused beam propagation through inhomogeneous biological tissues. A fifth order Gaussian beam profile was assumed for computational simplicity and the full electric and magnetic fields were determined in the focal region. The geometry consisted of a $50 \times 50 \times 50\ \mu m^3$ volume containing 27 cells (Fig. 4). Each cell contained a nucleus as well as a heterogeneous distribution of smaller organelles ranging in size from 0.25 to 1.5 μm. Refractive index values for each cell component were assumed from literature values (see Table 1) and most organelles contained internal variations in refractive index. Although this tissue geometry is not specific to neural tissue, these results still provide insight into the effects of optical properties on focused beam propagation.

The two-photon excitation point spread function of the focused beam was determined by converting the steady state field distribution to intensity and squaring the resulting intensity. Figure 4 illustrates the spatial distribution of intensity when the focal plane is located approximately 42 μm into the sample. The beam appears distorted due to scattering but the beam profile through the focus remains mostly intact. This modeling study also illustrates that the central width of the excitation point spread function remains mostly constant as a function of focal plane depth, at least within the first 50 μm that were modeled. In addition, the intensity in the beam focus decayed exponentially with focal plane depth. Therefore, these results are similar to previous

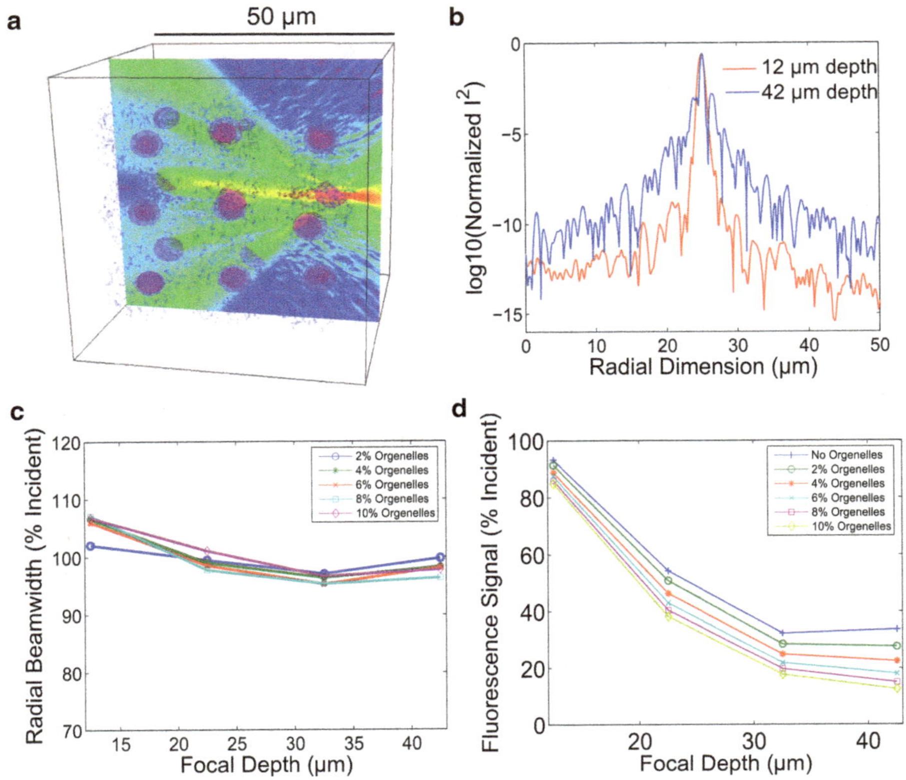

Fig. 4 Electromagnetic modeling of the effects of light scattering on beam profile of a focused laser using FDTD [38]. (**a**) The spatial intensity of a beam focused approximately 42 μm into a heterogeneous collection of cells. (**b**) The radial beam profile for two focal planes at 12 and 42 μm depth demonstrates that the width of the central lobe of the beam remains intact with increasing depth, but the intensity in the side lobes increases significantly. (**c**) The percent change in radial beam width shows little change with depth even as the fraction of subcellular organelles increases. (**d**) The integrated intensity within the focal volume, however, decreases exponentially with depth, consistent with the experimental finding that the limitation on imaging depth in two-photon fluorescence microscopy is loss of signal due to scattering rather than loss of resolution.

Monte Carlo and experimental results that scattering degrades signal strength but does not significantly affect spatial resolution. Although these FDTD studies were limited to relatively small volumes, further advances in computing power should enable full electromagnetic simulation of focused beam propagation over larger length scales.

6 Conclusions

The optical properties of neural tissue play important roles in almost all methods for imaging cortical function and structure. In some cases the optical properties provide the source of contrast, while in other cases they limit the imaging penetration depth or resolution. In both situations it is important to understand how optical properties influence imaging data. Modeling techniques can offer important insights into the role of optical properties. However, one of the greatest limitations of modeling methods is often the uncertainty in the optical properties of neural tissue, particularly scattering. Therefore, further effort is required to obtain more conclusive values for scattering coefficients at multiple wavelengths as well as refractive index values.

References

1. Villringer A, Chance B (1997) Non-invasive optical spectroscopy and imaging of human brain function. Trends Neurosci 20(10):435–442
2. Bevilacqua F, Piguet D, Marquet P, Gross G, Tromberg B, Depeursinge C (1999) In vivo local determination of tissue optical properties: applications to human brain. Appl Optics 38(22):4939–4950
3. Palmer KF, Williams D (1974) Optical properties of water in the near infrared. J Opt Soc Am 64(8):1107
4. Beuthan J, Minet O, Helfman J, Muller G (1996) The spatial variation of the refractive index in biological cells. Phys Med Biol 41:369–382
5. Kohl M, Cope M (1994) Influence of glucose concentration on light scattering in tissue. Opt Lett 17:2170–2172
6. Brunsting A, Mullaney P (1974) Differential light scattering from spherical mammalian cells. Biophys J 14:439–453
7. Liu H, Beauvoit B, Kimura M, Chance B (1996) Dependence of tissue optical properties on solute-induced changes in refractive index and osmolarity. J Biomed Opt 1:200–211
8. Vitkin I, Woolsey J, Wilson B, Anderson R (1994) Optical and thermal characterization of natural (Sepia oficinalis) melanin. Photochem Photobiol 59:455–462
9. Lanni F, Waggoner A, Taylor D (1985) Internal reflection fluorescence microscopy. J Cell Biol 100:1091
10. Bereiter-Han J, Fox C, Thorell B (1979) Quantitative reflection contrast microscopy of living cells. J Cell Biol 82:767–779
11. Barer R (1957) Refractometry and interferometry of living cells. J Opt Soc Am 47:545–556
12. Barer R, Joseph S (1954) Refractometry of living cells. Q J Microsc Sci 95:399–423
13. Dunn A, Richards-Kortum R (1996) Three-dimensional computation of light scattering from cells. IEEE J Sel Top Quantum Electron 2:898–905
14. Starosta MS, Dunn AK (2010) Far-field superposition method for three-dimensional computation of light scattering from multiple cells. J Biomed Opt 15(5):055006
15. Arridge SR (1999) Optical tomography in medical imaging. Inverse Problems 15(2):R41–R93
16. Barnett AH, Culver JP, Sorensen AG, Dale AM, Boas DA (2003) Robust inference of baseline optical properties from the human head with 3D segmentation from magnetic resonance imaging. Appl Optics 42:3095–3108
17. Prahl SA, van Gemert MJC, Welch AJ (1993) Determining the optical properties of turbid mediaby using the adding–doubling method. Appl Optics 32(4):559
18. Matcher SJ, Cope M, Delpy DT (1997) In vivo measurements of the wavelength dependence of tissue-scattering coefficients between 760 and 900 nm measured with time-resolved spectroscopy. Appl Optics 36(1):386–396
19. Stepnoski R, LaPorta A, Raccuia-Behling F, Blonder G, Slusher R, Kleinfeld D (1991) Noninvasive detection of light changes in membrane potential in cultured neurons by light scattering. Proc Natl Acad Sci USA 88:9382–9386
20. Cohen LB, Keynes RD, Hille B (1968) Light scattering and birefringence changes during nerve activity. Nature 218(5140):438–441
21. Rector DM, Poe GR, Kristensen MP, Harper RM (1997) Light Scattering Changes Follow Evoked Potentials From Hippocampal Schaeffer Collateral Stimulation. J Neurophysiol 78(3): 1707–1713
22. Gratton G, Fabiani M, Friedman D, Franceschini MA, Fantini S, Corballis P, Gratton E (1995) Rapid changes of optical parameters in the human brain during a tapping task. J Cogn Neurosci 7(4):446–456

23. Franceschini MA, Boas DA (2004) Noninvasive measurement of neuronal activity with near-infrared optical imaging. Neuroimage 21(1):372–386
24. Grinvald A, Lieke E, Frostig R, Gilbert C, Wiesel T (1986) Functional architecture of cortex revealed by optical imaging of intrinsic signals. Nature 324:361–364
25. Malonek D, Grinvald A (1996) Interactions between electrical activity and cortical microcirculation revealed by imaging spectroscopy: implications for functional brain mapping. Science 272(5261):551–554
26. Mayhew J, Johnston D, Berwick J, Jones M, Coffey P, Zheng Y (2000) Spectroscopic analysis of neural activity in brain: increased oxygen consumption following activation of barrel cortex. Neuroimage 12(6):664–675
27. Kohl M, Lindauer U, Royl G, Kuhl M, Gold L, Villringer A, Dirnagl U (2000) Physical model for the spectroscopic analysis of cortical intrinsic optical signals. Phys Med Biol 45(12):3749–3764
28. Dunn AK, Devor A, Bolay H, Andermann ML, Moskowitz MA, Dale AM, Boas DA (2003) Simultaneous imaging of total cerebral hemoglobin concentration, oxygenation, and blood flow during functional activation. Opt Lett 28:28–30
29. Drew PJ, Shih AY, Driscoll JD, Knutsen PM, Blinder P, Davalos D, Akassoglou K, Tsai PS, Kleinfeld D (2010) Chronic optical access through a polished and reinforced thinned skull. Nat Methods 7(12):5–10
30. Prahl S, Jacques S, Welch AJ (1989) A Monte Carlo model of light propagation in tissue. Proc SPIE 5:102–111
31. Wang L, Jacques SL, Zheng L (1995) MCML—Monte Carlo modeling of light transport in multi-layered tissues. Comput Methods Programs Biomed 47(2):131–146
32. Jacques S, Wang L (1995) Monte Carlo modeling of light transport in tissue. In: Welch AJ, Gemert MV (eds) Optical-thermal response of laser irradiated tissue. Plenum, New York
33. Boas DA, Culver JP, Stott JJ, Dunn AK (2002) Three dimensional Monte Carlo code for photon migration through complex heterogeneous media including the adult human head. Opt Express 10:159–170
34. Kumar AT, Skoch J, Bacskai BJ, Boas DA, Dunn AK (2005) Fluorescence-lifetime-based tomography for turbid media. Opt Lett 30(24):3347–3349
35. Davis MA, Shams Kazmi SM, Ponticorvo A, Dunn AK (2011) Depth dependence of vascular fluorescence imaging. Biomed Opt Express 2(12):3349–3362
36. Pfefer TJ, Barton JK, Smithies D, Milner TE, van Gemert MJC, Nelson JS, Welch AJ (1999) Modeling laser treatment of port wine stains using a computer-reconstructed biopsy. Lasers Surg Med 24:151–166
37. Barton JK, Pfefer TJ, Welch AJ, Smithies DJ, Nelson J, Gemert MJV (1998) Optical Monte Carlo modeling of a true portwine stain anatomy. Opt Express 2:391–396
38. Starosta MS, Dunn AK (2009) Three-dimensional computation of focused beam propagation through multiple biological cells. Opt Express 17(15):12455–12469
39. Yee K (1966) Numerical solutions of initial boundary value problems involving Maxwell's equations in isotropic media. IEEE Trans Antennas Propagat AP-14:302–307
40. Taflove A (1995) Computational electrodynamics: the finite-difference time-domain method. Artech House, Norwood
41. Drezek R, Dunn A, Richards-Kortum R (1999) Light scattering from cells: finite-difference time-domain simulations and goniometric measurements. Appl Optics 38(16):3651–3661
42. Tian P, Devor A, Sakadžić S, Dale AM, Boas DA (2011) Monte Carlo simulation of the spatial resolution and depth sensitivity of two-dimensional optical imaging of the brain. J Biomed Opt 16(1):016006
43. Tian P, Teng IC, May LD, Kurz R, Lu K, Scadeng M, Hillman EM, De Crespigny AJ, D'Arceuil HE, Mandeville JB, Marota JJ, Rosen BR, Liu TT, Boas DA, Buxton RB, Dale AM, Devor A (2010) Cortical depth-specific microvascular dilation underlies laminar differences in blood oxygenation level-dependent functional MRI signal. Proc Natl Acad Sci USA 107:15246–15251
44. Sakadzić S, Yuan S, Dilekoz E, Ruvinskaya S, Vinogradov SA, Ayata C, Boas DA (2009) Simultaneous imaging of cerebral partial pressure of oxygen and blood flow during functional activation and cortical spreading depression. Appl Optics 48(10): D169–D177
45. Ponticorvo A, Dunn AK (2010) Simultaneous imaging of oxygen tension and blood flow in animals using a digital micromirror device. Opt Express 18(8):8160–8170
46. Shonat RD, Wachman ES, Niu W, Koretsky AP, Farkas DL (1997) Near-simultaneous hemoglobin saturation and oxygen tension maps in mouse brain using an AOTF microscope. Biophys J 73(3):1223–1231
47. Raabe A, Beck J, Gerlach R, Zimmermann M, Seifert V (2003) Near-infrared indocyanine green

video angiography: a new method for intraoperative assessment of vascular flow. Neurosurgery 52(1):132–139; discussion 139

48. Dunn AK, Wallace VP, Coleno M, Berns MW, Tromberg BJ (2000) Influence of optical properties on two-photon fluorescence imaging in turbid samples. Appl Optics 39:1194–1201
49. Oheim M, Beaurepaire E, Chaigneau E, Mertz J, Charpak S (2001) Two-photon microscopy in brain tissue: parameters influencing the imaging depth. J Neurosci Methods 111(1):29–37
50. Gan X, Gu M (1999) Effective point-spread function for fast image modeling and processing in microscopic imaging through turbid media. Opt Lett 24:741–743
51. Theer P, Denk W (2006) On the fundamental imaging-depth limit in two-photon microscopy. J Opt Soc Am 23(12):3139–3149

Chapter 4

Fluorescent Indicators for Functional Optical Imaging

Walther Akemann, Hiroki Mutoh, and Thomas Knöpfel

Abstract

Optical imaging is developing towards the method of choice for monitoring the activities of large numbers of cells in the cortex of living rodents. During the last decade this approach was fueled by methodological progress in both optical instrumentation and fluorescent indicators of neuronal activities. In the first part of this chapter, we provide information required to reach an informed decision regarding the most suitable fluorescent indicators for the different imaging modalities and the experimental approaches for imaging cortical circuit dynamics in vivo. In particular, an overview is given of the available palettes of synthetic and genetically encoded calcium- and voltage-sensitive indicators. In the second half of the chapter we describe the status of voltage imaging of cortical activity using genetically encoded voltage-sensitive fluorescent proteins.

Key words Optical imaging, Calcium, Voltage, Genetically encoded activity indicators, In vivo, Signal-to-noise ratio, Neuronal circuit dynamics

1 Introduction

Functional optical imaging for measuring cortical circuit dynamics is motivated by the need to move forward from classical single-cell microelectrode recordings to an experimental approach that reveals how the activity of large populations of neurons is orchestrated and assembled into activity patterns. The ultimate goal of these experiments is to shed light on the rules the brain uses to encode, store, and process data by what is known as the neuronal code. It is generally accepted that information about the activities of large number of neurons during physiologically relevant circuit operations is necessary to experimentally decipher (and theoretically conceptualize) the neuronal code. While pooling of independent single-cell recordings allows for an understanding of what a population of neurons does "on the average," cellular interactions and population phenomena such as synchronicity can only be extracted from datasets where many neurons are recorded simultaneously.

Bruno Weber and Fritjof Helmchen (eds.), *Optical Imaging of Neocortical Dynamics*, Neuromethods, vol. 85,
DOI 10.1007/978-1-62703-785-3_4, © Springer Science+Business Media New York 2014

The recording of interactions between neighboring neurons requires a spatial resolution on the scale of a few micrometers. This requirement limits the use of microelectrode arrays because even the finest microelectrodes will mechanically lesion and disintegrate cortical circuits when applied at such high density. Moreover, since action potential firing and some types of synaptic connections are rather sparse in the cortex, the number of neurons that need to be recorded for an accurate assessment of circuit operations is probably higher than the number of simultaneous recording sites provided by the most advanced microelectrode techniques.

Following the discovery of activity-dependent changes of endogenous fluorescence and absorption or scattering of light in neural tissue [1, 2] several decades ago, optical methods were developed to provide the spatial resolution needed to solve the multisite recording challenge in a relatively noninvasive manner. The use of exogenously applied dyes for reporting neuronal activities has been pioneered by Lawrence B. Cohen and co-workers who demonstrated changes in dye absorption during action potentials in giant axons from squid followed by the first multisite optical recordings of network activities by means of voltage-sensitive dyes (VSDs) [3, 4]. While the search for better voltage indicators was initially focused on absorption dyes, fluorescent dyes became popular later on as they provide signals with higher contrast due to lower background. More importantly, fluorescent dyes enable an epi-illumination approach as required for the imaging of cortical circuits in living animals. Imaging of cortical circuit dynamics has been done most successfully with VSDs using photodiode arrays, CCD, or CMOS type image sensors. However, conventional CCD/CMOS-based wide field imaging of the cortex in living mice does not provide single-cell resolution as opposed to two-photon laser scanning microscopy (see Chapter 9 in this volume). The drawback of laser scanning microscopy, on the other hand, is a stricter limitation in terms of shot noise (see Box 1) because photons emitted from different locations are sampled serially and not in parallel as with CCD/CMOS image sensors. A larger photon noise level (see Box 1) is a more serious problem with voltage imaging that typically achieves smaller $\Delta F/F$ values than calcium imaging. To overcome the signal-to-noise issue while aiming at cellular resolution, many researchers during the last decade moved not only from wide field imaging to two-photon laser scanning microscopy but also from voltage to calcium imaging. Another trend in methodological development concerns the fluorescent indicators. With the quest to record from identified cell types, organic dyes are more and more replaced by genetically encoded probes. Following the mainstream of these trends might be a safe strategy for researchers that like to focus on biological questions while lacking expertise or resources to contribute to methodological developments. Yet, choices have to be made when selecting the

Box 1 Noise in Optical Functional Imaging

The time course of light intensity emitted from fluorescent indicators reflects not only the signals of interest (representations of neuronal activity) but also random statistical fluctuations (noise). The main technical challenge of optical imaging is to achieve an acceptable signal-to-noise ratio (SNR). Photon shot noise (a consequence of the quantum nature of light) defines a theoretical lower limit of the total noise. Other sources of noise (caused by mechanical vibrations, instabilities of the light source, or electronic interference) may, under optimal conditions, be sufficiently minimized to a level below the unavoidable photon shot noise. In fact, the shot noise limit is routinely achieved in many whole field and laser scanning imaging experiments. Photon shot noise equals to the square root of the total number of photons (n_T) that are sampled per time bin. A signal expressed as the change of fluorescence (ΔS) normalized to baseline (S_0) is reported at best with an SNR:

$$SNR \leq \frac{\Delta S}{S_0}(1 - f_B)\sqrt{n_T} \tag{1}$$

with f_B, the fraction of background fluorescence. The factor $(1 - f_B)$ in the above equation accounts for the fact that the detected photons may include not only signal photons but also photons from nonresponsive sources, e.g., stained glia, and neurons that do not contribute to the signal [5, 6]. Importantly, background photons reduce the SNR because they both reduce the apparent amplitude of the response and increase the noise. The equation also explains why an SNR better than two for a probe response amounting 1 % $\Delta S/S_0$ in the presence of 50 % background requires the detection of at least 10^5 photons per time bin. The necessity of a minimal photon yield for the detection of optical signals has several notable implications. Obviously, there would be no issue if light intensities could be cranked up with no limit. However, a single dye molecule can on average emit maximally one photon per fluorescence lifetime. With fluorescence lifetimes of common dyes or fluorescent proteins of 1–3 ns this results in a maximal photon yield per probe molecule of 10^5–10^6 photons/ms. While this limit may be reached with saturating excitation intensities in laser scanning microscopy (during the dwell time), wide field excitation and detection with a camera is performed routinely at an order of magnitude lower to prevent rapid photobleaching of the dye. On the other hand, since CCD or CMOS cameras sample photons at many points almost simultaneously and continuously, their effective photon yield can easily exceed that of laser scanning devices.

best experimental strategy. In the first part of this chapter, we describe the methodological background for guiding an informed decision on which indicator to use. In the second part, we describe the status of our own ("neo-classical") approach based on wide field imaging using genetically encoded calcium indicators (GECIs) and voltage indicators (GEVIs).

1.1 Calcium Versus Voltage Imaging

After settling for optical imaging as the method to study electrical signaling in cortical circuits, a choice has to be made whether electrical activity should be imaged either directly using voltage indicators or indirectly using calcium indicators. Calcium imaging can be used to detect action potentials because a brief influx of calcium ions is triggered during an action potential, leading to a transient increase of intracellular Ca^{2+} concentration (a "calcium transient") in many types of neurons including cortical pyramidal cells and interneurons [7–11]. The perhaps strongest argument for using calcium indicators to detect action potential firing of cortical neurons is that they generally provide a better signal-to-noise ratio (SNR) than voltage indicators. A second reason in favor of calcium imaging is its relative insensitivity to subthreshold voltage transients. This might be a paradoxical argument but in fact this feature is particularly useful when action potentials are the focus of interest (with the premise that "it is only the action potential that matters"). Many other features of calcium imaging would actually argue against the use of calcium indicators. These drawbacks include the usually slow decay of the signals provided by calcium indicators and the disturbances the indicators may exert on neuronal excitability, calcium homeostasis, synaptic integration, and plasticity. The slow dynamics provided by calcium indicators is partly caused by the delayed kinetics of calcium unbinding but also due to the fact that cytosolic calcium transients are usually much slower than electrical signals. Therefore, timing information with a precision of a few milliseconds (but usually not at higher precision) can be extracted from calcium imaging data using signal deconvolution and similar procedures [12]. Finally, the probably most important concern related to the interpretation of calcium imaging data (obtained either from large numbers of individual cells or as population bulk responses) is the assumption that "only action potentials matter." This may be true if it was feasible to record from very large numbers of cells simultaneously in which case synaptic interactions might be derived from correlations between single-cell spike patterns. However, in practice imaging technologies are far away from monitoring the very large numbers (e.g., thousands) of cells that would be required to reconstruct even a limited connectome.

Voltage imaging, on the other hand, would be ideal because the neuronal code is more likely represented by voltage rather than by calcium signals and because tracking voltage signals can provide more complete information about all electrical activity (action potentials and synaptic potentials). In the ideal case of perfect voltage imaging the main challenge would be related to the analysis of large-scale datasets. Unfortunately, voltage imaging methods are still far from ideal because the signals delivered from currently available indicators have a small SNR. An inherent reason for the small SNR of voltage imaging as compared to calcium imaging lies in the much smaller number of photons that can be collected from the plasma membrane area as compared to the cytosolic volume. Moreover, there are concerns related to pharmacological side effects and toxicity (e.g., photo-toxicity) of classical VSDs [13]. However, as argued in the second half of this chapter, voltage imaging recently gained a new momentum by the development of genetically encoded voltage probes that overcome many of the classical problems associated with voltage imaging, promising to fulfill long standing expectations in the field.

1.2 Organic Versus Genetically Encoded Indicators

After more than 20 years of development and steady improvement of calcium-sensitive dyes and VSDs, their performance probably reached the "as good as it gets" level and further improvements in performance are rather unlikely. Since these dye indicators are produced with the methods of organic chemistry, they are commonly referred to as "organic" indicators. These classical organic indicators share the drawback that they stain, when applied into the extracellular space of brain tissue, all cells unselectively. This approach therefore does not overcome the classical system physiological dilemma of "single unit" recordings from cells of undefined identity. In the context of circuit properties it is, however, obviously important to know if a cell that fires action potentials inhibits or excites the cells to which it connects synaptically. A first step towards resolving this issue was the use of transgenic mice in which specific cell populations (e.g., GABAergic cells) are marked by a fluorescent protein (FP) so that they can be identified in one color channel while a fluorescent indicator reports the activity from cells unselectively labeled with an indicator. This approach was nicely used to compare the tuning curves of excitatory versus inhibitory neurons in the visual system of mice [14, 15].

The more direct way to associate activity signals with cellular identity is through the expression of genetically encoded indicators under cell type-specific promoters. There are several advantages of genetically encoded indicators as compared to organic dyes. Since they can be targeted to specific cell populations, they allow for recording of population ("bulk") signals in a cell type-specific manner [5, 16–18]. In other words, it is possible to record responses averaged over a specific cell population, much like summary data from single-cell recordings but much more efficiently. Another advantage of genetically encoded indicators is the possibility to generate transgenic animals with stable labeling of cells, resulting in preparations ready for acute and chronic in vitro or in vivo imaging without the need for time-consuming and invasive staining procedures. In addition to expediency, genetic labeling is more reliable than conventional organic dye staining due to fewer variations between individual cells within or between preparations. Finally, genetically encoded indicators facilitate observations over long periods of time (weeks and months) as required for developmental and regeneration studies.

Based on this reasoning, it is becoming increasingly acknowledged that genetically encoded optical indicators, combined with advanced optical instrumentation, will be the tool of choice for monitoring neural activity in the intact animal. Since the initial descriptions of genetically encoded indicators of neuronal activity [19–23] intense efforts were made to further develop and improve such indicators so that they are applicable under in vivo conditions. Serious technical challenges in generating optical reporters, with sufficient signal in the intact brain, have however made probe development a lengthy process. Those challenges included correct folding and targeting of the engineered proteins in neurons of living animals and their expression at sufficient levels in the desired neurons without significant side effects. Many of these hurdles have successfully been overcome in recent years and several classes of genetically encoded activity indicators are now working robustly when expressed in living animals. Accordingly, GECIs have already been proven to successfully record calcium transients associated with action potential firing in living rodents [17, 24–27]. More recently, GEVIs have been shown to report sensory-induced responses in living mice [28].

1.3 Single-Cell Versus Population Imaging

Imaging methodologies are motivated by the need (or at least desire) to monitor the activities of many cells simultaneously. The golden standard is population imaging with single-cell resolution, because brain tissue consists of different subpopulations of neurons, with specific functions in the circuit. In the cerebral cortex, these different cell types cannot be identified simply by their localization but rather through genetic markers. In addition, analyzing well-defined individual cells avoids errors resulting from averaging over inhomogeneous populations. However, genetic targeting by means of cell type-specific promoters enhances population imaging with features that in the past were limited to single-cell studies.

2 Calcium Indicators

Optical imaging of neuronal activity using fluorescent probes became an established method following the development of the calcium indicator fura-2 [29]. Its design principle provided the basis for practically all calcium indicators available today. Fura-2 combined the properties of a fast calcium binding molecule (BAPTA) with a bright fluorescent dye (stilbene chromophore) in a manner that caused a change in fluorescence output upon binding of calcium. Nowadays, Oregon Green BAPTA-1 that has a chromophore that fluoresces in green is more widely used for acute in vivo two-photon imaging (see Chapter 9 in this volume). GECIs followed this inspiration. They involve the molecular fusion of a calcium binding protein (calmodulin, troponin) that undergoes a conformational transition upon binding of calcium, with a conformation-sensitive FP-based reporter protein component

[21, 22, 30] (see Chapter 10 in this volume). The reporter component involves either a single or two fluorescent proteins (Fig. 1a). The GECIs with two FPs exploit Förster resonance energy transfer (FRET, also referred to as Fluorescence Resonance Energy transfer) as a mechanism to translate changes in protein structure into modulation of fluorescence output. In FRET, a donor FP in its excited state can transfer energy by a dipole–dipole coupling mechanism to an acceptor FP. For efficient FRET, the two chromophores need to show sufficient overlap of donor emission and acceptor absorption spectra; they need to be in close proximity (<10 nm) and oriented with their transition dipoles close to parallel. Conformation of the indicator protein, that is its calcium binding state, modulates FRET efficacy via the latter two parameters. The first GECI based on FRET used the calcium-dependent conformation of the protein calmodulin in combination with the M13 peptide as the sensor, and pairs of blue/green or cyan/yellow FPs as reporters [21, 22]. Since modulation of FRET efficacy results in opposing changes of donor and acceptor fluorescence, the ratio of these two intensities provides a relative measure of FRET efficacy independent of indicator concentration, excitation light intensity, and absorption in the optical path. The ratio is therefore also insensitive to fluctuations of these parameters. Importantly, this includes out-of-focus movements that are often a troublesome complication in in vivo experiments. It should be noted, however, that simple rationing does not compensate for artifacts mediated by indicator-independent components of the measured light intensities or changes in tissue absorption spectra. A more recent, but similar, design used the skeletal muscle calcium sensor troponin C (TnC) which, unlike calmodulin, is not expressed at high levels in neurons. The potential advantage is that TnC-based sensors are proposed to have less interaction with endogenous cellular mechanisms than calmodulin-based sensors. An alternative approach to address this issue was explored with D2 and D3 cameleon-based calcium indicators in which the calcium sensing moiety was reengineered to prevent interaction with endogenous calmodulin [31]. A second general principle to make a FP sensitive to conformational alterations is circular permutation (cp). In cpFPs the sequential order of the N- and C-terminal parts of the amino sequence is swapped, with the effect that the amino acids where the initial sequence is divided become the new N- and C-terminals. Importantly, positions have been identified in the amino acid sequence of several FPs where circular permutation conserves the ability of the protein to form a fluorescent chromophore. The principle behind using cpFPs is that the fluorescent properties of an FP are more sensitive to the impact of a fused sensor, following structural weakening by circular permutation [32–34]. Original cpFP-based Ca^{2+} sensors suffered from pH sensitivity and inefficient protein folding [32, 34]. However, optimizing

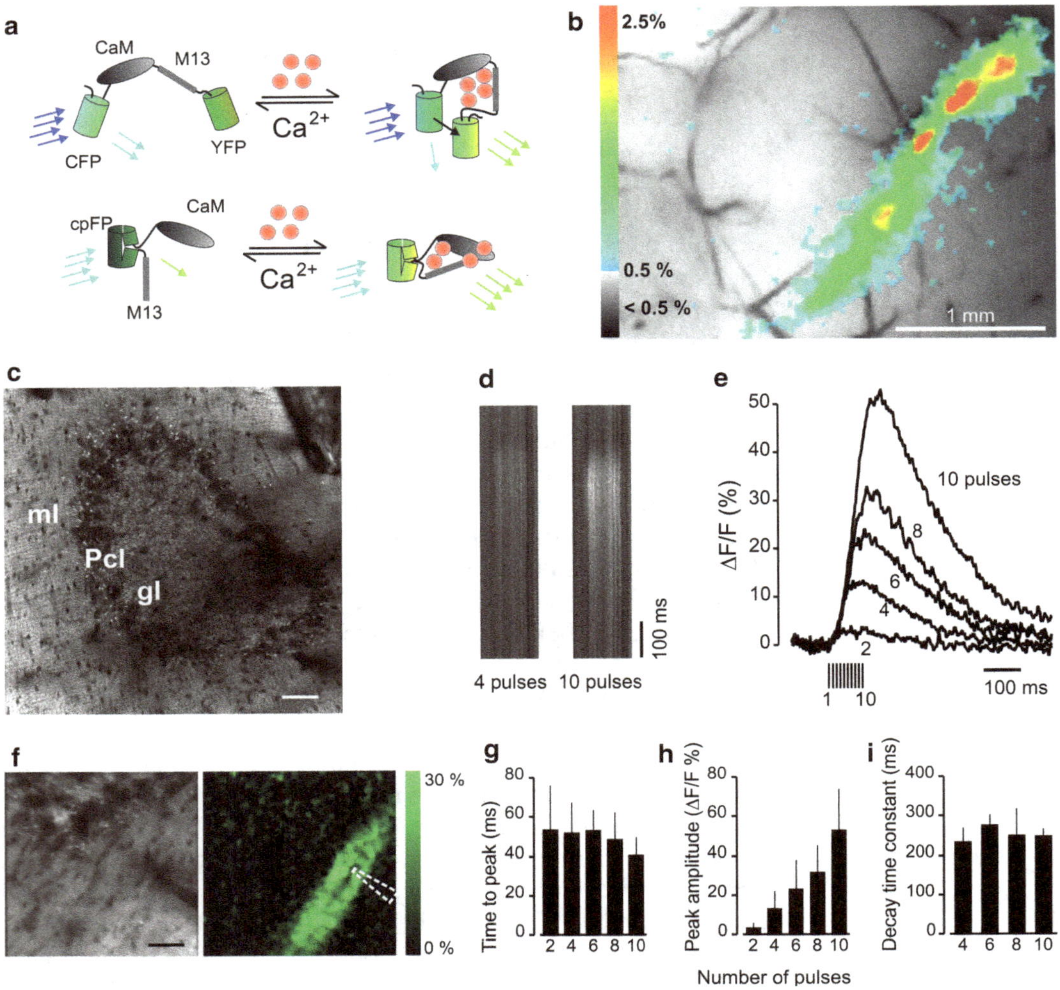

Fig. 1 Genetically encoded Ca^{2+} indicators and their application in vivo. (**a**) Schematic depiction of Ca^{2+} indicators using FRET (fluorescence resonance energy transfer) or single fluorescent protein reporters. *Upper row*: FRET-based calcium indicators such as cameleons often use cyan (CFP) and yellow (YFP) fluorescent proteins. Excitation of CFP with 440 nm light (*blue arrows*) results in cyan (*cyan arrows*) and (via FRET) yellow fluorescence (*yellow arrows*). Binding of Ca^{2+} to the calmodulin (CaM)-M13 complex increases the efficacy of energy transfer between cyan and yellow fluorescent proteins. *Lower row*: Single fluorescent protein indicators of the GCaMP family incorporate a circularly permuted fluorescent protein (cpFP). Binding of Ca^{2+} to CaM-M13 increases the quantum yield of cpFP fluorescence. (**b**) Ca^{2+} transients in cerebellar parallel fibers of a transgenic mouse expressing GCaMP2 obtained by transcranial whole-field epifluorescence imaging. The Ca^{2+} signal (color-coded map of $\Delta F/F_0$) was induced by electrical micro-stimulation of parallel fibers. (**c**) In vivo two-photon image of a GCaMP-transgenic mouse of same lineage showing the three principal cortical layers (*gl* granule cell layer, *Pcl* Purkinje cell layer, *ml* molecular layer) 120 μm below the cortical surface. Note that the fluorescence is generated by GCaMP2 expressed in the granule cells and their axons, whereas other cell types are non-expressing and therefore appear dark. (**d**) Two-photon line scans across the beam-shaped responsive areas induced by electrical stimulation in the molecular layer of area crus I/II in a GCaMP2-transgenic mouse. (**e**) Time course of fluorescence transients induced by delivering 2–10 stimulation pulses (0.2 ms duration) at 100 Hz as derived from two-photon line scans. (**f**) Two-photon fluorescence imaging of

folding efficacy of the fluorescent proteins as well as the Ca^{2+} affinity and dynamic range of the sensing mechanisms has subsequently yielded improved variants such as GCaMP2 and GCaMP3 and variants covering different spectral bands [16, 26, 35]. Figure 1b illustrates the characterization of signals than can be obtained from a transgenic mouse line that expresses GCaMP2 [16, 17, 36].

There are important differences in the reporting characteristics of organic indicators and GECIs. All commonly used organic indicators have a single Ca^{2+} binding site. Their response kinetics can therefore easily be derived from a simple reversible binding reaction with an equilibrium dissociation constant (K_d = unbinding rate/binding rate). Since the binding rate is diffusion-limited for most dyes, variations in K_d across different indicators are paralleled by proportional variations in unbinding rates. Therefore, high affinity dyes produce slow signals while low affinity dyes give fast signals. Unfortunately, the fast kinetics of low affinity dyes goes with smaller optical signals (because only a small fraction of the dye molecules are driven in the calcium-bound state if calcium concentration is much below the K_d). Moreover, this cannot be compensated by increasing dye concentration because the amount of calcium available for binding is limited. Finally, the fluorescence changes of organic calcium indicators follow approximately a linear relationship with Ca^{2+} concentration for values up to K_d. GECIs do not overcome these biochemical principles but add several complications. Firstly, most GECIs have several Ca^{2+} binding sites (Hill coefficient >1, see Table 1) and therefore the fluorescence–[Ca^{2+}] relationship is highly nonlinear. Moreover, Ca^{2+} binding to GECIs is not diffusion-limited, causing additional dampening of the kinetics of the optical signal. Therefore, closely spaced action potentials (>10 Hz) are difficult to resolve directly from the shape of the calcium transient and the number of action potentials in a burst response needs to be deduced from the amplitude of the cumulative response.

Fluorescence at resting calcium concentration (determined by the fluorescence in Ca^{2+}-free state and occupancy at resting [Ca^{2+}]) is another important selection criterion for calcium indicators. Low baseline fluorescence is advantageous because it increases the contrast of activity-related signals against background fluorescence and decreases undesirable out-of-focus light absorption. As a drawback, low resting fluorescence causes difficulties in quantifying sig-

Fig. 1 (continued) Ca^{2+} transients in parallel fibers. Electrical stimulation (10 pulses of 0.2-ms duration at 100 Hz) was delivered via an electrode placed in the molecular layer in proximity to the Purkinje cell layer. *Left*: image of basal fluorescence indicating the position of the stimulus electrode. *Right*: fluorescence response ($\Delta F/F_0$, average of six trials recorded at 30-s intervals) at the time of the response maximum. Position of electrode tip is indicated as dotted line. (**g–i**) Histograms summarizing time to peak (**g**, measured from the time of the last stimulation pulse), peak amplitude (**h**), and decay time constants (**i**, mean ± SD from two animals). Remounted data in panels (**b–i**) taken from the work of Diez-Garcia et al. [17].

Table 1
Overview of calcium-sensitive probes for in vivo detection of action potentials in neocortical neurons.

	Biochemical characterization				Optical signals associated with neuronal action potentials in brain slices or living rodents at typical concentrations of the indicator		
Indicator	**Components**	**K_d (μM)**	**Hill slope**	**Max($\Delta F/F$) or Max($\Delta R/R$)**	**Signal size ($\Delta F/F$ or $\Delta R/R$)**	**Rise and decay times**	**References**
Organic calcium probes							
Fura-2	Stilbene-BAPTA	0.14[a]	~1[a]	>30[a]	~5 % (1 s AP train) 2–3 %/1 AP *b.sl.*		[10, 29, 37]
Fluo-4	Fluorescein-BAPTA	0.34[a]	~1[a]	>30[a]	16 ± 5 %/1 AP	On: <25 ms Off: ~600 ms	[38]
Oregon green 488 BAPTA-1	Fluorescein-BAPTA	0.17[a]	0.97[a]	~13[a]	10 ± 1 %/1 AP	On: <25 ms Off: 700 ± 210 ms	[39]
Genetically encoded probes							
GCaMP2	M13-cpGFP-CaM	0.66	3.8	~4	3.2 ± 2.2 %/2 AP 3.6 ± 2.3 %/1AP *b.sl.*	On: 30 ± 10 ms Off: 160 ± 40 ms	[17, 26, 40]
GCaMP3	M13-cpGFP-CaM	0.84	n/a	~11	7.9 ± 2.8 %/1 AP 14 %/1 AP *b.sl.*	On: <100 ms Off: 385 ± 80 ms	[26, 27]
Yellow Cameleon 3.6	CFP-CaM-M13-cpYFP	0.25	3.6	~14	2.0 ± 0.1 %/1 AP ~4 %/1 AP *b.sl.*	On: < 100 ms Off: 540 ± 40 ms	[18, 41, 42]
D3cpv	CFP-CaM-M13-cpYFP	0.6	0.76	~5	3.5 ± 0.2 %/1 AP 5.0 ± 3 %/1 AP *b.sl.*	On: <100 ms Off: 1,200 ± 110 ms	[25, 31] [26]
TN-XXL	CFP-Trop-YFP	0.8	1.5	~2.3	~7.5 % (4 s AP train) 4.2 ± 2 %/1 AP *b.sl.*	On: n/a Off: ~610 ms	[26, 43]

AP action potential, *On* rise time from half maximal to maximal signal amplitude, *Off* decay time from maximal to half maximal signal amplitude, *BAPTA* bis-aminophenoxyethane-tetraacetic acid, *Trop* chicken skeletal muscle troponin C, *CFP, GFP, and YFP* cyan, green, and yellow variants of jellyfish green fluorescent protein, *cpGFP* circularly permuted GFP, *cpYFP* circularly permuted YFP, *CaM* calmodulin, *M13* CaM-binding peptide M13, *b.sl.*, brain slice.
[a]Taken from: The MolecularProbes® Handbook, 11th Edition, Invitrogen, 2011.

nals that are normalized with regard to a noisy baseline. This issue can, however, be resolved by adding a reference dye with a color different from that of the indicator dye. Moreover, low baseline fluorescence implies that cells cannot be visualized and identified unless they show activity.

Table 1 summarizes key parameters of widely used organic indicators and GECIs.

3 Voltage Indicators

Since the discovery of dyes that transduce membrane potential changes into optical signals in the mid-1970s [44], efforts by several groups led to a series of fluorescent potentiometric dyes that have been proven powerful tools to image cortical circuit dynamics. The most advanced and widely used VSDs for in vivo imaging are "blue dyes." They are named "blue" because they absorb red light (hence they are blue when observed in transmission or reflectance). Their emission in the far red solved a major issue with earlier shorter wavelength "red" dyes whose absorption spectrum overlapped with that of hemoglobin, causing large heartbeat and oxygenation level-dependent components in the optical signals. Blue dyes include oxonols (RH-1691, RH-1692, and RH-1838; see Chapter 8 in this volume), [45–47] while red dyes comprise many dyes of the hemicyanine and styryl family of dyes [48–51]. VSDs partition into membranes of living cells and rapidly (within a microsecond) alter the intensity and/or wavelengths of their fluorescence as a function of membrane potential. Their fluorescence–voltage relationship is linear. While baseline-normalized fluorescence changes ($\Delta F/F$ values) amounting to several ten percent per 100 mV have been reported in isolated membranes, the responses observed in vivo are one or two orders of magnitude smaller because only a minor fraction of the dye molecules associate with activated cells while the additional staining produces large background and shot noise (see term $(1 - f_B)$ in Eq. 1 of Box 1).

Genetically encoded optical voltage reporter proteins have considerably evolved over the last 15 years. Initial design concepts were based on the exploitation of voltage-dependent conformational changes associated with voltage-gated ion channels (FlaSh [23]; SPARC [52]) or their isolated voltage sensor domain (VSFP1 [53]), resulting in modulation of the reporter protein fluorescence intensity. The term VSFP (Voltage Sensing Fluorescent Protein) is now commonly used for the GEVIs that are based on an isolated voltage sensor domain. More recently, use of the voltage sensing domain of the non-ion channel protein, *Ciona intestinalis* voltage sensor-containing phosphatase (Ci-VSP), gave rise to VSFPs with improved targeting to the plasma membrane and reliable responsiveness to membrane potential signaling of cortical pyramidal cells in living mice [27, 51–55]. Figure 2 gives an overview on the different ways FPs are combined with the voltage sensor domain.

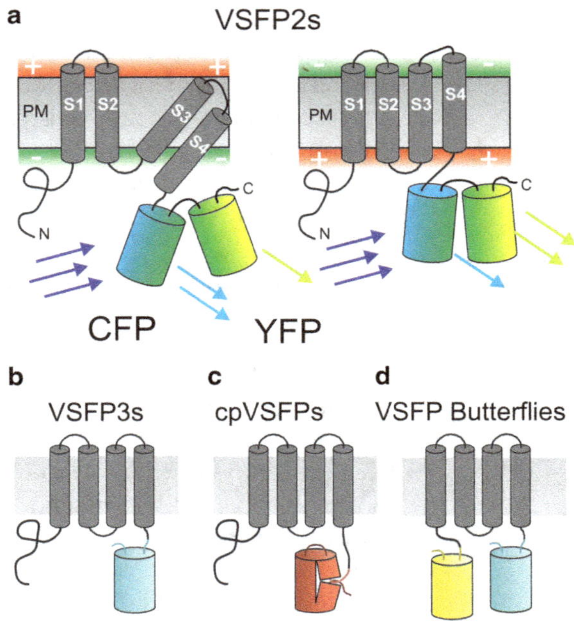

Fig. 2 Schematic design of VSFP probes derived from a combination of a voltage sensor domain and fluorescent proteins. (**a**) VSFP2s are FRET-based voltage sensors comprising an acceptor (e.g., Citrine) and donor (e.g., Cerulean) fluorophore fused to the fourth transmembrane segment (S4) of the voltage sensing domain (S1–S4) of Ci-VSP. The efficacy of FRET increases upon membrane depolarization as the voltage sensor adopts its depolarized state. (**b**) VSFP3s involve single FPs. (**c**) cpVSFPs are based on circularly permuted fluorescent proteins (e.g., mKate). (**d**) In VSFP Butterflies, the voltage sensing domain is sandwiched between a fluorescent protein FRET pair.

FRET-based VSFPs are at present the only GEVIs that have shown to provide reliable signals from cortical cells in vivo [28]. A detailed description of the development and characterization of VSFPs has been presented elsewhere [54]. Here we describe the features that are most relevant for voltage imaging in vivo. The kinetics of voltage sensing domain-based indicators can be explained by voltage-dependent state transitions using formalisms developed for voltage-gated ion channels. It is therefore not surprising that the fluorescence–voltage curves resemble the sigmoid activation curves of ion channels and that response time constants are voltage-dependent. In analogy to the functional specialization of ion channels, VSFPs can be tuned to suit a specific range of neurophysiological applications. The sigmoid shape of fluorescence–voltage relationship of VSFPs implies that the sensitivity is maximal at the half-activation voltage ($V_{1/2}$) [55]. Therefore, VSFPs with $V_{1/2}$ values close to the resting membrane potential of the cells of interest are particularly sensitive to subthreshold synaptic potentials. Conversely, VSFPs with $V_{1/2}$ values at more depolarized voltages and fast response kinetics are better suited to report action potentials. Table 2 summarizes key parameters of widely used organic indicators and GEVIs.

Table 2
Overview of voltage-sensitive probes for in vivo population imaging in neocortex.

Indicators	Chromophore	Biophysical properties				Neuronal responses to sensory stimuli in living rodents	References
Organic dyes (VSDs)		*Sensitivity*[a] *($\Delta F/F$ per 100 mV)*	*Linearity*		*Response time*	*$\Delta F/F$ or $\Delta R/R$*	
RH-1691 RH-1692 RH-1838	Oxonol	10…25 %	Linear		<10 µs	~0.02…1 %	[45, 47, 56, 57]
RH-795 JPW-3028 JPW-5029	Styryl-pyridinium	10…25 %	Linear		<10 µs	~0.005…0.05 %	[50, 58, 59]
Annine-6plus	Annellated benzene-pyridinium	15…35 %	Weakly nonlinear		<10 µs	~0.2…0.7 %	[60–62]
Genetically encoded probes (VSFPs)		*Sensitivity*[a] *($\Delta R/R$ per 100 mV at $V_{1/2}$)*	*$V_{1/2}$*	*On time constant*	*Off time constant*	*$\Delta R/R$*	
VSFP2.3	mCerulean mCitrine	~19 %	−50 mV	3.3 (56)[b] ms	n/a (77)[c] ms	~0.2…0.5 %	[28, 63, 64]
VSFP2.42	mCitrine mKate2	~18 %	−55 mV	2.5 (57)[b] ms	n/a (75)[c] ms	~0.2…0.5 %	[28, 64]
Mermaid	mUKG mKO	~18 %	−44 mV	2.5 (57)[b] ms	n/a (73)[c] ms	None detected	[64, 65]
VSFP Butterfly 1.2	mCitrine mKate2	~22 %	−79 mV	2.6 (44)[b] ms	7.2 (64)[c] ms	~0.5…1 %	[76]

[a]Measured in cultured cells with minimal fluorescence background.
[b]Fast (slow) ON time constants at −20 mV (35 °C).
[c]Fast (slow) OFF time constant at −70 mV (35 °C).

4 Cortical Circuit Dynamics Imaged with VSFPs In Vivo

Over the last 20 years, voltage imaging facilitated the analysis of neuronal population dynamics in the cerebral cortex at higher temporal and spatial resolution than other methods available at the mesoscopic (covering millimeters of cortical space) scale [45, 47, 58, 59, 66–69]. To establish the performance of VSFP2s for this experimental approach, we selected the mouse somatosensory cortex as a model following earlier work using organic VSD-based functional imaging [56, 57, 68–71].

VSFPs can be expressed in specific cell types using a variety of gene transfer approaches including viral gene delivery, transgenesis, and in utero electroporation of VSFP plasmids (unpublished data). In our lab, we have mainly used the latter approach as it allows for a relatively fast generation of experimental preparations expressing new indicator versions. For in vivo imaging experiments, we prepared an optically transparent brain window over the primary somatosensory cortex of in utero-electroporated mice. The procedure essentially follows the description published by the Kleinfeld

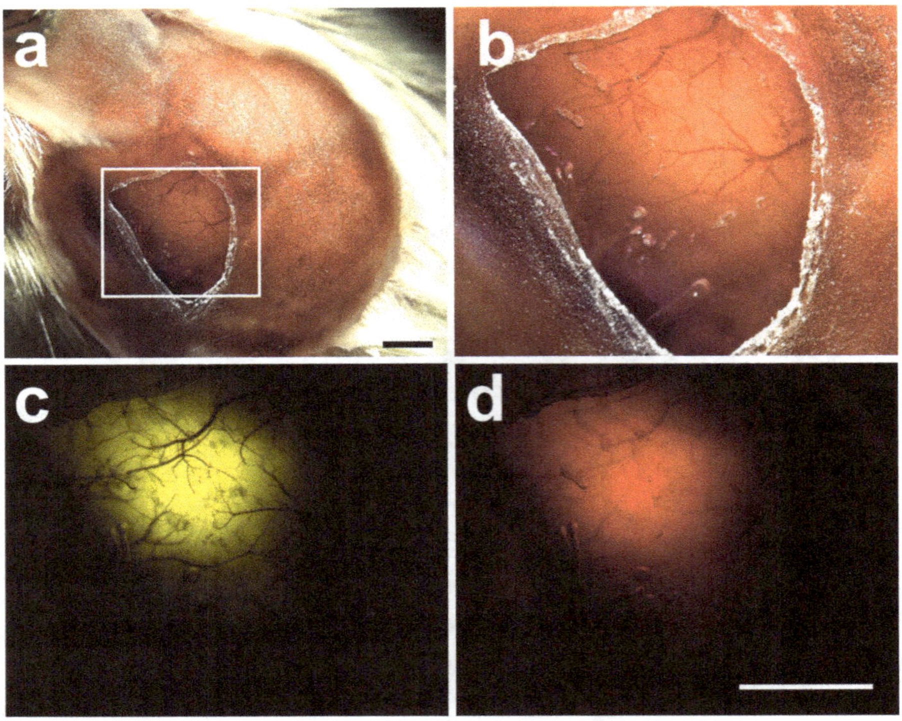

Fig. 3 VSFP imaging through a thinned-skull cranial window in a living mouse. (**a**) View of a mouse head after mounting a cover glass onto the thinned cranial bone. The boxed area is shown at higher magnification in (**b**). (**c**, **d**) Fluorescence images captured through the glass window showing the mouse cortex expressing a VSFP variant with a *yellow–red* (mCitrine/mkate2) FRET reporter. Scale bars, 2 mm.

group [72] that includes mounting a coverslip on the thinned cranial bone (Fig. 3). As this approach does not necessitate craniotomy, it maintains the intracranial pressure that stabilizes the brain, thereby helping to minimize motion-induced imaging artifacts.

In deeply anesthetized mice, a single deflection of the C1 whisker induced ratiometric responses in the donor and acceptor channels (Fig. 4) over an area of the approximate size of the C1 barrel field as indicated by optical imaging of intrinsic signals. Notably, these signals can be resolved at the level of single trials. Depending on the level of anesthesia, these signals expanded over a larger region or triggered a wave of excitation that can spread over considerable distances.

From the methodological viewpoint, it should be noted that the emission wavelengths of current VSFPs overlap with hemoglobin absorption. Therefore, tissue absorption of VSFP fluorescence fluctuates with each heartbeat and this causes a modulation of the fluorescence signal. Fortunately, this effect can be compensated for in dual emission recordings using FRET-based VSFPs because the

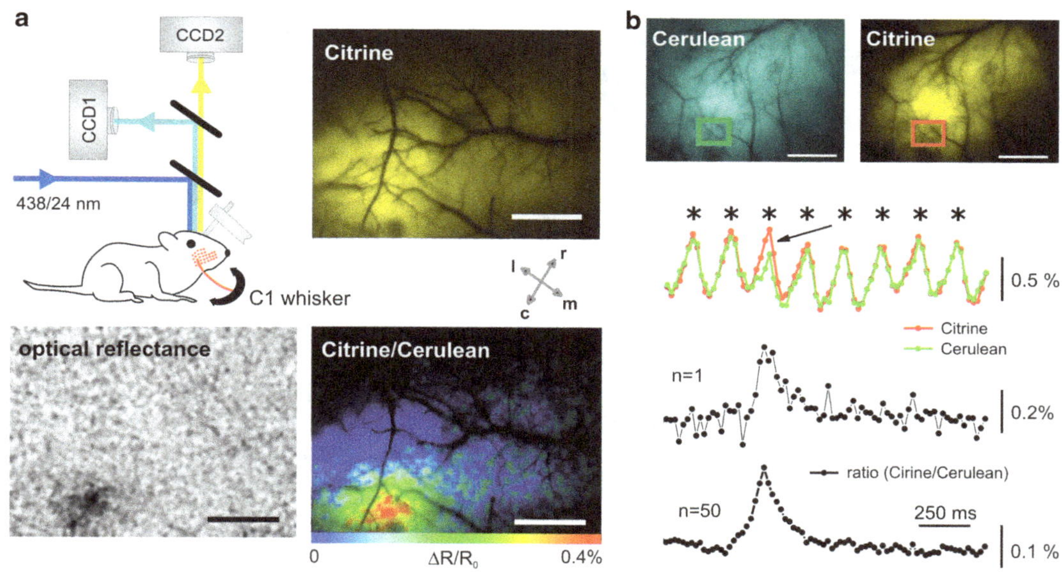

Fig. 4 VSFP2.3 imaging in the somatosensory cortex in vivo. (**a**) *Upper left*: schematic diagram of the experimental configuration for imaging donor and acceptor fluorescence with two synchronized CCD cameras in living animals. *Upper right*: example of Citrine fluorescence image captured with CCD2. *Lower left*: optical reflectance signal (optical imaging of intrinsic signal) evoked by repetitive C1 whisker stimulation. *Lower right*: ratiometric VSFP2.3 response ($\Delta R/R_0$) to a single whisker deflection at the time of the maximal response. *Arrows* indicate the rostral (R), caudal (C), lateral (L), and medial (M) directions. Scale bars, 1 mm. Adapted from [28]. (**b**) Modulation of VSFP signals at heartbeat frequency is strongly reduced in the ratio trace. *Upper row* shows fluorescence images in the cyan and yellow channels. The signal in both emission channels (obtained from the region of interest indicated by *rectangles* in the fluorescence images) is modulated in the rhythm of the heartbeat (*asterisk*). The voltage signal (induced as in (**a**)) is seen as a small deviation from this correlated modulation (*arrow*) while the ratio of yellow over cyan fluorescence is little affected by the heartbeat (middle trace shows the ratio of a single trial, and lower trace the mean ratio of 50 trials).

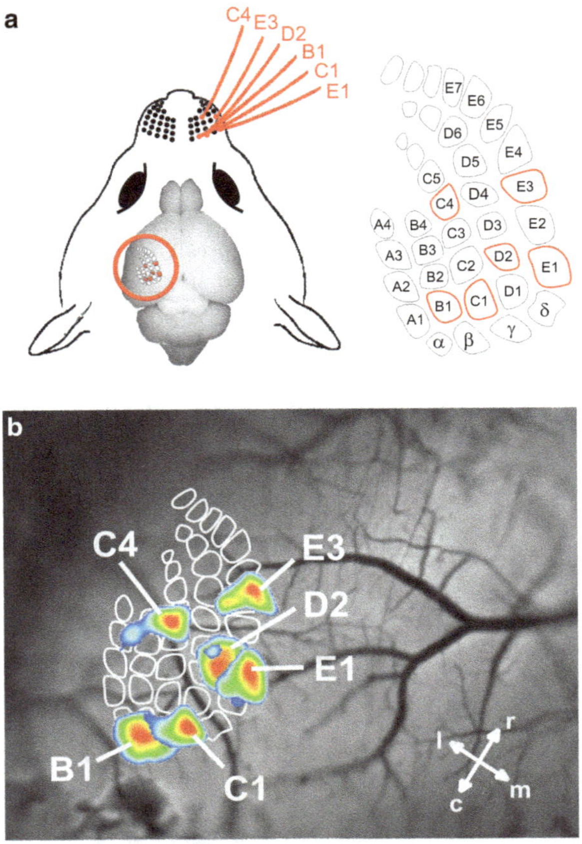

Fig. 5 Construction of receptive field maps with VSFP2.3 imaging in living mice. (**a**) *Left panel*: selection of six whiskers (C4, E3, D2, E1, B1, and C1) for sequential stimulation and receptive field imaging. *Right panel*: identity of selected whiskers in a schematic of the cortical barrel field of mouse somatosensory cortex. (**b**) Composite image showing the barrel map together with the responsive areas from six stimulated whiskers acquired from the same VSFP2.3-expressing mouse. Responsive areas are shown in color. Arrows indicate the rostral (R), caudal (C), lateral (L), and medial (M) directions. Scale bar, 1 mm. Data are from [28].

acceptor and donor fluorescence change in opposite direction in response to membrane voltage changes (i.e., the acceptor emission increases when the donor emission decreases) whereas the emission of the donor and the acceptor changes in the same direction in response to variations of tissue absorption. Movement and heartbeat-related artifacts therefore largely cancel out when forming the ratio between the two color channels (Fig. 4b) [28]. This approach, in turn, requires splitting the VSFP fluorescence emission onto two imaging devices, one for each color channel. While there are optical devices that can split and project two color channels on a single imaging sensor, we found that this approach results in large aperture losses in combination with low magnification objectives used for imaging over millimeters of cortical space. A configuration that is more efficient in light sampling is an expansion of the classical tandem lens configuration into a dual emission port configuration.

VSFP2 imaging enabled the construction of maps of cortical representation of different whiskers in the same mouse with receptive areas mainly overlapping with the cortical barrel field representation (Fig. 5). Notably, VSFP imaging along with a preparation as illustrated in Fig. 3 can be conducted over weeks and even months. It is therefore suitable for studying developmental and experience-dependent plasticity in living animals.

5 Concluding Remarks

Both calcium and voltage imaging using either organic or genetically encoded fluorescent indicators have provided fruitful approaches to foster our understanding of cortical circuit dynamics. In this chapter, we first provided an overview of different calcium and voltage indicators for an informed choice of the most suitable methodological approach. We concluded the chapter with a description of voltage imaging using genetically encoded probes that is well suited for imaging circuit dynamics over millimeters of cortical space while taking into account cell type specificity. In the future, this type of voltage imaging might be expanded to the scale of single-cell resolution using two-photon laser microscopy either in scanning [60, 73] or in whole-field mode [74, 75]. We like to argue that along with the ability to target specific genetically defined cell populations, VSFPs open new avenues for the study of interaction dynamics of neuronal assemblies. This will facilitate the investigation of fundamental questions related to the information-processing mechanisms of the brain, such as the circuit operations involved in sensing our environment and generation of body movements, but will also be applicable to directly visualize cognitive functions.

References

1. Hill DK, Keynes RD (1949) Opacity changes in stimulated nerve. J Physiol 108:278–281
2. Cohen LB, Keynes RD, Hille B (1968) Light scattering and birefringence changes during nerve activity. Nature 218:438–441
3. Ross WN, Salzberg BM, Cohen LB, Davila HV (1974) A large change in dye absorption during the action potential. Biophys J 14:983–986
4. Grinvald A, Salzberg BM, Cohen LB (1977) Simultaneous recording from several neurones in an invertebrate central nervous system. Nature 268:140–142
5. Knöpfel T, Diez-Garcia J, Akemann W (2006) Optical probing of neuronal circuit dynamics: genetically encoded versus classical fluorescent sensors. Trends Neurosci 29:160–166

6. Knöpfel T, Fromherz P (1987) Fluorescence monitoring of membrane potentials: the spatio-temporal resolution in isolated neurons of *Helix pomatia*. Z Naturforsch C 42:986–990
7. Cossart R, Ikegaya Y, Yuste R (2005) Calcium imaging of cortical networks dynamics. Cell Calcium 37:451–457
8. Ikegaya Y, Aaron G, Cossart R, Aronov D, Lampl I, Ferster D, Yuste R (2004) Synfire chains and cortical songs: temporal modules of cortical activity. Science 304:559–564
9. Mao BQ, Hamzei-Sichani F, Aronov D, Froemke RC, Yuste R (2001) Dynamics of spontaneous activity in neocortical slices. Neuron 32:883–898
10. Smetters D, Majewska A, Yuste R (1999) Detecting action potentials in neuronal populations with calcium imaging. Methods 18: 215–221
11. Knöpfel T, Gähwiler BH (1992) Activity-induced elevations of intracellular calcium concentration in pyramidal and nonpyramidal cells of the CA3 region of rat hippocampal slice cultures. J Neurophysiol 68:961–963
12. Grewe BF, Langer D, Kasper H, Kampa BM, Helmchen F (2010) High-speed in vivo calcium imaging reveals neuronal network activity with near-millisecond precision. Nat Methods 7:399–405
13. Mennerick S, Chisari M, Shu HJ, Taylor A, Vasek M, Eisenman LN, Zorumski CF (2010) Diverse voltage-sensitive dyes modulate GABAA receptor function. J Neurosci 30:2871–2879
14. Hofer SB, Ko H, Pichler B, Vogelstein J, Ros H, Zeng H, Lein E, Lesica NA, Mrsic-Flogel TD (2011) Differential connectivity and response dynamics of excitatory and inhibitory neurons in visual cortex. Nat Neurosci 14:1045–1052
15. Zariwala HA, Madisen L, Ahrens KF, Bernard A, Lein ES, Jones AR, Zeng H (2011) Visual tuning properties of genetically identified layer 2/3 neuronal types in the primary visual cortex of cre-transgenic mice. Front Syst Neurosci 4:162
16. Diez-Garcia J, Matsushita S, Mutoh H, Nakai J, Ohkura M, Yokoyama J, Dimitrov D, Knöpfel T (2005) Activation of cerebellar parallel fibers monitored in transgenic mice expressing a fluorescent Ca^{2+} indicator protein. Eur J Neurosci 22:627–635
17. Diez-Garcia J, Akemann W, Knöpfel T (2007) In vivo calcium imaging from genetically specified target cells in mouse cerebellum. Neuroimage 34:859–869
18. Lütcke H, Murayama M, Hahn T, Margolis DJ, Astori S, Zum Alten Borgloh SM, Göbel W, Yang Y, Tang W, Kügler S, Sprengel R, Nagai T, Miyawaki A, Larkum ME, Helmchen F, Hasan MT (2010) Optical recording of neuronal activity with a genetically-encoded calcium indicator in anesthetized and freely moving mice. Front Neural Circuits 4:9
19. Miesenböck G, Rothman JE (1997) Patterns of synaptic activity in neural networks recorded by light emission from synaptolucins. Proc Natl Acad Sci USA 94:3402–3407
20. Miesenböck G, De Angelis DA, Rothman JE (1998) Visualizing secretion and synaptic transmission with pH-sensitive green fluorescent proteins. Nature 394:192–195
21. Miyawaki A, Llopis J, Heim R, McCaffery JM, Adams JA, Ikura M, Tsien RY (1997) Fluorescent indicators for Ca^{2+} based on green fluorescent proteins and calmodulin. Nature 388:882–887
22. Romoser VA, Hinkle PM, Persechini A (1997) Novel fluorescent indicator proteins for monitoring free intracellular Ca^{2+}. J Biol Chem 272:13270–13274
23. Siegel MS, Isacoff EY (1997) A genetically encoded optical probe of membrane voltage. Neuron 19:735–741
24. Hasan MT, Friedrich RW, Euler T, Larkum ME, Giese G, Both M, Duebel J, Waters J, Bujard H, Griesbeck O, Tsien RY, Nagai T, Miyawaki A, Denk W (2004) Functional fluorescent Ca^{2+} indicator proteins in transgenic mice under TET control. PLoS Biol 2:e163
25. Wallace DJ, Meyer zum Alten BS, Astori S, Yang Y, Bausen M, Kügler S, Palmer AE, Tsien RY, Sprengel R, Kerr JN, Denk W, Hasan MT (2008) Single-spike detection in vitro and in vivo with a genetic Ca^{2+} sensor. Nat Methods 5:797–804
26. Tian L, Hires SA, Mao T, Huber D, Chiappe ME, Chalasani SH, Petreanu L, Akerboom J, McKinney SA, Schreiter ER, Bargmann CI, Jayaraman V, Svoboda K, Looger LL (2009) Imaging neural activity in worms, flies and mice with improved GCaMP calcium indicators. Nat Methods 6:875–881
27. Dombeck DA, Harvey CD, Tian L, Looger LL, Tank DW (2010) Functional imaging of hippocampal place cells at cellular resolution during virtual navigation. Nat Neurosci 13:1433–1440
28. Akemann W, Mutoh H, Perron A, Rossier J, Knöpfel T (2010) Imaging brain electric signals with genetically targeted voltage-sensitive fluorescent proteins. Nat Methods 7:643–649
29. Grynkiewicz G, Poenie M, Tsien RY (1985) A new generation of Ca^{2+} indicators with greatly improved fluorescence properties. J Biol Chem 260:3440–3450

30. Heim N, Griesbeck O (2004) Genetically encoded indicators of cellular calcium dynamics based on troponin C and green fluorescent protein. J Biol Chem 279:14280–14286
31. Palmer AE, Giacomello M, Kortemme T, Hires SA, Lev-Ram V, Baker D, Tsien RY (2006) Ca^{2+} indicators based on computationally redesigned calmodulin-peptide pairs. Chem Biol 13:521–530
32. Baird GS, Zacharias DA, Tsien RY (1999) Circular permutation and receptor insertion within green fluorescent proteins. Proc Natl Acad Sci USA 96:11241–11246
33. Akemann W, Raj CD, Knöpfel T (2001) Functional characterization of permuted enhanced green fluorescent proteins comprising varying linker peptides. Photochem Photobiol 74:356–363
34. Nakai J, Ohkura M, Imoto K (2001) A high signal-to-noise Ca^{2+} probe composed of a single green fluorescent protein. Nat Biotechnol 19:137–141
35. Zhao Y, Araki S, Wu J, Teramoto T, Chang YF, Nakano M, Abdelfattah AS, Fujiwara M, Ishihara T, Nagai T, Campbell RE (2011) An expanded palette of genetically encoded Ca^{2+} indicators. Science 333:1888–1891
36. Fletcher ML, Masurkar AV, Xing J, Imamura F, Xiong W, Nagayama S, Mutoh H, Greer CA, Knöpfel T, Chen WR (2009) Optical imaging of postsynaptic odor representation in the glomerular layer of the mouse olfactory bulb. J Neurophysiol 102:817–830
37. Sobel EC, Tank DW (1994) In vivo Ca^{2+} dynamics in a cricket auditory neuron: an example of chemical computation. Science 263:823–826
38. Sato TR, Gray NW, Mainen ZF, Svoboda K (2007) The functional microarchitecture of the mouse barrel cortex. PLoS Biol 5:e189
39. Kerr JN, Greenberg D, Helmchen F (2005) Imaging input and output of neocortical networks in vivo. Proc Natl Acad Sci USA 102:14063–14068
40. Tallini YN, Ohkura M, Choi BR, Ji G, Imoto K, Doran R, Lee J, Plan P, Wilson J, Xin HB, Sanbe A, Gulick J, Mathai J, Robbins J, Salama G, Nakai J, Kotlikoff MI (2006) Imaging cellular signals in the heart in vivo: cardiac expression of the high-signal Ca^{2+} indicator GCaMP2. Proc Natl Acad Sci USA 103:4753–4758
41. Nagai T, Yamada S, Tominaga T, Ichikawa M, Miyawaki A (2004) Expanded dynamic range of fluorescent indicators for Ca^{2+} by circularly permuted yellow fluorescent proteins. Proc Natl Acad Sci USA 101: 10554–10559
42. Horikawa K, Yamada Y, Matsuda T, Kobayashi K, Hashimoto M, Matsu-ura T, Miyawaki A, Michikawa T, Mikoshiba K, Nagai T (2010) Spontaneous network activity visualized by ultrasensitive Ca(2+) indicators, yellow Cameleon-Nano. Nat Methods 7:729–732
43. Mank M, Santos AF, Direnberger S, Mrsic-Flogel TD, Hofer SB, Stein V, Hendel T, Reiff DF, Levelt C, Borst A, Bonhoeffer T, Hubener M, Griesbeck O (2008) A genetically encoded calcium indicator for chronic in vivo two-photon imaging. Nat Methods 5:805–811
44. Cohen LB, Salzberg BM, Davila HV, Ross WN, Landowne D, Waggoner AS, Wang CH (1974) Changes in axon fluorescence during activity: molecular probes of membrane potential. J Membr Biol 19:1–36
45. Shoham D, Glaser DE, Arieli A, Kenet T, Wijnbergen C, Toledo Y, Hildesheim R, Grinvald A (1999) Imaging cortical dynamics at high spatial and temporal resolution with novel blue voltage-sensitive dyes. Neuron 24:791–802
46. Spors H, Grinvald A (2002) Spatio-temporal dynamics of odor representations in the mammalian olfactory bulb. Neuron 34:301–315
47. Slovin H, Arieli A, Hildesheim R, Grinvald A (2002) Long-term voltage-sensitive dye imaging reveals cortical dynamics in behaving monkeys. J Neurophysiol 88:3421–3438
48. Gupta RK, Salzberg BM, Grinvald A, Cohen LB, Kamino K, Lesher S, Boyle MB, Waggoner AS, Wang CH (1981) Improvements in optical methods for measuring rapid changes in membrane potential. J Membr Biol 58:123–137
49. Grinvald A, Hildesheim R, Farber IC, Anglister L (1982) Improved fluorescent probes for the measurement of rapid changes in membrane potential. Biophys J 39:301–308
50. Fluhler E, Burnham VG, Loew LM (1985) Spectra, membrane binding, and potentiometric responses of new charge shift probes. Biochemistry 24:5749–5755
51. Wuskell JP, Boudreau D, Wei MD, Jin L, Engl R, Chebolu R, Bullen A, Hoffacker KD, Kerimo J, Cohen LB, Zochowski MR, Loew LM (2006) Synthesis, spectra, delivery and potentiometric responses of new styryl dyes with extended spectral ranges. J Neurosci Methods 151:200–215
52. Ataka K, Pieribone VA (2002) A genetically targetable fluorescent probe of channel gating with rapid kinetics. Biophys J 82:509–516
53. Sakai R, Repunte-Canonigo V, Raj CD, Knöpfel T (2001) Design and characterization of a DNA-encoded, voltage-sensitive fluorescent protein. Eur J Neurosci 13:2314–2318
54. Perron A, Akemann W, Mutoh H, Knöpfel T (2012) Genetically encoded probes for optical imaging of brain electrical activity. Prog Brain Res 196:63–77

55. Dimitrov D, He Y, Mutoh H, Baker BJ, Cohen L, Akemann W, Knöpfel T (2007) Engineering and characterization of an enhanced fluorescent protein voltage sensor. PLoS One 2:e440
56. Berger T, Borgdorff A, Crochet S, Neubauer FB, Lefort S, Fauvet B, Ferezou I, Carleton A, Luscher HR, Petersen CC (2007) Combined voltage and calcium epifluorescence imaging in vitro and in vivo reveals subthreshold and suprathreshold dynamics of mouse barrel cortex. J Neurophysiol 97:3751–3762
57. Ferezou I, Bolea S, Petersen CC (2006) Visualizing the cortical representation of whisker touch: voltage-sensitive dye imaging in freely moving mice. Neuron 50:617–629
58. Arieli A, Shoham D, Hildesheim R, Grinvald A (1995) Coherent spatiotemporal patterns of ongoing activity revealed by real-time optical imaging coupled with single-unit recording in the cat visual cortex. J Neurophysiol 73:2072–2093
59. Orbach HS, Cohen LB, Grinvald A (1985) Optical mapping of electrical activity in rat somatosensory and visual cortex. J Neurosci 5:1886–1895
60. Kuhn B, Denk W, Bruno RM (2008) In vivo two-photon voltage-sensitive dye imaging reveals top-down control of cortical layers 1 and 2 during wakefulness. Proc Natl Acad Sci USA 105:7588–7593
61. Kuhn B, Fromherz P, Denk W (2004) High sensitivity of Stark-shift voltage-sensing dyes by one- or two-photon excitation near the red spectral edge. Biophys J 87:631–639
62. Fromherz P, Hubener G, Kuhn B, Hinner MJ (2008) ANNINE-6plus, a voltage-sensitive dye with good solubility, strong membrane binding and high sensitivity. Eur Biophys J 37: 509–514
63. Lundby A, Mutoh H, Dimitrov D, Akemann W, Knöpfel T (2008) Engineering of a genetically encodable fluorescent voltage sensor exploiting fast Ci-VSP voltage-sensing movements. PLoS One 3:e2514
64. Mutoh H, Perron A, Dimitrov D, Iwamoto Y, Akemann W, Chudakov DM, Knopfel T (2009) Spectrally-resolved response properties of the three most advanced FRET based fluorescent protein voltage probes. PLoS One 4: e4555
65. Tsutsui H, Karasawa S, Okamura Y, Miyawaki A (2008) Improving membrane voltage measurements using FRET with new fluorescent proteins. Nat Methods 5:683–685
66. Arieli A, Sterkin A, Grinvald A, Aertsen A (1996) Dynamics of ongoing activity: explanation of the large variability in evoked cortical responses. Science 273:1868–1871
67. Grinvald A, Hildesheim R (2004) VSDI: a new era in functional imaging of cortical dynamics. Nat Rev Neurosci 5:874–885
68. Petersen CC, Hahn TT, Mehta M, Grinvald A, Sakmann B (2003) Interaction of sensory responses with spontaneous depolarization in layer 2/3 barrel cortex. Proc Natl Acad Sci USA 100:13638–13643
69. Petersen CC, Grinvald A, Sakmann B (2003) Spatiotemporal dynamics of sensory responses in layer 2/3 of rat barrel cortex measured in vivo by voltage-sensitive dye imaging combined with whole-cell voltage recordings and neuron reconstructions. J Neurosci 23:1298–1309
70. Ferezou I, Haiss F, Gentet LJ, Aronoff R, Weber B, Petersen CC (2007) Spatiotemporal dynamics of cortical sensorimotor integration in behaving mice. Neuron 56:907–923
71. Poulet JF, Petersen CC (2008) Internal brain state regulates membrane potential synchrony in barrel cortex of behaving mice. Nature 454:881–885
72. Drew PJ, Shih AY, Driscoll JD, Knutsen PM, Blinder P, Davalos D, Akassoglou K, Tsai PS, Kleinfeld D (2010) Chronic optical access through a polished and reinforced thinned skull. Nat Methods 7:981–984
73. Helmchen F, Denk W (2005) Deep tissue two-photon microscopy. Nat Methods 2:932–940
74. Therrien OD, Aube B, Pages S, Koninck PD, Cote D (2011) Wide-field multiphoton imaging of cellular dynamics in thick tissue by temporal focusing and patterned illumination. Biomed Opt Express 2:696–704
75. Oron D, Papagiakoumou E, Anselmi F, Emiliani V (2012) Two-photon optogenetics. Prog Brain Res 196:119–143
76. Akemann W, Mutoh H, Perron A, Park YK, Iwamoto Y, Knöpfel T (2012) Imaging neural circuit dynamics with a voltage-sensitive fluorescent protein. J Neurophysiol 108: 2323–2337

Chapter 5

Optogenetic Tools for Control of Neural Activity

Lief E. Fenno and Karl Deisseroth

Abstract

Understanding the contribution of defined elements to the dynamics of neural circuitry requires not only control on a time scale relevant to neural activity, but also specificity such that the experimental manipulation leaves other components of the circuit unaltered [1]. Electrical stimulation, while providing precise temporal control, modulates neurons based only on location, regardless of identity. Conversely, pharmacological control manipulates the activity of neurons based on specific properties, but in the absence of precise temporal control and with limited spatial control. The adaptation of light-activated ion channels and pumps to neuroscience has enabled temporally precise manipulation of genetically defined circuit elements and allows collection of causal, rather than correlative, data in describing the function of neural elements in acute slice preparation and anesthetized or awake, behaving animals [2]. The ready integration of these "optogenetic" tools [3] into the standard neuroscience toolbox has opened the door to experimental paradigms and insights that were not previously available. Here, we provide an overview of optogenetic principles followed by a description of currently available tools for control of depolarization, hyperpolarization, and biochemical signaling cascades.

Key words Optogenetics, Channelrhodopsin, Halorhodopsin, Opsin, Neural targeting

1 Introduction

Light is the major source of energy for life on earth, utilized directly or indirectly by all known living organisms in a remarkable diversity of ways. For example, single-cell organisms can utilize light-activated ion pumps and channels for transduction of photons into ionic gradients to generate energy, maintain osmotic homeostasis in high salinity environments, and tune flagella-driven phototaxis or direction to avoid unfavorable situations or seek symbiotic or nutrient-rich situations, among other purposes. These pumps and channels covalently bind a retinal moiety (related to vitamin A, and the same molecule used by human rhodopsin but in a different conformation); this light-sensitive molecule changes shape upon absorbing energy from a photon of the correct wavelength [4]. This physical change in the shape of retinal initiates a molecular

Bruno Weber and Fritjof Helmchen (eds.), *Optical Imaging of Neocortical Dynamics*, Neuromethods, vol. 85,
DOI 10.1007/978-1-62703-785-3_5,

cascade of events within the rhodopsin protein, in which amino acids shift to accommodate the new conformation of retinal. When the opsin protein is a channel (a "channelrhodopsin"), a conducting pore is opened and ions (in this case cations) are able to flow through the channel, until the retinal returns to its ground state via either spontaneous relaxation or is driven there by the absorption of a separate photon of different energy (wavelength) [5–7]. If the opsin protein is a pump (for instance, a "halorhodopsin"), a single photocycle entails the piecemeal movement of ions (in this case, chloride) one position further from the entrance pore and closer to the exit. In contrast to the channelrhodopsins, no conducting pore is opened, and the ion that exits the pump on one side of the membrane is not thought to be the same ion that was taken up on the other side during that particular pumping cycle. It can therefore be said that these pumps move through a photocycle that entails a spatially discontinuous set of ion displacements from position to position within the protein [4, 8, 9]. This intricate choreography of light-sensing and ion transduction unified into a single protein enables the central operating principle of optogenetics [3].

The description of a light-driven proton pump from the archaea *Halobacterium salinarum* in 1971 by Oesterhelt and Stoeckenius was the first report of such a protein [10]. This pump was termed "bacteriorhodopsin" (BR) and is utilized by haloarchaea to drive ATP synthesis via maintenance of a proton gradient, especially in low oxygen [10–12]. The BR structure and photocycle have been extensively characterized over the ensuing decades in many dozens of laboratories and thousands of papers, and this wealth of mutagenic analyses later informed and guided the diversification of optogenetic tools [13–16]. The first halorhodopsin (HR) was characterized from halophilic archaea in 1977 [17] and the first channelrhodopsin (ChR1) was characterized from the single-celled algae *Chlamydomonas reinhardtii* in 2002 [18] followed thereafter by ChR2 in 2003 [19] (Fig. 1).

But it was not until 2005 (3 decades after the characterization of BR) that a microbial opsin was shown to be useful in controlling neurons with light, driving individual action potentials with millisecond resolution (first in cultured mammalian hippocampal neurons; [22]). Though it was known that embryonic and optical tissues contained retinal, a year passed until it was shown that the intact postnatal mammalian brain does not require exogenous retinal supplementation for the function of ChR2 [23], a surprising finding central to the widespread adaptation of optogenetics by neuroscience. Indeed in 2007 the chloride-pumping halorhodopsin from *Natronomonas pharaonis* (NpHR; [20]; Fig. 1) was shown to hyperpolarize membrane potential sufficiently to precisely silence action potentials in response to yellow light, in intact tissue and without added retinoids. That there is sufficient endogenous retinal within mature brain tissues has been consistently

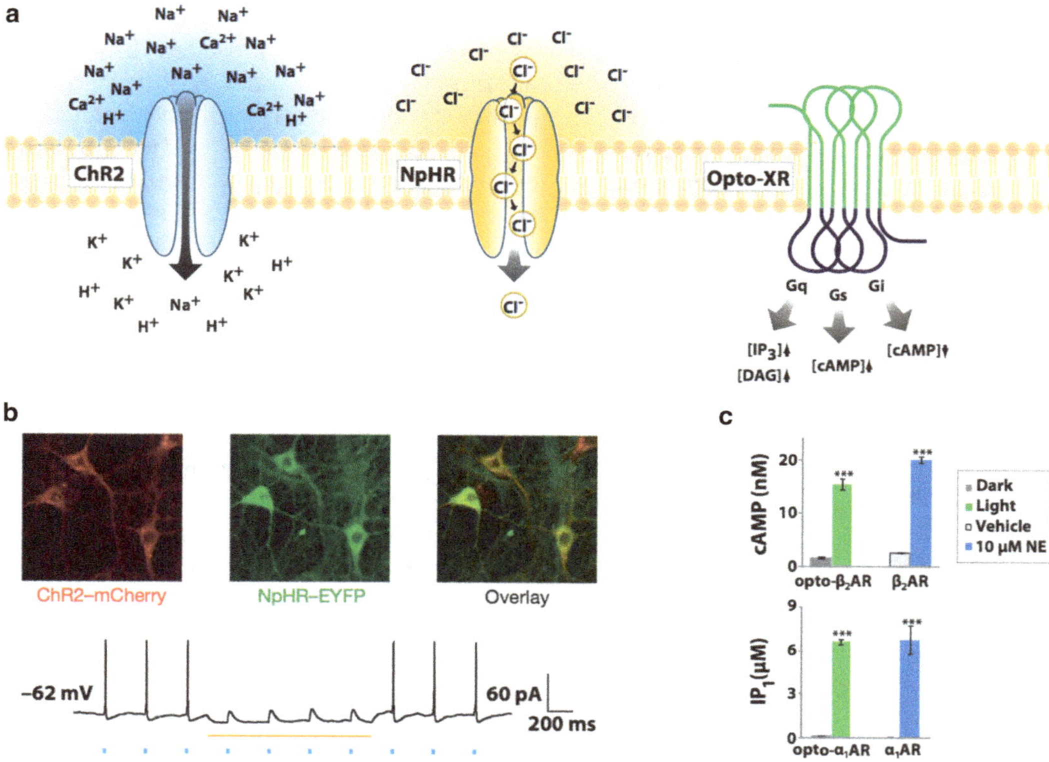

Fig. 1 Optogenetic tool families. (**a**) Channelrhodopsins (*left*), such as ChR2, induce cation photocurrents in response to illumination, resulting in a net depolarization neurons. Halorhodopsins (*middle*) conduct chloride (in the case of NpHR) and bacteriorhodopsin and related opsins conduct protons (not shown) in response to illumination, both hyperpolarizing neurons. The OptoXRs (*right*) are chimeric proteins that elicit biochemical signal cascades in response to illumination. The specific biochemical cascade is determined by the source of the intracellular loops. (**b**) Whole-cell patch clamp response of a neuron expressing both ChR2 and NpHR. Pulses of blue light activate ChR2 and induce sufficient depolarizing photocurrent to produce action potentials whereas yellow light induces an inward chloride current that hyperpolarizes the neuron and prevents firing. (**c**) Response of HEK cells expressing Opto-beta-2 or native beta-2 adrenergic receptor (*top*) or Opto-alpha-1 or native alpha-1 adrenergic receptor (*bottom*). Production of downstream biochemical effectors is similar in magnitude from optical stimulation of Opto-XRs (*left*) and agonist-induced response in native receptors (*right*). Panel (**a**) adapted from [2], panel (**b**) adopted from [20], panel (**c**) adapted from [21].

observed in all but two model organisms (*D. melanogaster* and *C. elegans*, although these animals are easily supplied with sufficient retinal by dietary supplementation; non-neuronal cell types, such as cultured HEK cells also require supplementation [15, 21]).

Mammalian (type II) rhodopsins, like the prokaryotic (type I) relatives, require a retinal derivative to act as a photon antenna for diverse light-modulated roles in vision, the circadian rhythm, and regulation of pigmentation [24, 25]. The similarities between type I and type II rhodopsins are limited, however; while both are 7-transmembrane domain proteins, mammalian rhodopsins are

G protein-coupled receptors (GPCRs) not ion channels and pumps (although type I opsins do function to modulate biochemical signaling cascades; for example, light-activated sensory rhodopsins I and II from *H. salinarum* modulate a phosphorylation cascade that controls flagellar direction [26, 27]). Sequence homology between the two rhodopsin families is virtually nonexistent, although there is significant similarity within families. In type I rhodopsins this homology is particularly interesting in residues defining the ionic environment in the vicinity of bound retinal, while type II rhodopsins follow the general homology patterns of GPCRs. This similarity has been exploited in both families of rhodopsins to engineer chimeras using a modular approach (C1V1 for type I opsins [15] and the optoXRs for type II opsins [21]; Fig. 1).

The advent of optogenetic tools has filled a hole in the neuroscience toolbox [1], allowing, for the first time, neuroscientists to obtain with high temporal precision direct evidence of the sufficiency or necessity of specific events within genetically defined populations of neurons as they relate to the function of neural circuitry and behavior. Below we touch on key principles governing the use of optogenetic tools; while the specifics of many topics are beyond the purview of this chapter, applicable reviews and papers may provide fuller context [2, 28–30].

2 Channelrhodopsins

Since the identification of ChR1 and ChR2 from *Chlamydomonas*, a handful of other light-activated cation channels have been described; these include VChR1 and VChR2 from colonial algae *Volvox carteri* [31], and MChR1 from *Mesostigma viride* [32]. These proteins vary widely in magnitude of induced photocurrent and peak activation wavelength, with activation wavelength profiles generally divided between the blue (~470 nm; ChR1, ChR2, VChR2) and red-shifted (540 nm; VChR1, MChR1; see Fig. 2 for summary). In some cases currents may be augmented by formation of chimeras between these opsins (such as C1V1 [15], ChIEF [33], Green Receiver, and Wide Receiver [34]) or by mutation of a histidine residue homologous to the proton donor in BR (Asp-96) to arginine (in ChR2, this is H134R [35, 36]). For H134R, larger photocurrents are offset by reduced temporal resolution, lowering the maximum rate at which photocurrent-induced action potentials may be generated [28, 35, 36].

The relatively small number of initial genomically identified excitatory channelrhodopsins has now been dwarfed by an ever-increasing portfolio of derivatives designed to manipulate peak wavelength, current magnitude, and inactivation time constant (τ_{off}). τ_{off} is an intrinsic property that describes the time taken for an opsin to spontaneously relax from the conducting state to the nonconducting state after the cessation of light. The τ_{off} of ChR2 is

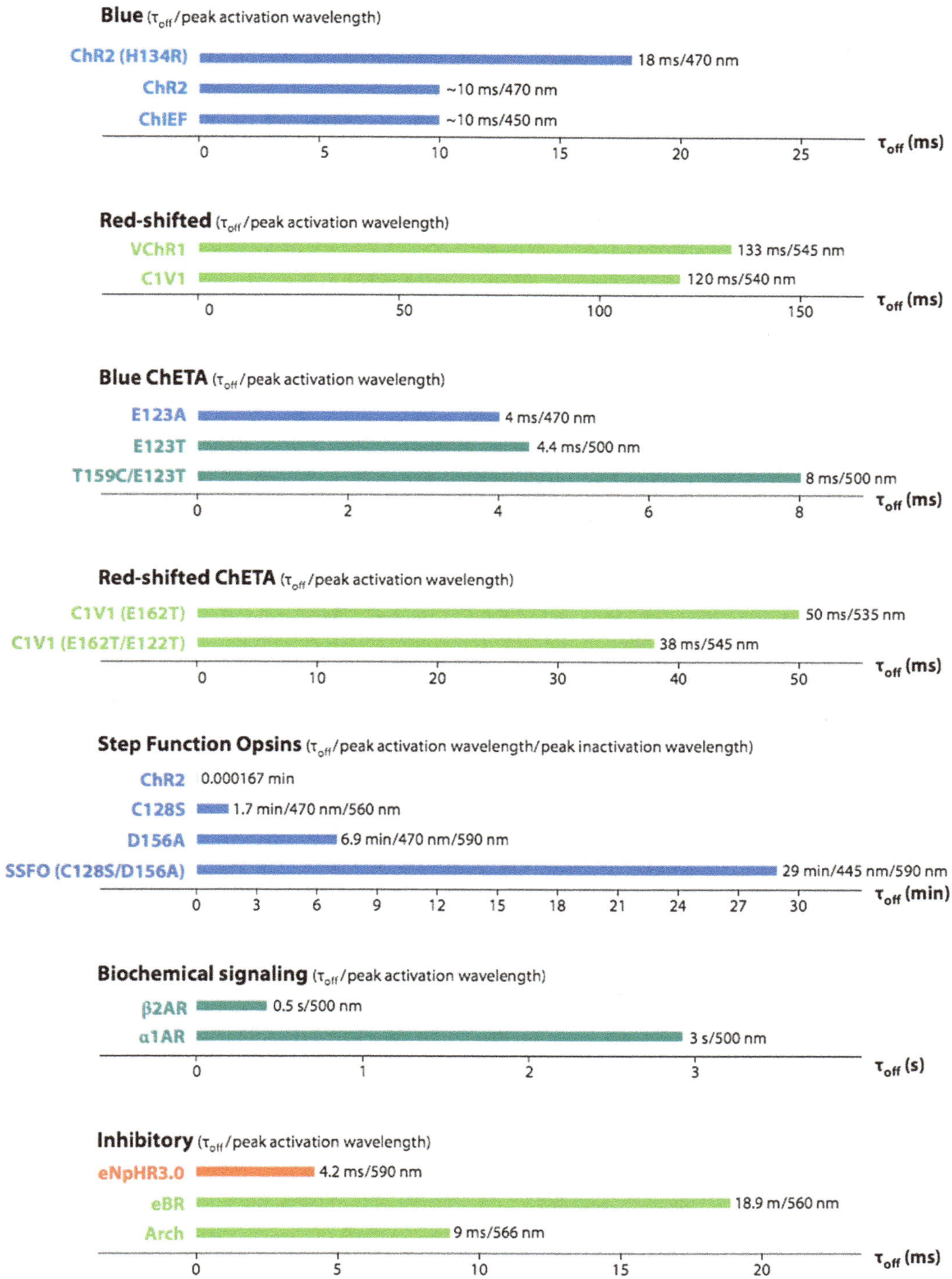

Fig. 2 Properties of optogenetic tools. Deactivation time constants (τ_{off}) and activation/inactivation wavelengths are illustrated for comparison purposes and as a look-up table. ChR2 is grouped with step function opsins for scale purposes. Decay kinetics were measured where precise published values were not available. Values from room temperature except for OptoXRs (37 °C). Adapted from [2].

~12 ms whereas (for example) that of ChR2-H134R is approximately doubled [28] (Fig. 2); this property in part determines the maximum frequency at which a neuron may produce precisely timed light-initiated action potentials [14, 28]. Indeed, an opportunity was seen to speed up optogenetic control function by modulating τ_{off}; destabilizing the open state of the channel could lead to a more rapid spontaneous reversion to the nonconducting state of ChR2 following light cessation, thereby increasing the frequency ceiling for driving neurons—an especially important consideration for some classes of fast-spiking inhibitory interneurons, including those expressing parvalbumin (PV). By applying observations in the bacteriorhodopsin literature to ChR2 based on the high degree of homology among type I rhodopsins, it was found that shifting a single amino acid residue was sufficient to approximately halve the τ_{off} (ChETA; [14, 28]), thereby increasing the top high-fidelity frequency to at least 200 Hz and increasing overall performance. A separate creative approach to creating a channelrhodopsin with a rapid τ_{off} arose from chimeras of ChR1 and ChR2 (ChEF/ChIEF) [33] (Fig. 2).

Opsins optimized for rapid closure and high frequency stimulation are complemented by ChR2 variants designed to induce prolonged depolarization after light cessation. The engineering approach to design these proteins also took advantage of homology between ChR2 and BR to produce variants that stabilize, as opposed to destabilize, the active retinal conformer. These "step-function opsins" (SFO) allow precise induction and inactivation of a depolarizing current "step" into the population of neurons, the duration of which is dependent upon the particular mutation (C128S: 100 s [13]; D156A: 6.9 min [37]; C128S/D156A: 29 min (Stabilized Step Function Opsin; SSFO [15])) (Fig. 2). As these are derivatives of ChR2, they are activated by blue (470 nm) light. This depolarizing step can then be terminated with yellow light (590 nm), which drives the opsin to its resting state. In the case of SSFO, the most stable of the SFOs, the depolarization step occurs without crossing the sodium channel activation threshold, which has the useful effect of modestly sensitizing a population of neurons to incoming excitatory input.

One additional benefit of an extended τ_{off} is that the cell expressing the protein may effectively act as a photon integrator: in the case of SSFO, once a channel has been activated, it remains open for a period approximately 150,000 times longer than native ChR2. The implication of cellular photon integration is that the total number of photons needed to induce a maximal photocurrent in the cell may be spread over a longer period of time (in contrast to stimulation parameters of ChR2 that require a high density of photons in order to activate enough channels to move the neuron beyond action potential threshold). In the context of behavioral neuroscience, this property has been exploited to allow for extracerebral light delivery, obviating the need to implant fiber optic light delivery devices into animals for behavioral trials at depths of at least 2.8 mm [15].

A second dimension into which considerable engineering resources have been invested is "red-shifting" the maximum excitation wavelength of channelrhodopsins. A tool that is sufficiently red-activated might be able to be modulated independently of ChR2, thus allowing for the manipulation of multiple, independent populations of neurons within the same space [15, 31]. While both VChR1 and recently described MChR1 are significantly red-shifted relative to ChR2, VChR1 does not produce sufficient photocurrents to reliably drive supra-threshold events in neural populations and the function of MChR1 in neurons has yet to be described (although data from other heterologous systems suggests that its currents will be similar in magnitude to VChR1; [32]). The most potent red-shifted channelrhodopsins currently available are not derived from any one organism but are chimeras between ChR1 and VChR1 (C1V1); resulting currents are significantly larger than ChR2 [15, 28] and action potentials may be elicited by even 630 nm red light [15]. In a recent series of experiments, C1V1 was shown to independently modulate cell bodies and terminals in the same tissue as neurons expressing ChR2 (experiments conducted in brain slice, in anesthetized ("optrode") recordings in vivo, and in awake, behaving animals [15]). Independent modulation of multiple populations is of immediate utility in understanding the dynamics of circuits and, indeed, has been used to examine independent inputs into thalamus as well as the interplay between pyramidal neurons and parvalbumin-expressing (PV) fast-spiking interneurons in medial prefrontal cortex (mPFC; [15]).

In summary, the explosion of channelrhodopsins, mutational variants, and chimeras since the original introduction of ChRs into neurons in 2005 has included the engineering of tools differently specialized for increased photocurrent, for driving action potentials at high frequencies, for depolarizing neurons for extended periods in the absence of direct action potential generation, and for combinatorial control via red-shifting of excitation wavelength. Used in concert with molecular techniques to be discussed later, currently available channelrhodopsin tools are able to probe the sufficiency of multiple components of neural circuits simultaneously in awake, behaving animals, in anesthetized recordings, or in acute slice preparations. As these tools continue to be refined and enhanced, designing the ideal experimental protocol becomes more difficult; recent reviews by our lab are available that compare available excitatory and inhibitory tools under standardized conditions [28], describe technical details of experimental design [29], and give an overview of the myriad systems and experimental designs undertaken with optogenetics [2].

With the immense diversity of nature and ongoing efforts to characterize the vast number of microbial organisms, it is likely that the number of primary sequences encoding channelrhodopsins will grow, leading to further diversity among the described parameters (excitation wavelength and temporal dynamics); it is also possible that channels with ionic conduction preferences that may be exploited for specific neuroscience applications will be discovered or engineered (for instance, potassium- or calcium-selective channels). Indeed, preliminary work in this domain has already begun and will accelerate with the long-awaited determination of the crystal structure of a channelrhodopsin [38].

3 Light-Activated Ion Pumps

Using channelrhodopsins to control the excitability of a population of neurons with fine temporal resolution allows neuroscientists to ask if the activity of a genetically and/or anatomically specified population of neurons is sufficient to cause a given outcome. The reverse side of this experimental coin is to ask if the activity of these elements is necessary for the outcome. To obtain direct evidence of necessity, a tool is required that is able to silence the activity of a targeted population of neurons while leaving others unaffected [38].

One candidate to fill this role would be the light-activated halorhodopsin (HR) from *Halobacterium salinarum*. HR is a type I rhodopsin that functions as an electrogenic pump producing inward chloride translocation. Although highly homologous to BR, HR does not pump protons; this has been shown to be due to a single amino acid change of an Asp that acts as a proton acceptor during the BR proton pumping cycle: during active pumping, instead of a proton being translocated out of the cell, as is the case with BR, an extracellular chloride ion is shuttled inward, past the protonated Retinal Schiff Base (RSBH+). Subsequent steps release this chloride ion into the intracellular milieu; the extracellular Cl^--binding site is refilled, and the cycle continues. In theory, expressing HR in neurons would increase the intracellular concentration of chloride while light was applied, thus hyperpolarizing the neuron and preventing action potentials.

Heterologous expression of the *H. salinarum*, the most well-known and characterized HR, was found to elicit unstable photocurrents that were not likely to be useful in the context of neuron silencing [20]. A broader experimental screen found that the *Natronomonas pharaonis* HR (NpHR), in contrast, induced sufficiently strong and stable hyperpolarizing photocurrents in response to yellow light (590 nm; Fig. 2) even in intact mammalian brain tissue and behaving animals [20]. NpHR has since been used in vivo in a number of experimental systems [20, 39–42], even over long (e.g., tens of minutes) time scales [43–45] and in systems as complex as behaving primates [46, 47].

Optogenetic function of pumps from other organisms that can hyperpolarize neurons have since also been described. These have all been proton efflux pumps, and include the original BR [48] and a Geminigeraceae member *Guillardia theta* rhodopsin-3 (Gtr3), both of which are activated by green light. Although able to silence action potentials in cultured neurons, the use of these two opsins has been hampered by unimpressive photocurrents [48]; larger proton currents were obtained from an opsin nicknamed Mac, from the fungus *Leptosphaeria maculans* and an opsin nicknamed Arch, from the Halobacteriaceae member *Halorubrum sodomense* [49]. Arch has also benefited greatly from a general membrane trafficking improvement approach first outlined for NpHR [48, 50, 51] (detailed below) that has enhanced peak photocurrents for proton pumps as well [28], enabling robust use for silencing neurons. One potential drawback of using these proton efflux pumps is local environment acidification; the consequence to neighboring neurons of their prolonged or repeated use warrants further investigation [28].

As true ion pumps, these tools require constant light to move through the steps of the cycle and continue translocating ions; therefore, lengthening of temporal dynamics is not likely to be amenable to strategies that produced a continuum of off-kinetics ranging from 6 ms to 30 min seen with channelrhodopsins. Nor is combinatorial inhibition possible at present in intact mammalian tissue, in contrast to the situation for excitation, since available spectra are not sufficiently separated. Finally, since these pumps cannot shunt or change input resistance as native inhibitory channels can, efficacy is suboptimal in intact systems. Together these challenges point the way to the need for further tool development and refinement.

4 Opsin Optimization

The latest generation of optogenetic tools [15, 28] display high expression, large photocurrents in transduced cells, and safe and effective membrane trafficking from translation to the membrane in mammalian neurons. The feat of adapting proteins from a different kingdom of life, for expression in the soma and processes of a neuron, has required multiple engineering iterations. As described [48, 50, 51], these challenges may be met by meticulously optimizing each step in the process of protein production. Initially, opsin gene transcription levels are increased through codon optimization (the process of recapitulating the preference of a given organism to encode a given amino acid with a specific three-nucleotide codon; for instance, arginine may be encoded by six separate codons). The translated amino acid sequence in codon-optimized sequences is identical, however, the use of a codon-optimized primary transcript increases the total functional expression [20, 23, 52, 53]. After translation, the protein must be properly trafficked out of the endoplasmic reticulum (ER) and to the membrane. Early NpHR versions aggregated within the ER at high expression levels [51]; the addition of a trafficking peptide and targeting sequence to NpHR increased photocurrents and significantly reduced ER accumulations [48, 50, 51]. These simple steps were successfully applied even to BR, and when applied to NpHR as eNpHR3.0 [48] finally enabled the first optogenetic behavioral loss-of-function experiments in mammals [45, 54, 55], complementing the initial *C. elegans* behavioral loss-of-function [20]. It is likely that this strategy will be successfully applied to other heterologously expressed membrane proteins; indeed the same process has now been applied to Arch and Mac, robustly increasing photocurrents and reducing intracellular aggregations [28].

5 Subcellular Targeting

Intersectional strategies available for selectively manipulating less than an entire neuron include (1) restricting expression of the protein to a subcompartment, which allows the entire neuron to be bathed in light while retaining spatial specificity, or (2) to express the construct throughout the neuron and to restrict the activating light to a specific region of the neuron. Such efforts could lead to local induction of vesicle release in the absence of an action potential, or to selective stimulation of a class of cells defined by specific axonal projections from a brain region [35, 54, 56].

Restricting the expression of a high copy-number protein to a specific compartment is a difficult task; our work in optimizing the expression of various opsins to overcome aggregation in the endoplasmic reticulum [48, 51], however, constitutes proof-of-principle for applying native targeting motifs to heterologous proteins. Work done to add targeting motifs to control axonal or dendritic localization have met with some (though incomplete) success [40, 57, 58]; incomplete exclusion from nontargeted regions of the cell is still typically seen. Alternatively, spatial specificity may be obtained by restricting light to only portion of the neuron or projections. This technique has been applied using tools to shape light, including guided two-photon or single-photon light [54, 59], beveled metal cannulas to restrict the exit angle of light emitted from a fiber optic cable [54], and digital micro-mirror devices [60].

6 Optical Control of Cell Signaling

In addition to optically modulated mediators of neuronal membrane potential, a class of tools for optical control of biochemical signaling cascades (the Opto-XRs) now exists for temporally precise manipulation of GPCR signaling cascades (Fig. 1). These chimeric proteins are composed of the light-sensing extracellular domain of the type II mammalian rhodopsin and the intracellular domain of a given GPCR, for instance, alpha-1 and beta-2 adrenergic receptors (G_q- and G_s-coupled, respectively [21]) and 5-HT1a ($G_{i/o}$-coupled [61]) and activate native signaling pathways in response to light. These receptors are activated by 500-nm light and useful for control of biochemical signaling both in vivo [21] and in vitro [21, 61].

7 Conclusion

Optogenetics allows for manipulation of genetically defined populations of neurons with millisecond-scale temporal resolution. This technology is based on light-activated ion channels and pumps that vary in their ionic selectivity, temporal dynamics, and activation wavelength. These tools have been of great utility in analyzing the contribution of specific neurons to neural circuit behavior, and indeed now multiple optogenetic tools may be used in concert to analyze the causal contributions of activity of neurons alone or in combination, in acute slice preparations, anesthetized animals, or during behavioral experiments across the animal kingdom (see Chap. 11 in this volume for applications of optogenetics to studies of neural circuitry). As the optogenetic toolset continues to be expanded and diversified, circuit-level experiments will be conducted with progressively more refinement and resolution, elucidating the necessity and sufficiency of precisely defined events in specified cells to circuit function and behavior.

References

1. Crick FH (1979) Thinking about the brain. Sci Am 241:219–232
2. Fenno L, Yizhar O, Deisseroth K (2011) The development and application of optogenetics. Annu Rev Neurosci 34:389–412
3. Deisseroth K, Feng G, Majewska AK, Miesenbock G, Ting A, Schnitzer MJ (2006) Next-generation optical technologies for illuminating genetically targeted brain circuits. J Neurosci 26:10380–10386
4. Haupts U, Tittor J, Bamberg E, Oesterhelt D (1997) General concept for ion translocation by halobacterial retinal proteins: the isomerization/switch/transfer (IST) model. Biochemistry 36:2–7
5. Ernst OP, Sanchez Murcia PA, Daldrop P, Tsunoda SP, Kateriya S, Hegemann P (2008) Photoactivation of channelrhodopsin. J Biol Chem 283:1637–1643
6. Hegemann P, Ehlenbeck S, Gradmann D (2005) Multiple photocycles of channelrhodopsin. Biophys J 89:3911–3918
7. Ritter E, Stehfest K, Berndt A, Hegemann P, Bartl FJ (2008) Monitoring light-induced

structural changes of Channelrhodopsin-2 by UV-visible and Fourier transform infrared spectroscopy. J Biol Chem 283:35033–35041
8. Essen LO (2002) Halorhodopsin: light-driven ion pumping made simple? Curr Opin Struct Biol 12:516–522
9. Schobert B, Lanyi JK (1982) Halorhodopsin is a light-driven chloride pump. J Biol Chem 257:10306–10313
10. Oesterhelt D, Stoeckenius W (1971) Rhodopsin-like protein from the purple membrane of Halobacterium halobium. Nat New Biol 233:149–152
11. Racker E, Stoeckenius W (1974) Reconstitution of purple membrane vesicles catalyzing light-driven proton uptake and adenosine triphosphate formation. J Biol Chem 249:662–663
12. Michel H, Oesterhelt D (1976) Light-induced changes of the pH gradient and the membrane potential in H. halobium. FEBS Lett 65:175–178
13. Berndt A, Yizhar O, Gunaydin LA, Hegemann P, Deisseroth K (2009) Bi-stable neural state switches. Nat Neurosci 12:229–234
14. Gunaydin LA, Yizhar O, Berndt A, Sohal VS, Deisseroth K, Hegemann P (2010) Ultrafast optogenetic control. Nat Neurosci 13:387–392
15. Yizhar O, Fenno LE, Prigge M, Schneider F, Davidson TJ, O'Shea DJ, Sohal VS, Goshen I, Finkelstein J, Paz JT, Stehfest K, Fudim R, Ramakrishnan C, Huguenard JR, Hegemann P, Deisseroth K (2011) Neocortical excitation/inhibition balance in information processing and social dysfunction. Nature 477:171–178
16. Bamann C, Gueta R, Kleinlogel S, Nagel G, Bamberg E (2010) Structural guidance of the photocycle of channelrhodopsin-2 by an interhelical hydrogen bond. Biochemistry 49:267–278
17. Matsuno-Yagi A, Mukohata Y (1977) Two possible roles of bacteriorhodopsin; a comparative study of strains of Halobacterium halobium differing in pigmentation. Biochem Biophys Res Commun 78:237–243
18. Nagel G, Ollig D, Fuhrmann M, Kateriya S, Musti AM, Bamberg E, Hegemann P (2002) Channelrhodopsin-1: a light-gated proton channel in green algae. Science 296:2395–2398
19. Nagel G, Szellas T, Huhn W, Kateriya S, Adeishvili N, Berthold P, Ollig D, Hegemann P, Bamberg E (2003) Channelrhodopsin-2, a directly light-gated cation-selective membrane channel. Proc Natl Acad Sci U S A 100:13940–13945
20. Zhang F, Wang LP, Brauner M, Liewald JF, Kay K, Watzke N, Wood PG, Bamberg E, Nagel G, Gottschalk A, Deisseroth K (2007) Multimodal fast optical interrogation of neural circuitry. Nature 446:633–639
21. Airan RD, Thompson KR, Fenno LE, Bernstein H, Deisseroth K (2009) Temporally precise in vivo control of intracellular signalling. Nature 458:1025–1029
22. Boyden ES, Zhang F, Bamberg E, Nagel G, Deisseroth K (2005) Millisecond-timescale, genetically targeted optical control of neural activity. Nat Neurosci 8:1263–1268
23. Zhang F, Wang LP, Boyden ES, Deisseroth K (2006) Channelrhodopsin-2 and optical control of excitable cells. Nat Methods 3:785–792
24. Sakmar TP (2002) Structure of rhodopsin and the superfamily of seven-helical receptors: the same and not the same. Curr Opin Cell Biol 14:189–195
25. Shichida Y, Yamashita T (2003) Diversity of visual pigments from the viewpoint of G protein activation—comparison with other G protein-coupled receptors. Photochem Photobiol Sci 2:1237–1246
26. Buldt G, Heberle J, Dencher NA, Sass HJ (1998) Structure, dynamics, and function of bacteriorhodopsin. J Protein Chem 17:536–538
27. Spudich JL, Luecke H (2002) Sensory rhodopsin II: functional insights from structure. Curr Opin Struct Biol 12:540–546
28. Mattis J, Tye KM, Ferenczi EA, Ramakrishnan C, O'Shea DJ, Prakash R, Gunaydin LA, Hyun M, Fenno LE, Gradinaru V, Yizhar O, Deisseroth K (2011) Principles for applying optogenetic tools derived from direct comparative analysis of microbial opsins. Nat Methods 9:159–172
29. Yizhar O, Fenno LE, Davidson TJ, Mogri M, Deisseroth K (2011) Optogenetics in neural systems. Neuron 71:9–34
30. Zhang F, Gradinaru V, Adamantidis AR, Durand R, Airan RD, de Lecea L, Deisseroth K (2010) Optogenetic interrogation of neural circuits: technology for probing mammalian brain structures. Nat Protoc 5:439–456
31. Zhang F, Prigge M, Beyriere F, Tsunoda SP, Mattis J, Yizhar O, Hegemann P, Deisseroth K (2008) Red-shifted optogenetic excitation: a tool for fast neural control derived from Volvox carteri. Nat Neurosci 11:631–633
32. Govorunova EG, Spudich EN, Lane CE, Sineshchekov OA, Spudich JL (2011) New channelrhodopsin with a red-shifted spectrum and rapid kinetics from Mesostigma viride. MBio 2:e00115–e00111
33. Lin JY, Lin MZ, Steinbach P, Tsien RY (2009) Characterization of engineered channelrhodopsin variants with improved properties and kinetics. Biophys J 96:1803–1814

34. Wang H, Sugiyama Y, Hikima T, Sugano E, Tomita H, Takahashi T, Ishizuka T, Yawo H (2009) Molecular determinants differentiating photocurrent properties of two channelrhodopsins from chlamydomonas. J Biol Chem 284:5685–5696
35. Gradinaru V, Thompson KR, Zhang F, Mogri M, Kay K, Schneider MB, Deisseroth K (2007) Targeting and readout strategies for fast optical neural control in vitro and in vivo. J Neurosci 27:14231–14238
36. Nagel G, Szellas T, Kateriya S, Adeishvili N, Hegemann P, Bamberg E (2005) Channelrhodopsins: directly light-gated cation channels. Biochem Soc Trans 33:863–866
37. Bamann C, Kirsch T, Nagel G, Bamberg E (2008) Spectral characteristics of the photocycle of channelrhodopsin-2 and its implication for channel function. J Mol Biol 375:686–694
38. Kato HE, Zhang F, Yizhar O, Ramakrishnan C, Nishizawa T, Hirata K, Ito J, Aita Y, Tsukazaki T, Hayashi S, Hegemann P, Maturana AD, Ishitani R, Deisseroth K, Nureki O (2012) Crystal structure of the channelrhodopsin light-gated cation channel. Nature 482:369–374
39. Arrenberg AB, Del Bene F, Baier H (2009) Optical control of zebrafish behavior with halorhodopsin. Proc Natl Acad Sci U S A 106:17968–17973
40. Greenberg KP, Pham A, Werblin FS (2011) Differential targeting of optical neuromodulators to ganglion cell soma and dendrites allows dynamic control of center-surround antagonism. Neuron 69:713–720
41. Schoonheim PJ, Arrenberg AB, Del Bene F, Baier H (2010) Optogenetic localization and genetic perturbation of saccade-generating neurons in zebrafish. J Neurosci 30:7111–7120
42. Tonnesen J, Sorensen AT, Deisseroth K, Lundberg C, Kokaia M (2009) Optogenetic control of epileptiform activity. Proc Natl Acad Sci U S A 106:12162–12167
43. Goshen I, Brodsky M, Prakash R, Wallace J, Gradinaru V, Ramakrishnan C, Deisseroth K (2011) Dynamics of retrieval strategies for remote memories. Cell 147:678–689
44. Gradinaru V, Mogri M, Thompson KR, Henderson JM, Deisseroth K (2009) Optical deconstruction of parkinsonian neural circuitry. Science 324:354–359
45. Witten IB, Lin SC, Brodsky M, Prakash R, Diester I, Anikeeva P, Gradinaru V, Ramakrishnan C, Deisseroth K (2010) Cholinergic interneurons control local circuit activity and cocaine conditioning. Science 330:1677–1681
46. Diester I, Kaufman MT, Mogri M, Pashaie R, Goo W, Yizhar O, Ramakrishnan C, Deisseroth K, Shenoy KV (2011) An optogenetic toolbox designed for primates. Nat Neurosci 14:387–397
47. Han X, Qian X, Bernstein JG, Zhou HH, Franzesi GT, Stern P, Bronson RT, Graybiel AM, Desimone R, Boyden ES (2009) Millisecond-timescale optical control of neural dynamics in the nonhuman primate brain. Neuron 62:191–198
48. Gradinaru V, Zhang F, Ramakrishnan C, Mattis J, Prakash R, Diester I, Goshen I, Thompson KR, Deisseroth K (2010) Molecular and cellular approaches for diversifying and extending optogenetics. Cell 141:154–165
49. Chow BY, Han X, Dobry AS, Qian X, Chuong AS, Li M, Henninger MA, Belfort GM, Lin Y, Monahan PE, Boyden ES (2010) High-performance genetically targetable optical neural silencing by light-driven proton pumps. Nature 463:98–102
50. Zhao S, Cunha C, Zhang F, Liu Q, Gloss B, Deisseroth K, Augustine GJ, Feng G (2008) Improved expression of halorhodopsin for light-induced silencing of neuronal activity. Brain Cell Biol 36:141–154
51. Gradinaru V, Thompson KR, Deisseroth K (2008) eNpHR: a Natronomonas halorhodopsin enhanced for optogenetic applications. Brain Cell Biol 36:129–139
52. Aravanis AM, Wang LP, Zhang F, Meltzer LA, Mogri MZ, Schneider MB, Deisseroth K (2007) An optical neural interface: in vivo control of rodent motor cortex with integrated fiberoptic and optogenetic technology. J Neural Eng 4:S143–S156
53. Adamantidis AR, Zhang F, Aravanis AM, Deisseroth K, de Lecea L (2007) Neural substrates of awakening probed with optogenetic control of hypocretin neurons. Nature 450:420–424
54. Tye K, Prakash R, Kim S-Y, Fenno LE, Grosenick L, Zarabi H, Thompson KR, Gradinaru V, Ramakrishnan C, Deisseroth K (2011) Amygdala circuitry mediating reversible and bidirectional control of anxiety. Nature 471(7338):358–362
55. Stuber GD, Sparta DR, Stamatakis AM, van Leeuwen WA, Hardjoprajitno JE, Cho S, Tye KM, Kempadoo KA, Zhang F, Deisseroth K, Bonci A (2011) Excitatory transmission from the amygdala to nucleus accumbens facilitates reward seeking. Nature 475:377–380

56. Petreanu L, Huber D, Sobczyk A, Svoboda K (2007) Channelrhodopsin-2-assisted circuit mapping of long-range callosal projections. Nat Neurosci 10:663–668
57. Grubb MS, Burrone J (2010) Channelrhodopsin-2 localised to the axon initial segment. PLoS One 5:e13761
58. Lewis TL Jr, Mao T, Svoboda K, Arnold DB (2009) Myosin-dependent targeting of transmembrane proteins to neuronal dendrites. Nat Neurosci 12:568–576
59. Rickgauer JP, Tank DW (2009) Two-photon excitation of channelrhodopsin-2 at saturation. Proc Natl Acad Sci U S A 106:15025–15030
60. Leifer AM, Fang-Yen C, Gershow M, Alkema MJ, Samuel AD (2011) Optogenetic manipulation of neural activity in freely moving Caenorhabditis elegans. Nat Methods 8:147–152
61. Oh E, Maejima T, Liu C, Deneris E, Herlitze S (2010) Substitution of 5-HT1A receptor signaling by a light-activated G protein-coupled receptor. J Biol Chem 285:30825–30836

Part II

Shedding Light on Cortical Activity States

Chapter 6

In Vivo Dendritic Mapping of Sensory Inputs in Cortical Neurons

Hongbo Jia and Arthur Konnerth

Abstract

Mammalian cortical neurons integrate sensory information that arrives through numerous synaptic inputs located on their dendrites. Here we introduce an approach to identify sensory-evoked dendritic input sites in cortical neurons in vivo involving the use of two-photon calcium imaging combined with targeted whole-cell recordings. We provide basic technical descriptions as well as experimental procedures of this method. First, we discuss various scanning modes for two-photon imaging of cortical neuron dendrites in vivo. Second, we focus on practical aspects of dye-loading by means of whole-cell recordings in vivo. Third, we provide a step-by-step experimental protocol and a data analysis algorithm for dendritic imaging in cortical neurons. Fourth, we show, as an example, results obtained from dendrites of orientation-tuned neurons in layer 2/3 of mouse visual cortex. We conclude that the combination of whole-cell recording and two-photon calcium imaging is a useful tool for the functional mapping of input sites in dendrites that can be readily applied to other cortical areas and layers, helping to reveal the fine organization of sensory information representation in the dendrites of mammalian cortical neurons in vivo.

Key words Cortical neuron, Dendrite, Synapse, Sensory stimulation, Two-photon microscopy, Calcium imaging, Whole-cell recording

1 Introduction

Brain function relies on signaling in networks of interconnected neurons. Neurons of the mammalian brain communicate with each other mostly through chemical synapses. For example, excitatory synapses using glutamate as a transmitter form many of their contacts with fine dendritic protrusions, the so-called spines (for review see [1]). In the mammalian cortex, single pyramidal neurons receive thousands of synaptic inputs, most of them located on such spines. An understanding of the dendritic organization of synaptic inputs is important because of their great potential for signal integration and computation (for reviews see [2, 3]). In recent years there has been substantial progress in our understanding of the local connectivity in selected brain regions [4–6]. However, despite

Bruno Weber and Fritjof Helmchen (eds.), *Optical Imaging of Neocortical Dynamics*, Neuromethods, vol. 85,
DOI 10.1007/978-1-62703-785-3_6, © Springer Science+Business Media New York 2014

many years of detailed anatomical studies [7] and more recent efforts in revealing the brain's connectome [8], the functional organization of synaptic connectivity in vivo, in relation to defined sensory stimuli, remains largely unknown. For example, in visual cortex neurons it is largely unknown how and where synaptic input sites that are activated by distinct orientations of visual stimuli are organized on different dendrites (Fig. 1).

One approach to studying synaptic function in dendrites is the analysis of postsynaptic signals, for example, calcium transients that are evoked by synaptic activity (for review see ref. [9]). In the mammalian cortex, pioneering two-photon imaging studies of dendritic calcium signaling in vivo have initially characterized calcium transients induced by action potentials that are triggered by either sensory stimulation or local electrical stimulation [10–12]. As in brain slices [13, 14], such action potentials back-propagate in the dendrites of cortical neurons also in vivo, recruiting most of the dendritic tree. The action potential-evoked calcium transients interact with the local calcium signals induced by the afferent synaptic activity, resulting in complex dendritic calcium signaling patterns [15, 16]. This complexity combined with the difficulties of imaging in the intact mammalian brain posed major challenges in the identification of the precise locations of the synaptically evoked calcium signals as indicators of afferent synaptic inputs. Therefore, the identification of synaptic input sites in the dendrites of spiking neurons had been largely restricted to more transparent brain tissues of lower vertebrates [17].

One method to overcome these challenges consists of the recording of synaptically evoked subthreshold dendritic calcium signals by combining whole-cell recordings and two-photon calcium imaging. This approach has recently enabled the mapping of sensory-evoked dendritic input sites in layer 2/3 neurons of mouse visual cortex [18], auditory cortex [19], and vibrissal cortex [20]. The basic experimental protocol and analysis procedures can be

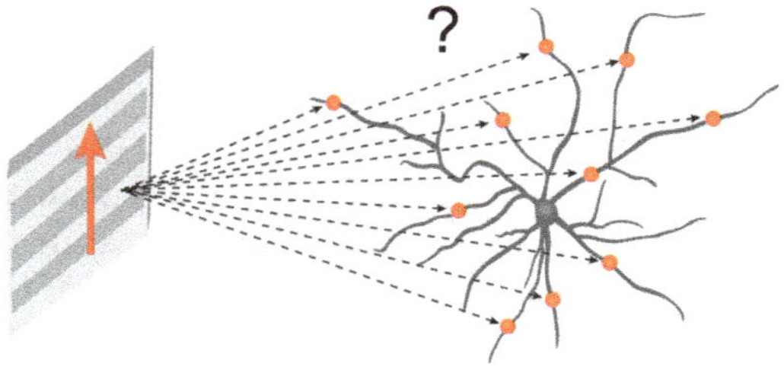

Fig. 1 Search for visual stimulation-activated synaptic input sites on dendrites of cortical neurons in vivo. This is a simplified experimental scheme. *Gray bars* indicate drifting gratings along the direction of the *red arrow*. *Red dots* indicate possible synaptic input sites that can be activated by this specific visual stimulus. *Question mark* illustrates that the locations of stimulation-activated input sites are unknown.

found in a protocol report [21]. In this chapter, we provide a detailed description and an example application of in vivo dendritic mapping of sensory inputs in cortical neurons.

2 Two-Photon Calcium Imaging in Dendrites of Cortical Neurons In Vivo

2.1 Considerations for Laser-Scanning Modes in Dendrite Imaging

Two-photon imaging typically involves scanning the probe with a laser beam [22]. In order to obtain a complete image, the laser beam must be steered such that the trajectory of focal points covers the desired field within the optical focal plane. Depending on the experimental goals (e.g., high spatial and temporal resolution, or high temporal resolution from large fields of view, or high spatial–temporal resolution from selected subregions) different modes of scanning may be used. Figure 2 illustrates four examples (termed as full raster scan, arbitrary line scan, random access scan, and

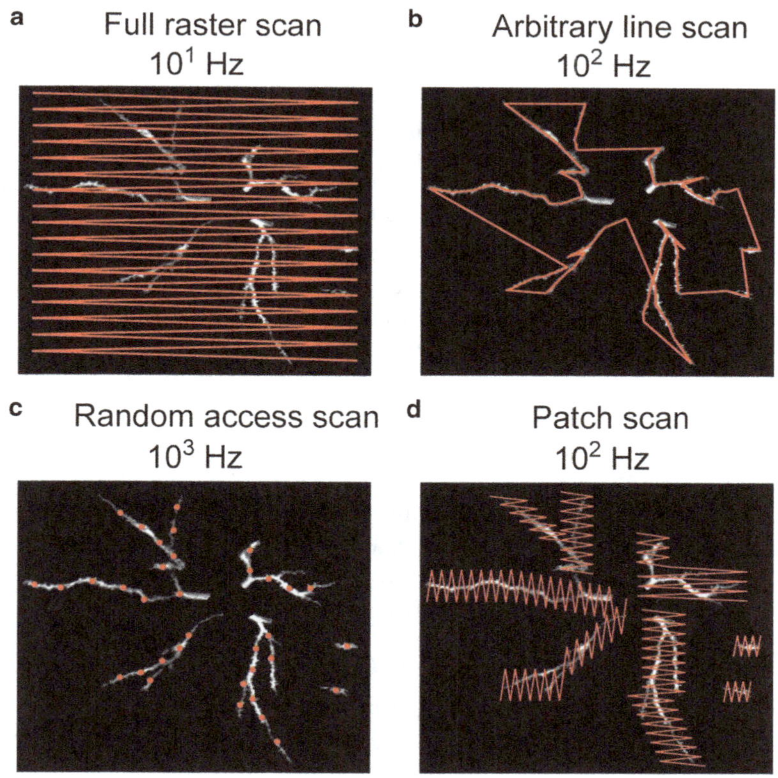

Fig. 2 Scanning modes for two-photon imaging of dendrites in cortical neurons. (**a–d**) *Red lines/dots* illustrate the scanning trajectory. The image is adapted from a real two-photon image in vivo taken from a layer 2/3 cortical neuron (filled with 100 μM OGB-1 and 25 μM Alexa-594), at focal depth of 165 μm below the cortical surface. The typical frame repetition rate for each scanning mode is expressed as an order of magnitude.

patch scan) of scanning modes for imaging dendrites of cortical neurons in vivo. The most common mode is the full raster scan that is frequently used for both visualization and recording with two-photon microscopes (Fig. 2a). However, for some applications the complete spatial information is less relevant than high temporal resolution in selected subregions. For such applications the repeated scanning of arbitrary lines represents an interesting alternative (Fig. 2b). Scanning along a manually selected line enables the visit of a given object at markedly higher repetition rates than that in raster scans (10^2 versus 10^1 Hz). Both scanning modes can be achieved by using standard galvanometric mirrors [23, 24]. A third mode is the random access scan mode (Fig. 2c). Here, instead of driving the focus point continuously along a scanning line, it is moved in a saltatory fashion to pre-specified locations and parked at each location of interest for a short time interval (10–100 μs). This mode can be only employed by inertia-free devices, such as those employing acoustic-optical-deflectors (AODs) (e.g., [25, 26]). This approach allows exceedingly high repetition rates (10^3 Hz) for the detection of signals in biological samples [27, 28] (see Chapter 9 in this volume). In practice, a potential source of errors for both line scan and random access scan in vivo is tissue motions that may disturb the alignment between the pre-specified scan lines/points and the object to be scanned. A possible solution to this problem is what we term as the "patch scan" mode (Fig. 2d). In this mode the scanning trajectory covers the entire object, for example, the dendrites of a neuron, by raster-scanning systematically over multiple selected regions of interest (ROIs) (Fig. 2d). Patch scan modes are not yet available commercially and their use for the detection of rapid biological signals may require implementation of inertia-free scanners, like AODs.

2.2 Using Resonant- and AOD-Based Raster Scans for Imaging Dendrites

As discussed in the previous section, the full raster scan is a practical and widely used scanning mode for imaging dendrites in vivo. It can involve either galvanometric mirror or AOD-based scanners. In either case the focus point is driven simultaneously along two orthogonal axes. Typically, the driving along the horizontal axis (x-axis) is fast, while the driving along the vertical axis (y-axis) is slower and uniformly spreads the horizontal lines into a square frame. In order to acquire an image with sufficient digital resolution, a minimum number of horizontal lines must be scanned; therefore, the repetition rate of imaging is largely dependent on the repetition rate of horizontal lines. The most commonly used scanner in confocal and multi-photon imaging is the linear galvanometric mirror, which changes angle linearly with the command voltage. The cycling frequency is typically as high as 1 kHz, which means 2 kHz repetition rate of horizontal scan lines, since one cycle consists of a forward and backward sweep and both can be used for image formation. When using 400 lines for the acquisition of an image, this means that the image frame rate is maximally 5 frames/s. However, for the detection of biological signals it is often necessary to use higher frame rates. For example, the accurate imaging of the fine structure of dendrites in vivo may require significantly higher frame rates, because of noise from the pulsation of the heart (at around 8 Hz for mice). According to the Nyquist–Shannon sampling theorem [29], in order to numerically filter this noise, the imaging frame rate must be at least double the frequency, i.e., 16 Hz in the case of imaging dendrites located near a pulsating blood vessel. One simple technical solution to increase the imaging speed is to reduce the number of scanned lines within the imaged frame. Note that because the last line cannot be instantly transitioned into the first line of the next frame, the gain in repetition rate is sub-proportional to the reduction in image size, resulting, in our example, in about 120 lines for images acquired at 16 frames/s.

Another solution for increasing the imaging frame rate is to replace the linear galvanometric mirror for the horizontal axis with a resonant mirror [30, 31]. The cycling frequency of a resonant mirror is typically 8 kHz, and thus the gain in frame rate is about eightfold and allows the acquisition of 30 video frames per second. Our group recently implemented a 12 kHz resonant scanner (configured for 40 frames/s with 512 lines/frame or 200 frames/s with 64 lines/frame) and used it successfully to image dendrites and spines in cortical neurons of the vibrissal cortex in vivo [20].

A further increase in imaging frame rate can be achieved by using AODs as scanning devices (e.g., [32–34]). AOD-based two-photon imaging is often used in the random access scanning mode (e.g., [25–27]), a mode which allows high-speed monitoring of selected recording sites (cells) within a large field of view but with a limited spatial resolution. By contrast, our group has recently implemented an AOD-based high-speed approach with a high spatial resolution in small fields of view [19]. In this device, we use a galvo-mirror for the vertical scan and an AOD for the horizontal scan (80 kHz line repetition rate). This configuration provides a high degree of flexibility concerning frame rates, frame shapes, and pixel resolution. For example, our group has recently used the mode of 1,000 frames/s with 80 lines/frame for the analysis of auditory stimulation-evoked single-spine responses in vivo [19]. Such a high frame repetition rate combined with a low laser power per frame allowed the implementation of an approach termed "low-power temporal oversampling" (LOTOS). Two different studies showed that LOTOS can be beneficial for recording calcium transients in spines of cortical neurons, because it allows prolonged recordings with high signal-to-noise ratio and low photodamage [19, 20].

In summary, the commonly used single-beam galvanometric mirror scanners provide a good control of the scanning beam, but at a limited speed. Because of this limitation, rapid biological signals such as dendritic calcium transients can be recorded typically only in the single line scan mode [10–12]. Resonant scanner-based devices at least partially overcome this limitation and are well suited for imaging local calcium transients simultaneously from multiple dendrites [18, 21] and with single-spine resolution [20]. High spatial resolution AOD-based scanning devices provide probably the best performance for imaging calcium signals in dendrites and spines in vivo, with high signal-to-noise ratio and low photodamage [19].

2.3 Calcium Dye-Loading Using "Shadow Patching" Whole-Cell Recordings In Vivo

The recording of calcium signals in neuronal dendrites most often involves the use of small molecule fluorescent calcium-sensitive dyes, for example, Oregon Green BAPTA-1 (OGB-1), Calcium Green-1, and Fluo-4 [18, 35–37] (see Chapter 4 in this volume). In recent years, genetically encoded calcium indicators (GECIs) have also been applied to calcium imaging of neuronal activities [38, 39] (see Chapter 10 in this volume) and, as these indicators continuously improve, there is good potential for their use for dendrite imaging in future studies. Here, we focus only on the use of small molecule dye-based calcium imaging, the presently best established approach for monitoring calcium signaling in dendrites and spines of cortical neurons in vivo (e.g., refs. [18–20]).

To perform dendritic calcium imaging with small molecule dyes in single cortical neurons in vivo, an important initial step is dye-loading, i.e., the delivery of membrane-impermeable calcium dyes to a target neuron. Early studies used calcium indicator dye delivery through sharp microelectrodes (Fig. 3a, e.g., ref. [12]). Such recordings leave the intracellular milieu intact, but they are

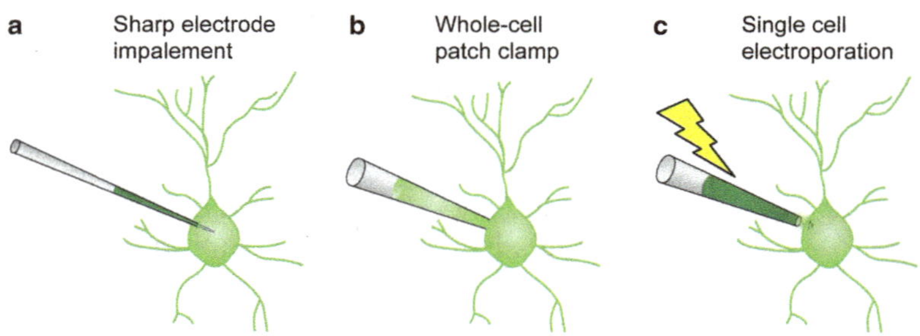

Fig. 3 Methods for delivery of small molecule intracellular calcium dye (e.g., OGB-1) to single neurons in vivo. (**a**) Sharp electrode impalement. The pipette contains a high concentration of dye (e.g., 10 mM) which diffuses into the cell via the small tip. (**b**) Whole-cell patch clamp. The pipette tip forms a GΩ-seal with the cell membrane and, after break-in, a direct access to the intracellular medium is established. Intracellular dye concentration eventually reaches equilibrium with the pipette solution dye concentration (e.g., 100 μM). (**c**) Single-cell electroporation. The pipette contains a high concentration of dye (e.g., 10 mM) but has no contact with intracellular medium. Applying short pulses of current (for example, each pulse at 0.5 μA, duration 30 ms) results in nanometer-sized pores forming in the membrane near the pipette tip, and thus allowing dye molecules to flow in.

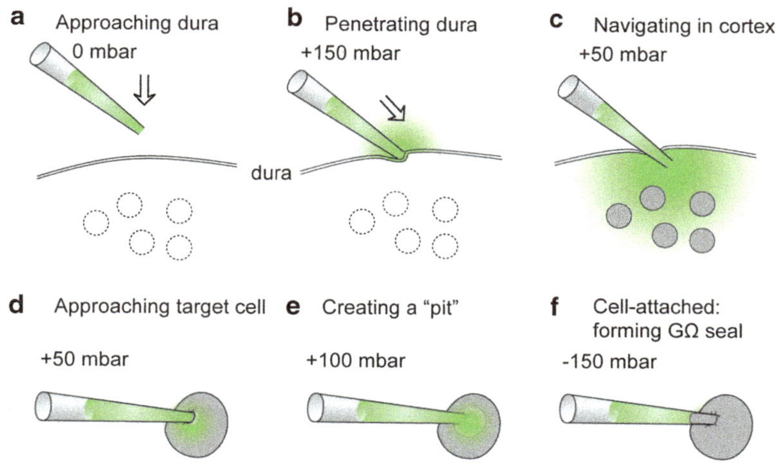

Fig. 4 "Shadow patching" of cortical neurons in vivo. (**a–c**) Penetrating the dura and navigating the pipette through cortex. (**d–f**) Approaching target neuron and establishing the GΩ-seal configuration. *Green shades* indicate spread of dye. *Empty circles* indicate neurons without extracellular dye nearby. *Filled circles* indicate neurons within the cloud of extracellular dye spreading out from the pipette.

tedious and it is difficult to control the intracellular concentration of the calcium dye. An effective alternative method, which overcomes some of these limitations and in addition allows the targeted delivery of the indicator dye to visually identified neurons, involves whole-cell recordings using the "shadow patching" procedure (Fig. 3b, e.g., ref. [21, 40]). Whole-cell recordings allow the monitoring of the electrical signals and a more accurate control of the intracellular dye concentration. However, there is the potential problem of washout of intracellular signaling molecules. Useful techniques that overcome this problem are targeted single-cell electroporation (Fig. 3c, e.g., ref. [41, 42]) or "bulk electroporation" of calcium indicator dyes [43]. Targeted single-cell electroporation can be readily combined with cell-attached recordings [44] and is a very useful technique for long-lasting, noninvasive dendrite calcium imaging recordings.

In this chapter we focus on the combination of targeted whole-cell recording and two-photon calcium imaging. Figure 4 illustrates the sequential steps of "shadow patching" [40]. When approaching the dura covering the cortex (Fig. 4a), a small puff of fluorescent dye is applied to prevent clogging of the pipette's tip (Fig. 4b). During the navigation through the cortex under visual guidance, fluorescent dye is continuously ejected by setting a mild positive pressure (50 mbar) to the patch pipette (Fig. 4c). The pipette is rapidly moved forward through the cortical tissue to minimize the amount of extracellular dye. The extracellular dye is in general quickly absorbed by the tissue, but the background

Box 1 Tips on Targeted Whole-Cell Recordings for Dye-Loading

1. If an obstacle such as a nontarget cell body or blood vessel lies directly in the path of the advancing pipette, the pipette should first be moved a bit laterally to avoid collision. When multiple lateral movements are necessary, they should not accumulate in one direction in order to avoid tissue deformation or pipette blockage. For large sideways movement first retract the pipette out of the tissue, reposition it, and then reinsert it into the tissue.
2. During navigation, when approaching the target cell, the amount of pipette solution blown out of the patch pipette should be as little as possible. As extracellular Ca^{2+} concentration is much higher than that inside the neuron, even a low concentration of dye in the extracellular tissue may produce adversely high background fluorescence.
3. After establishing a whole-cell recording, it is important to regularly check (about every 20 min) the position of the pipette tip relative to the cell body and to correct it if necessary. Tissue movements may cause the pipette to drift away from its initial position and thus the whole-cell recording condition may deteriorate, i.e., resulting in increased series resistance and/or leak current.

fluorescence resulting from residual dye may interfere with the recording of calcium signals in fine dendrites and spines. Fig. 4d–f illustrates the individual steps during GΩ seal formation in the neuron of choice (for more details see ref. [40]). Box 1 provides some practical tips for achieving stable and long-lasting dendritic imaging conditions in vivo.

After the whole-cell configuration has been established, it takes about 20–40 min for the calcium dye concentration to reach equilibrium in dendrites [21]. During this dye-loading period, it is recommended not to perform two-photon imaging except for checking pipette position, because the fluorescence signal is low such that more laser power is needed and photodamage may occur. Once a sufficient level of dye-loading is achieved, such that the baseline fluorescence signal is stable, combined whole-cell recordings and two-photon dendritic calcium imaging can begin. For the purpose of identifying sensory-evoked local dendritic calcium transients, it is practical to hyperpolarize the neuron to prevent action potential firing [18–20]. The hyperpolarization can be achieved by injecting negative current via the patch pipette in steps of −10 pA until action potential firing vanishes. The finally required level of hyperpolarizing current depends on the leak current, the series resistance, and the input resistance of the neuron. Therefore, the actual level of hyperpolarizing current should be readjusted if these conditions change over the duration of the recording. When taking into account the precautions mentioned above, one can perform a complete experiment by following the step-by-step protocol as described in Box 2. Note that the protocol was established for studying neurons in the visual cortex [21]; however, it can be readily adapted for studies in other cortical areas as well [19, 20].

2.4 Imaging Data Analysis

The combined whole-cell recording and two-photon calcium imaging experiments as described above allow the recording of sensory-evoked subthreshold dendritic calcium signals; however, post hoc data analysis is often needed to identify the exact dendritic sites of local calcium signaling. A simple and straightforward way to achieve this is to use small, consecutive ROIs covering all segments of the dendrites (Fig. 5). Then, the calcium signals from each ROI, i.e., the relative fluorescence changes over time, can be calculated (Box 3). The choice of ROI size is a compromise

Box 2 Experimental Protocol for Recording Visually Evoked Dendritic Calcium Signals in Cortical Neurons

This box contains a brief experimental protocol adapted from ref. [21].

1. Anesthetize the animal.
2. Clean the skull and fix the recording chamber (Fig. 6a).
3. Perform a small rectangular (1 mm × 0.8 mm) craniotomy above the left visual cortex (Fig. 6c).
4. Cover the exposed brain with agarose (20 mg/ml).
5. Transfer animal to imaging setup and prepare visual stimulation apparatus.
6. Prepare patch pipettes (6–7 MΩ) and pipette solution (112 mM K-gluconate, 8 mM KCl, 10 mM HEPES, 4 mM Mg-ATP, 0.375 mM Na2GTP, 10 mM Na-phosphocreatine, 100 μM OGB-1, and 25 μM Alexa-594; pH adjusted to 7.4 by KOH). For optional post hoc anatomical reconstruction of the dendritic tree, add 2 mg/ml Biocytin to the pipette solution.
7. Perform "shadow patching" to achieve whole-cell recording configuration of a layer 2/3 neuron (as illustrated in Fig. 4).
8. During the first 20–40 min for dye-loading, perform whole-cell recordings without two-photon calcium imaging, except for checking pipette positions.
9. Turn on two-photon imaging and search for a focal plane with many visible dendrites.
10. Hyperpolarize the cell in steps of −10 pA until action potential firing vanishes.
11. Start visual stimulation, whole-cell recording, and two-photon imaging simultaneously.
12. Repeat step 10 for multiple trials.
13. Repeat steps 10–11 for multiple focal planes.
14. Record *z*-stack images of the entire visible dendritic tree.
15. (Optional) Perfuse the animal with paraformaldehyde for post hoc anatomical reconstruction of the recorded neuron.

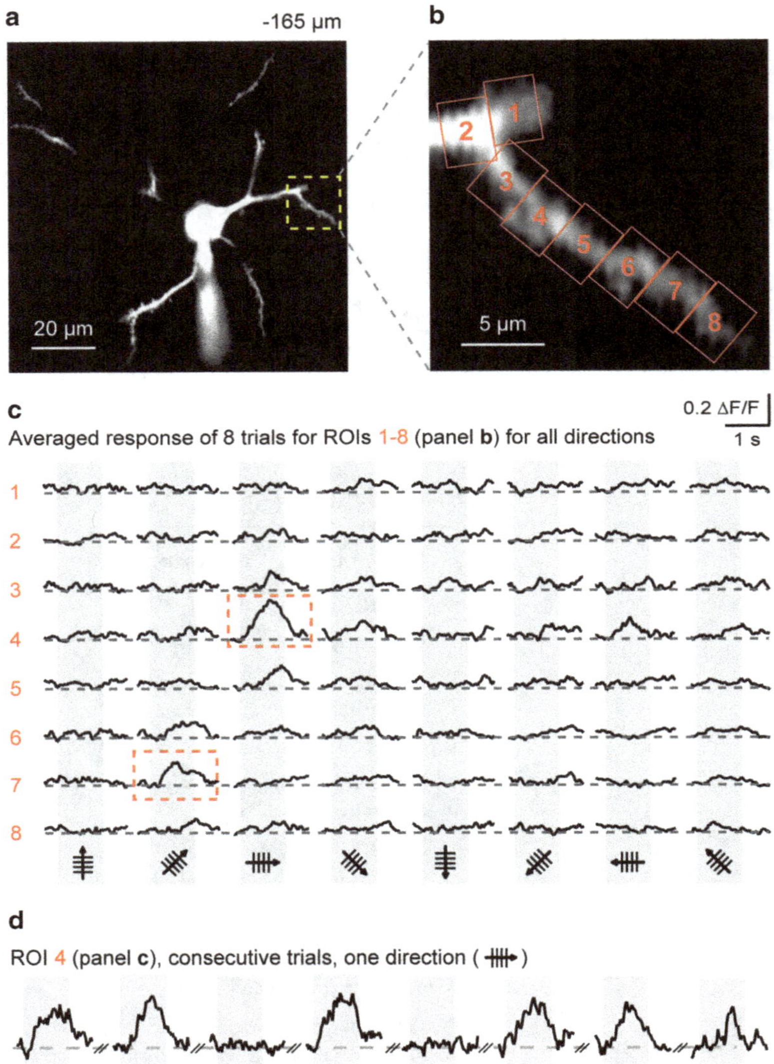

Fig. 5 Analysis of visual stimulation-evoked dendritic hotspots. (**a**) Image of layer 2/3 with a focal plane at 165 μm below the cortical surface. The image (OGB-1 fluorescence, average of 500 frames) shows a two-photon section through the cell body and dendrites of a neuron, the recording patch pipette, and a dashed frame containing a dendritic segment of interest. (**b**) Higher magnification image of the dendritic segment (indicated in (**a**)) covered by eight regions of interest (ROIs), each 3 μm × 4 μm, used to extract visually evoked local dendritic calcium transients. (**c**) Calcium transients (averages of eight consecutive trials) for all eight ROIs (marked in *red*) shown in (**b**), recorded during the presentation of eight directions of a drifting grating. Two local dendritic calcium transients are highlighted by *dashed red frames*. The vertical gray bands indicate the time of visual stimulation by the directions schematically indicated at the bottom. (**d**) Consecutive single calcium transients ($n = 8$) evoked by one direction (as indicated) and recorded in ROI 4 (see **b** and **c**). Scale bar same as (**c**). For all $\Delta F/F$ calculation the background fluorescence was negligible and thus not subtracted. Adapted from ref. [21].

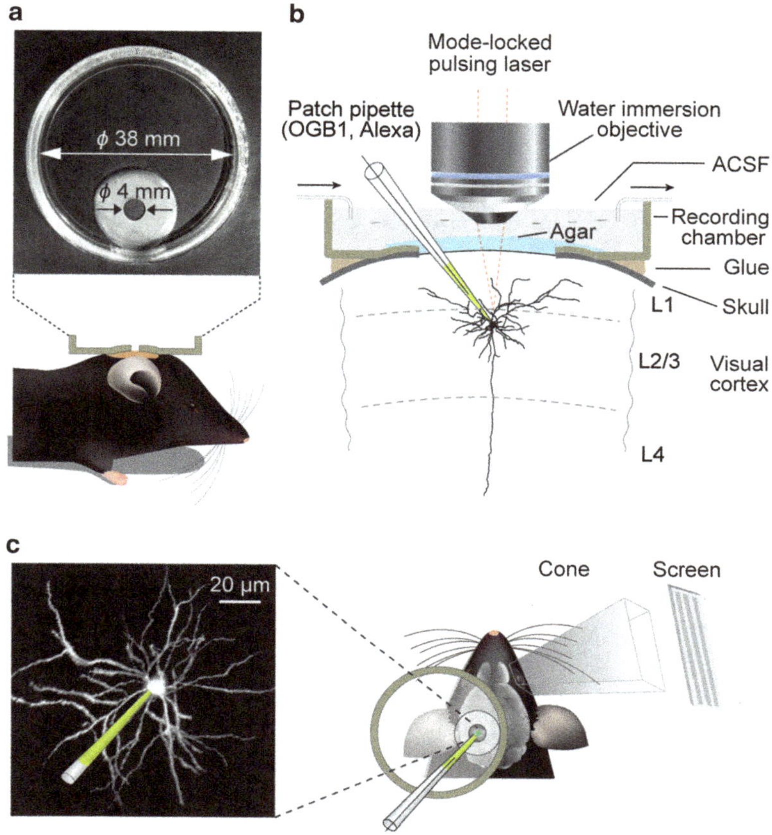

Fig. 6 Setup for two-photon imaging study of visual stimulation-evoked dendritic calcium signals. (**a**) Top view of the recording chamber (*upper panel*). The lower panel schematic shows the recording chamber (side view) fixed to the mouse skull. (**b**) Side view of the experimental arrangement. Note the positions of the dye-containing (OGB-1, Alexa) patch pipette, the water-immersion objective, the agar-covered exposed cortical surface, and the recording chamber glued on the skull. Cortical layers 1, 2/3, and 4 are indicated as L1, L2/3, and L4. The *arrows* indicate the flow direction of the ACSF superfusing the cortical surface. (**c**) *Left panel*, *z*-projection (along the dorsoventral axis) of in vivo two-photon images (from Alexa Fluor 594 fluorescence) of a visual cortical layer 2/3 neuron. The tip of the patch pipette (*green*) is in contact with the cell body. *Right panel*, top view of the recording chamber mounted on the skull, the patch pipette, and the screen displaying visual stimuli. A black paper cone that blocks stray light is placed between the eye and the screen. Adapted from ref. [21].

between the purpose of the study and the limit of the image resolution. On one hand, one would like to decrease ROI size to reveal the fine resolution of local calcium signals along the dendrites and possibly spines; on the other hand, a sufficient number of pixels within the ROI are required in order to reduce the

Box 3 Calculation of Calcium Signals ($\Delta F/F$) from Regions of Interest (ROIs)

1. Calculate the mean fluorescence $F(t)$ of an ROI for each time point t

$$F(t) = \frac{1}{N} \sum_{ROI} f_i(t)$$

where N is the number of pixels in the ROI and $f_i(t)$ is the value of pixel i at time t.

2. Calculate the time-dependent baseline $F_0(t)$
For each time point t, taken as the minimum value of smoothed $F(t)$ during a time window τ_2 before t.

$$F_0(t) = \min_{(t-\tau_2)<x<t} \overline{F(x)}$$

where

$$\overline{F(x)} = \frac{1}{\tau_1} \int_{x-\tau_1/2}^{x+\tau_1/2} F(\tau)\,d\tau$$

3. Calculate the relative change of fluorescence signal $R(t)$ from $F(t)$ and $F_0(t)$

$$R(t) = \frac{F(t) - F_0(t)}{F_0(t)}$$

Note that normally the background fluorescence is negligible; however, in case of high background fluorescence, a reference ROI should be taken in the area in the absence of any dendrites and its averaged pixel value should be subtracted from both $F(t)$ and $F_0(t)$.

4. Apply noise filtering (exponentially weighted moving average) to get final result

$$\Delta F / F(t) = \frac{\int_0^t R(t-\tau) \cdot w(\tau)\,d\tau}{\int_0^t w(\tau)\,d\tau}$$

where

$$w(\tau) = exp\left\{-\frac{|\tau|}{\tau_0}\right\}$$

Notes:

(a) Typically, effective filtering for 30 Hz imaging data is achieved by the following parameters: $\tau_0 = 0.2$ s, $\tau_1 = 0.75$ s, $\tau_2 = 3$ s

(b) This algorithm is robust against both the slow drifting of baseline signal and the fast oscillatory noise due to tissue pulsation, but it may slightly alter the shape of the calcium transient, depending largely on τ_0. In case the recording session is short (10 s or less) and baseline drifting can be neglected, set τ_1, τ_2 to be the same as the session duration to avoid this minor problem.

photonic–electronic noise. Based on our experience [21], when imaging multiple dendrites in a field of view of 100–200 μm size, the minimum size of ROI is about 2–4 μm. When using the LOTOS procedure for imaging a short segment of dendrite of

10–30 μm length, the size of ROI can be 0.5–1 μm [19]. For analyzing consecutive imaging sessions of the same dendrites, it is useful to keep the same set of ROIs while adjusting the positions of ROIs in case of relative drifts between sessions. After the calcium signal in each ROI is calculated, experiment-specific analyses can be performed with standard procedures (e.g., relation of dendritic signals and sensory features; see, for example, [18, 19]).

3 Example Application: Imaging Visual Stimulation-Evoked Local Dendritic Calcium Signals

Figure 7 illustrates an in vivo dendrite imaging experiment performed in a layer 2/3 neuron of mouse visual cortex [18]. As indicated in Fig. 7a, four focal planes were imaged consecutively at various cortical depths. Dendritic "hotspots" were defined as the dendritic sites which have reliable calcium transients (amplitudes at least threefold higher than baseline fluctuations) to at least one of the eight directions of moving gratings stimuli. 12 hotspots were identified in this particular neuron among all recorded dendrites (approx. 9 % of the whole dendritic tree). Figure 7b illustrates the calcium transients from three selected hotspots (number 4, 11, 7) for various stimulation directions. For each hotspot, polar plots of the orientation preference were calculated from the calcium transients (Fig. 7c). An overview of the polar plots revealed the presence of multiple direction and orientation preferences. The results demonstrate that the dendritic distribution of the orientation preferences is randomly distributed throughout the dendritic tree without any obvious rule. In more recent other two studies, the implementation of the LOTOS procedure allowed stable recordings at high zoom factors with minimal phototoxic damage, allowing the recording of calcium signals in individual dendritic spines of the auditory [19] and the vibrissal cortex [20]. While these experiments were performed in layer 2/3 cortical neurons, preliminary evidence indicates that similar recordings can be reliably performed also in layer 4 and layer 5 cortical neurons (Jia, Varga, Hill, Sakmann, Konnerth, in preparation).

4 Discussion and Outlook

The combination of whole-cell patch-clamp recording and two-photon calcium imaging has enabled the identification of sensory-evoked subthreshold local dendritic calcium transients in cortical neurons in vivo. These calcium transients are spatially restricted to specific dendritic segments representing synaptic inputs, whereas single spines are the major contributors of calcium

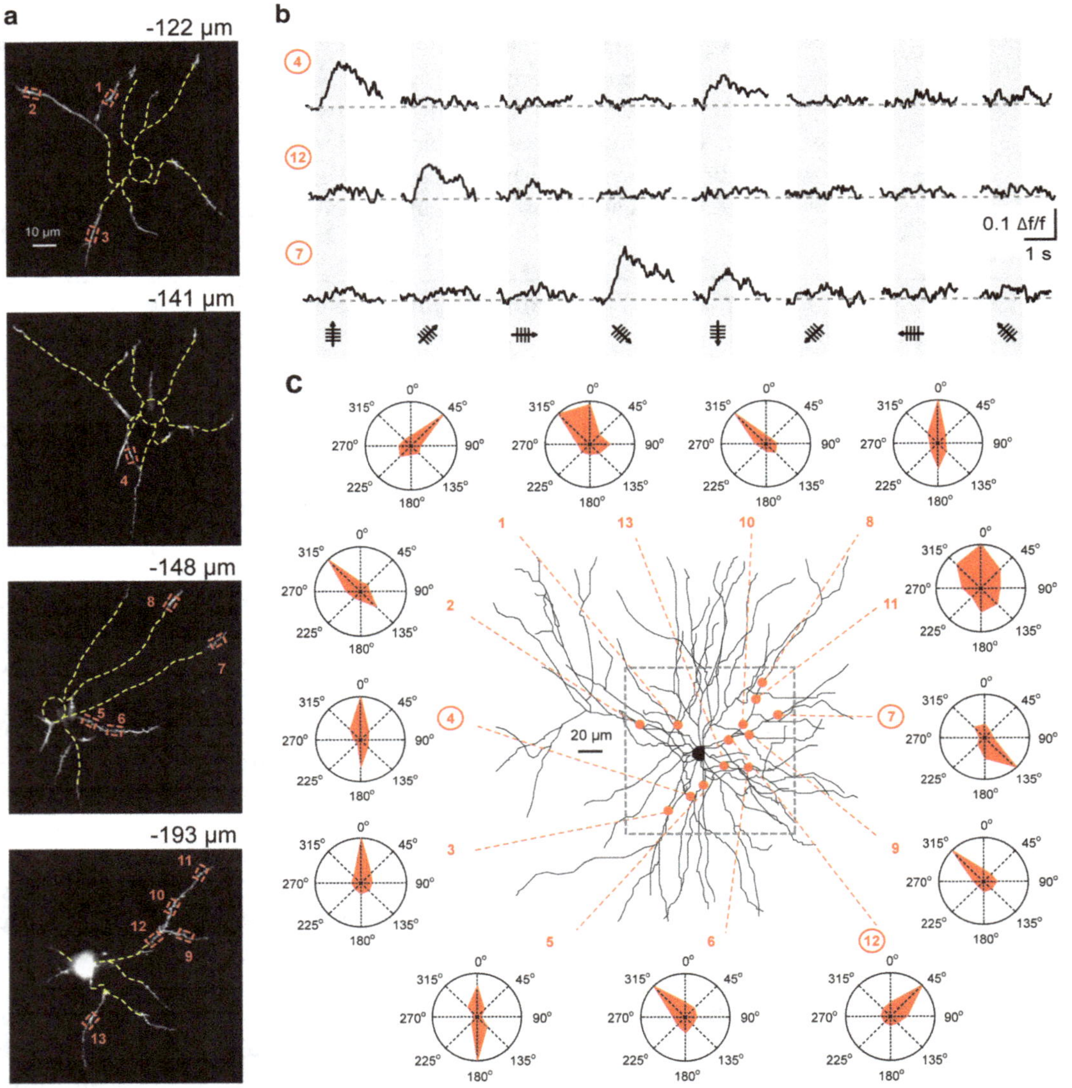

Fig. 7 Distribution pattern of orientation-tuned dendritic hotspots. (**a**) Four two-photon images (each an average of $n = 100$ frames) of a layer 2/3 neuron obtained at different depths under the cortical surface as indicated. *Red dashed boxes* indicate hotspots of local dendritic calcium signals. (**b**) Local dendritic calcium signals evoked by drifting gratings at different orientations (average of six trials) for three different dendritic sites indicated in (**a**). (**c**) Location of each hotspot is indicated as a *red dot* on the *z*-projection of the reconstructed dendritic tree. *Red dashed lines* point to the polar plot obtained for the corresponding local Ca^{2+} signals. The frame (*gray dashed line*) indicates the area of imaging. The output signal of the neuron was tuned for the vertical orientation. Adapted from ref. [18].

transients [19, 20]. A critical feature of this approach is the prevention of action potential back-propagation by hyperpolarizing the neuron. Under such condition, signaling involving the activation of voltage-gated calcium channels (VGCCs) has no major contribution to the localized dendritic Ca^{2+} influx upon synaptic activation and virtually all of the subthreshold synaptically evoked calcium signaling involves activation of NMDA-type glutamate receptor [19]. The revealed local dendritic calcium transients thus identify single synaptic sites that are activated by peripheral sensory stimuli [18–20] and allow a systematic mapping of sensory input sites in vivo.

The methods illustrated in this chapter rely on the use of chemically synthesized small molecule calcium dyes, which have been developed some years ago and have fluorescence properties well suited for sensing changes of intracellular Ca^{2+} concentration over a wide range of affinities [45]. Recently, GECIs have been used with increasing success for neuronal calcium imaging in vivo [46]. Although further improvements are still required, GECIs look promising for imaging dendritic activity, especially in awake animals and during chronic recordings [39, 47], because small molecule calcium dye imaging is largely restricted to acute experiments. One problem that remains to be solved is that dendrites of different neurons are likely to overlap each other, making it difficult to identify the parent neuron belonging to the active dendrite. One solution is to sparsely express GECIs to minimize such overlaps. Alternatively, brainbow mice [48, 49] expressing GECIs may be a useful animal model for large-scale dendritic analyses of interconnected neurons.

In conclusion, the combined whole-cell recording and in vivo two-photon dendritic calcium imaging provides a quite challenging but effective method for studying the fine-scale dendritic organization of the representation of sensory stimuli in single cortical neurons. This method offers a direct view of the excitatory circuitry from the perspective of a single neuron. The ensemble of input sites to a given neuron represents a subset of presynaptic neurons with functionally characterized features. The precise anatomical identification of the presynaptic neurons, the identification of inhibitory synaptic input sites, the analysis of dendritic input sites in other types of neurons, and the recording of dendritic signals in awake behaving animals represent only some of the challenges awaiting experimentalists in the future.

Acknowledgements

We thank Jia Lou for technical assistance with the figures. We thank Dr. Zsuzsanna Varga, Dr. Christine Grienberger, Dr. Xiaowei Chen, and Dr. Dan Hill for critically reading the chapter manuscript. The related works were supported by Deutsche Forschungsgemeinschaft (IRTG 1373), the ERA-Net Program, the Center for Integrated Protein Science Munich, and the Schiedel Foundation.

References

1. Nimchinsky EA, Sabatini BL, Svoboda K (2002) Structure and function of dendritic spines. Annu Rev Physiol 64:313–353
2. Silver RA (2010) Neuronal arithmetic. Nat Rev Neurosci 11:474–489
3. Branco T, Häusser M (2010) The single dendritic branch as a fundamental functional unit in the nervous system. Curr Opin Neurobiol 20:494–502
4. Katz LC, Dalva MB (1994) Scanning laser photostimulation: a new approach for analyzing brain circuits. J Neurosci Methods 54:205–218
5. Nikolenko V, Poskanzer KE, Yuste R (2007) Two-photon photostimulation and imaging of neural circuits. Nat Methods 4:943–950
6. Petreanu L, Mao T, Sternson SM, Svoboda K (2009) The subcellular organization of neocortical excitatory connections. Nature 457:1142–1145
7. Cajal SRY (1995) Histology of the nervous system of man and vertebrates. Oxford University Press, New York
8. Lichtman JW, Livet J, Sanes JR (2008) A technicolour approach to the connectome. Nat Rev Neurosci 9:417–422
9. Higley MJ, Sabatini BL (2008) Calcium signaling in dendrites and spines: practical and functional considerations. Neuron 59:902–913
10. Helmchen F, Svoboda K, Denk W, Tank DW (1999) In vivo dendritic calcium dynamics in deep-layer cortical pyramidal neurons. Nat Neurosci 2:989–996
11. Svoboda K, Helmchen F, Denk W, Tank DW (1999) Spread of dendritic excitation in layer 2/3 pyramidal neurons in rat barrel cortex in vivo. Nat Neurosci 2:65–73
12. Svoboda K, Denk W, Kleinfeld D, Tank DW (1997) In vivo dendritic calcium dynamics in neocortical pyramidal neurons. Nature 385:161–165
13. Stuart G, Spruston N, Sakmann B, Hausser M (1997) Action potential initiation and backpropagation in neurons of the mammalian CNS. Trends Neurosci 20:125–131
14. Spruston N, Schiller Y, Stuart G, Sakmann B (1995) Activity-dependent action potential invasion and calcium influx into hippocampal CA1 dendrites. Science 268:297–300
15. Waters J, Larkum M, Sakmann B, Helmchen F (2003) Supralinear Ca^{2+} influx into dendritic tufts of layer 2/3 neocortical pyramidal neurons in vitro and in vivo. J Neurosci 23:8558–8567
16. Eilers J, Konnerth A (1997) Dendritic signal integration. Curr Opin Neurobiol 7:385–390
17. Bollmann JH, Engert F (2009) Subcellular topography of visually driven dendritic activity in the vertebrate visual system. Neuron 61: 895–905
18. Jia H, Rochefort NL, Chen X, Konnerth A (2010) Dendritic organization of sensory input to cortical neurons in vivo. Nature 464:1307–1312
19. Chen X, Leischner U, Rochefort NL, Nelken I, Konnerth A (2011) Functional mapping of single spines in cortical neurons in vivo. Nature 475:501–505
20. Varga Z, Jia H, Sakmann B, Konnerth A (2011) Dendritic coding of multiple sensory inputs in single cortical neurons in vivo. Proc Natl Acad Sci USA 108:15420–15425
21. Jia H, Rochefort NL, Chen X, Konnerth A (2011) In vivo two-photon imaging of sensory-evoked dendritic calcium signals in cortical neurons. Nat Protoc 6:28–35
22. Denk W, Strickler JH, Webb WW (1990) Two-photon laser scanning fluorescence microscopy. Science 248:73–76
23. Katona G, Kaszas A, Turi GF, Hajos N, Tamas G, Vizi ES, Rozsa B (2011) Roller Coaster Scanning reveals spontaneous triggering of dendritic spikes in CA1 interneurons. Proc Natl Acad Sci USA 108:2148–2153
24. Göbel W, Helmchen F (2007) New angles on neuronal dendrites in vivo. J Neurophysiol 98:3770–3779
25. Iyer V, Hoogland TM, Saggau P (2006) Fast functional imaging of single neurons using random-access multiphoton (RAMP) microscopy. J Neurophysiol 95:535–545
26. Salome R, Kremer Y, Dieudonne S, Leger JF, Krichevsky O, Wyart C, Chatenay D, Bourdieu L (2006) Ultrafast random-access scanning in two-photon microscopy using acousto-optic deflectors. J Neurosci Methods 154:161–174
27. Grewe BF, Langer D, Kasper H, Kampa BM, Helmchen F (2010) High-speed in vivo calcium imaging reveals neuronal network activity with near-millisecond precision. Nat Methods 7:399–405
28. Katona G, Szalay G, Maak P, Kaszas A, Veress M, Hillier D, Chiovini B, Vizi ES, Roska B, Rozsa B (2012) Fast two-photon in vivo imaging with three-dimensional random-access scanning in large tissue volumes. Nat Methods 9:201–208
29. Shannon CE (1949) Communication in the presence of noise. Proc IRE 37:10–21
30. Fan GY, Fujisaki H, Miyawaki A, Tsay RK, Tsien RY, Ellisman MH (1999) Video-rate

scanning two-photon excitation fluorescence microscopy and ratio imaging with cameleons. Biophys J 76:2412–2420

31. Nguyen QT, Callamaras N, Hsieh C, Parker I (2001) Construction of a two-photon microscope for video-rate Ca^{2+} imaging. Cell Calcium 30:383–393
32. Lechleiter JD, Lin DT, Sieneart I (2002) Multiphoton laser scanning microscopy using an acoustic optical deflector. Biophys J 83:2292–2299
33. Kirkby PA, Srinivas Nadella KM, Silver RA (2010) A compact acousto-optic lens for 2D and 3D femtosecond based 2-photon microscopy. Opt Express 18:13721–13745
34. Roorda RD, Hohl TM, Toledo-Crow R, Miesenbock G (2004) Video-rate nonlinear microscopy of neuronal membrane dynamics with genetically encoded probes. J Neurophysiol 92:609–621
35. Eilers J, Augustine GJ, Konnerth A (1995) Subthreshold synaptic Ca^{2+} signalling in fine dendrites and spines of cerebellar Purkinje neurons. Nature 373:155–158
36. Kovalchuk Y, Eilers J, Lisman J, Konnerth A (2000) NMDA receptor-mediated subthreshold Ca^{2+} signals in spines of hippocampal neurons. J Neurosci 20:1791–1799
37. Carter AG, Sabatini BL (2004) State-dependent calcium signaling in dendritic spines of striatal medium spiny neurons. Neuron 44:483–493
38. Heim N, Garaschuk O, Friedrich MW, Mank M, Milos RI, Kovalchuk Y, Konnerth A, Griesbeck O (2007) Improved calcium imaging in transgenic mice expressing a troponin C-based biosensor. Nat Methods 4:127–129
39. Tian L, Hires SA, Mao T, Huber D, Chiappe ME, Chalasani SH, Petreanu L, Akerboom J, McKinney SA, Schreiter ER, Bargmann CI, Jayaraman V, Svoboda K, Looger LL (2009) Imaging neural activity in worms, flies and mice with improved GCaMP calcium indicators. Nat Methods 6:875–881
40. Kitamura K, Judkewitz B, Kano M, Denk W, Häusser M (2008) Targeted patch-clamp recordings and single-cell electroporation of unlabeled neurons in vivo. Nat Methods 5:61–67
41. Nevian T, Helmchen F (2007) Calcium indicator loading of neurons using single-cell electroporation. Pflügers Arch 454:675–688
42. Judkewitz B, Rizzi M, Kitamura K, Häusser M (2009) Targeted single-cell electroporation of mammalian neurons in vivo. Nat Protoc 4:862–869
43. Nagayama S, Zeng S, Xiong W, Fletcher ML, Masurkar AV, Davis DJ, Pieribone VA, Chen WR (2007) In vivo simultaneous tracing and Ca^{2+} imaging of local neuronal circuits. Neuron 53:789–803
44. Chen X, Kovalchuk Y, Adelsberger H, Henning HA, Sausbier M, Wietzorrek G, Ruth P, Yarom Y, Konnerth A (2010) Disruption of the olivo-cerebellar circuit by Purkinje neuron-specific ablation of BK channels. Proc Natl Acad Sci USA 107:12323–12328
45. Tsien RY (1989) Fluorescent probes of cell signaling. Annu Rev Neurosci 12:227–253
46. Rochefort NL, Konnerth A (2008) Genetically encoded Ca^{2+} sensors come of age. Nat Methods 5:761–762
47. Lütcke H, Murayama M, Hahn T, Margolis DJ, Astori S, Zum Alten Borgloh SM, Göbel W, Yang Y, Tang W, Kügler S, Sprengel R, Nagai T, Miyawaki A, Larkum ME, Helmchen F, Hasan MT (2010) Optical recording of neuronal activity with a genetically-encoded calcium indicator in anesthetized and freely moving mice. Front Neural Circuits 4:9
48. Livet J, Weissman TA, Kang H, Draft RW, Lu J, Bennis RA, Sanes JR, Lichtman JW (2007) Transgenic strategies for combinatorial expression of fluorescent proteins in the nervous system. Nature 450:56–62
49. Weissman TA, Sanes JR, Lichtman JW, Livet J (2011) Generating and imaging multicolor Brainbow mice. Cold Spring Harb Protoc 2011:763–769

Chapter 7

In Vivo Population Imaging of Dendritic Integration in Neocortex

Masanori Murayama and Matthew E. Larkum

Abstract

Translating the advances seen recently in recording from neurons in vivo to dendritic recordings presents special difficulties. In vivo two-photon imaging of dendrites was achieved over a decade ago and is still the method of choice for recording from small dendritic compartments in single neurons but has proven more difficult to apply to many dendrites simultaneously or to awake, freely moving preparations. An alternative that can be applied to layer 5 neocortical pyramidal neurons is the use of a fiber-optic method combined with bolus loading of fluorescent calcium indicator. This method takes advantage of the fact that the apical dendrites of these neurons are long (~1 mm) and always project in the same manner towards the cortical surface, thus allowing the fluorescence in these dendrites to be isolated from all other filled structures. Here we describe the details of this approach (also called the "periscope" method) and give examples of the kinds of recordings that can be made with this approach.

Key words Dendrites, Fiberoptic, Periscope, Calcium imaging, Pyramidal neuron, Neocortex

1 Introduction

Some of the first successes of two-photon microscopy for recording from neurons in vivo included imaging of dendritic calcium signals in pyramidal neurons [1, 2]. Despite the great promise of this method, far fewer studies on dendrites in vivo have appeared than from cell bodies [3–7]. Two-photon imaging can be used for simultaneous recordings from multiple cell bodies using membrane-permeant calcium indicators but this approach comes with the disadvantage of high background levels of fluorescence that make it virtually impossible to record from individual dendritic processes [8] (see Chapter 9 in this volume). Successful recordings of dendritic activity using two-photon calcium imaging require a high signal-to-background ratio and have tended therefore to be made from pyramidal neurons in layer 2/3 or layer 4 using intracellular indicator dye filling [2, 9–14]. Another limitation of high-resolution imaging of dendrites is that great stability is required.

Bruno Weber and Fritjof Helmchen (eds.), *Optical Imaging of Neocortical Dynamics*, Neuromethods, vol. 85, DOI 10.1007/978-1-62703-785-3_7,

Under in vivo conditions, both under anesthesia [3] and more strongly in awake animals with head fixation [15], tissue pulsations due to heartbeat, breathing, or animal motion cause movements on the order of ~2–5 μm [16], which is more than the width of a dendrite [3, 15]. More recently, fiber-optic-based miniaturized two-photon microscopes have been introduced as portable devices for imaging in freely moving animals [17–19]; however, they so far have not been successfully applied to dendritic recordings.

Over the past 2 decades, our knowledge of the active properties of dendrites has increased rapidly from direct electrical and optical recordings in vitro [20, 21]. Such recordings in brain slices have shown that the dendrites of pyramidal neurons are capable of very complex local processing [22–24]. However, the more we learn about the biophysical properties in vitro the more computational and functional hypotheses are generated [20, 25–28], which need to be tested in vivo, eventually in the behaving animal. In particular, it has been suggested that top-down inputs from higher cortical areas might have a particular influence on the distal apical dendrites of pyramidal neurons [29] because of the architecture of long-range projections in the cortex [30]. These hypotheses need to be addressed in awake animals because of the lack of top-down activity in the unconscious brain [31].

To enable such experiments, we have developed a method for recording from populations of layer 5 (L5) pyramidal cell dendrites, taking advantage of the fact that these neurons are the only cells in the infragranular layers extending their dendrites into the upper layers. This special spatial arrangement makes it possible to use the bolus-loading technique with a membrane-permeant calcium indicator dye injected into L5 in order to fill apical dendrites in the upper 500 μm of the cortex without background fluorescence from other sources [3]. Using a right-angled prism attached to a fiber-optic cable, a simple epifluorescence custom-built microscope can be used to cheaply and easily record from the dendrites [32]. We have dubbed this approach the "periscope" method, which makes reference to the appearance of the gradient-index (GRIN) lens and prism assembly attached to the end of the fiber-optic cable that is inserted into the cortex. In the following we provide methodological details on this new approach and give examples of its application for in vivo dendritic recording.

2 A "Periscope" Technique for Fiber-Optic Calcium Imaging of Populations of Dendrites

Fiber-optic imaging with the periscope technique depends on specific filling of dendrites with a membrane-permeant calcium indicator dye [3]. Successful filling can be difficult because of the thin caliber of dendrites and the long distance of the distal apical dendritic compartment from the cell body. We have had the greatest

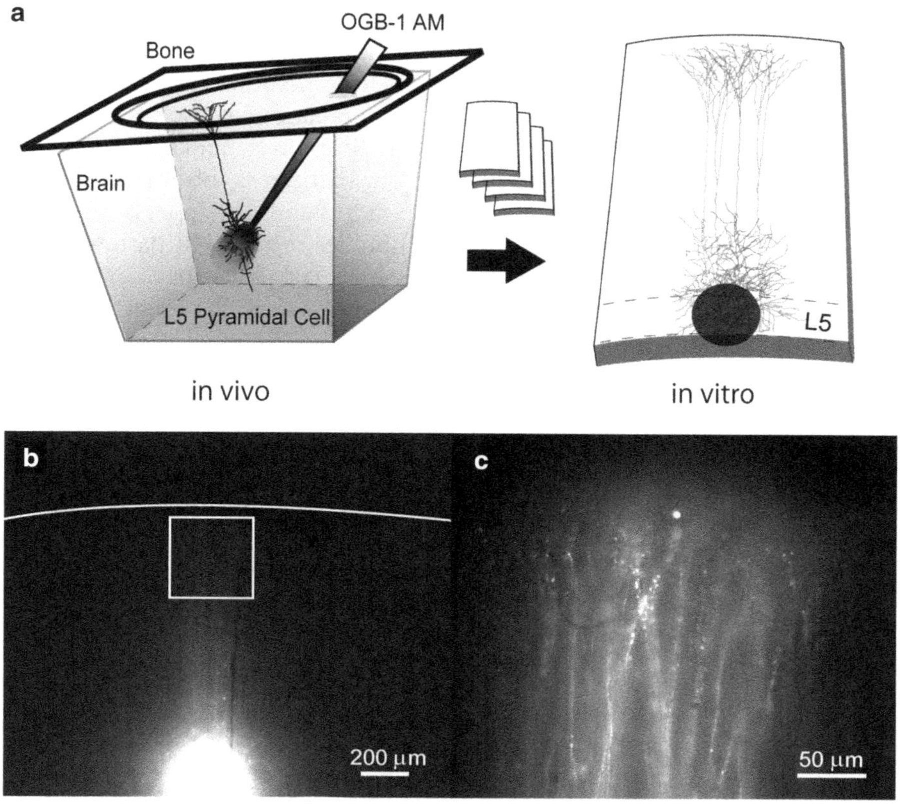

Fig. 1 Loading populations of L5 pyramidal cell dendrites with a calcium indicator. (**a**) Schematized pressure injection of a calcium indicator (OGB-1 AM) to layer 5 through a craniotomy in an anesthetized animal. Brain slices examined ex vivo that should reveal dendritic loading. (**b**) Fluorescence image from a neocortical brain slice following in vivo bolus loading of OGB-1 AM. (**c**) Higher magnification view of the area marked in (**b**), showing labeled vertically running dendrites in the superficial cortical layers.

success using Oregon Green BAPTA-1 AM (OGB-1 AM) which gave better signals than Fura-2 AM, Fluo-3 AM, and Calcium Orange AM [32]. We inject a small amount (<1 μl) of OGB-1 AM at a depth of about 1.2 mm from the pial surface in adult rats (>P28) and allow the dye to enter cells and redistribute by diffusion along the apical dendrites for at least 2 h (Fig. 1a). Since dye loading is critical, we advise optimizing this step with pilot experiments before applying the method. Injection of the bolus of OGB-1 AM must be in the right place in L5 to minimize direct filling of cells in the upper layers, which may occur via extracellular dye diffusion back up the injection pipette tract. It is important initially to perform control experiments checking the filling status of dendrites in brain slices (Fig. 1a–c). We find that this in vitro check following in vivo dye injection can be done quickly and routinely and is highly informative by confirming that signals are originating from filled dendrites. In principle, another method such as two-photon imaging can be used to perform similar control experiments [4] although our experience is that the concentration of dye in the apical dendrites is frequently too low for two-photon imaging [32].

In principle, other labeling methods could be useful for fiber-optic calcium imaging of populations of dendrites, e.g., electroporation [33, 34] or the use of genetically encoded calcium indicators [35]. Because our method involves recording aggregate light from multiple components and the signal is quickly swamped by any stray filling of cell bodies in the light path, bright and specific labeling of dendrites is crucial. Electroporation was not routinely successful in our hands, often filling not many (or any) neurons. Genetically encoded indicators currently are not optimal either, because it is difficult to guarantee that all fluorescence originates from dendritic structures. Perhaps in the future indicator expression can be better confined to a specific local subset of neurons (e.g., L5 pyramidal neurons) [36]. So far, bolus loading of calcium indicator dye has produced the most reliable signals.

The microscope required for recording calcium signals from dendrites can be relatively simply assembled (Fig. 2). The optical

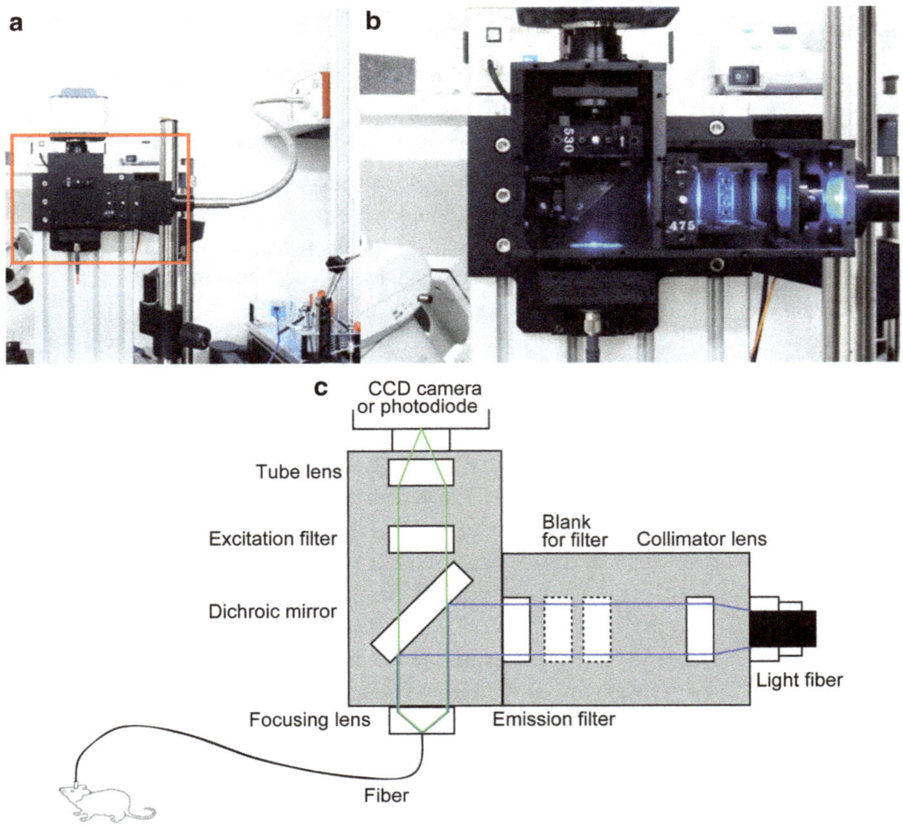

Fig. 2 Microscope. (**a**) Housing containing lens system, camera, and LED illumination via a light guide. (**b**) Higher magnification of (**a**) showing lens and mirror arrangement. (**c**) Schematic diagram of arrangement.

components consist of a 10× lens for coupling light to an optical fiber, a dichroic mirror and a tube lens in front of the CCD camera (or alternatively a photodiode if no spatial resolution is required). The fiber can be a bundle of several thousand fibers (providing spatial resolution) or a single-core fiber if only whole-field recording is required (see Section 4). An LED light source gives enough light, is cheap, and has very little noise although we also used a xenon–mercury lamp successfully [32, 37]. For the front end at the distal end of the optical fiber, we mainly use the GRIN lens and prism assembly approach shown in Fig. 3a. This approach ensures that light is directed horizontally lowering the possibility for light to illuminate the much brighter bolus of dye in L5. In addition, it can provide a read out of fluorescence versus depth, which can be important for determining the type of excitatory event that likely gives rise to the calcium signal (e.g., dendritic spikes versus back-

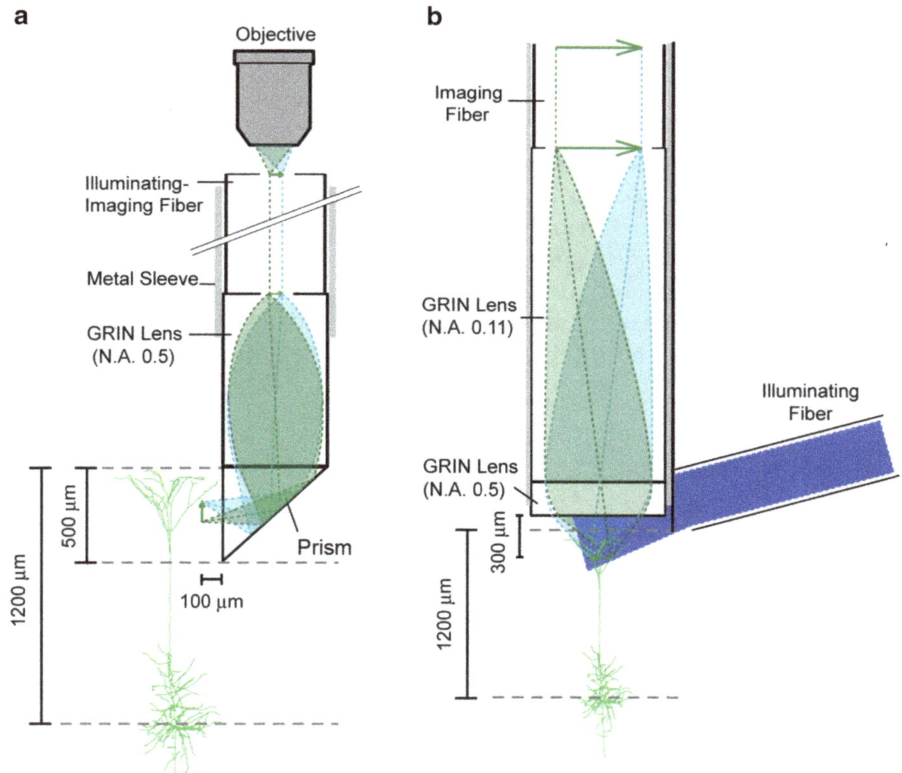

Fig. 3 Front end for fluorescence recordings from dendrites. (**a**) GRIN lens system with prism attachment for imaging horizontally. Here, illumination and light collection are made with the same system. The prism is invasively inserted into the cortex. This method is termed the "periscope system." Focus plane (*vertical green arrow*) can be transmitted to the end of the imaging fiber (*horizontal green arrow*). (**b**) Oblique illumination with separate fiber and vertical imaging GRIN lens system. This system is noninvasive. A single-core multi-mode fiber was used for the illumination fiber to maximally utilize light intensity.

propagating action potentials). Disadvantage of the prism approach is that the prism is invasively inserted into the cortical tissue. To circumvent this problem, we have also developed another approach involving two fibers, one arranged obliquely for illumination, the other for collecting fluorescence light (Fig. 3b). This alternative method is noninvasive and was shown to enable adequate signal recordings [32]. In addition, it is possible to make fluorescence recordings vertically without the prism (illuminating and collecting fluorescence through the same fiber). Whereas we found that there was too much contamination from the bolus of dye in L5 for dendritic recordings, this approach is well suited for recording population responses from L2/3 neurons expressing a genetically encoded indicator [38].

3 Example Applications of Fiber-Optic Dendritic Recordings

The studies of dendritic activity in vivo so far achieved with the periscope system have exceeded expectations. Clear signals can be recorded from a population of dendrites in the somatosensory cortex of rats and mice with evoked sensory stimulation or spontaneous signals in head-fixed, anesthetized, and freely moving preparations [32, 38–40]. Air-puff or cutaneous electrical stimulation of the hindlimb reliably evokes biphasic responses (Fig. 4). The first component of signal was highly susceptible to modulation by $GABA_A$ inhibition [39]. Gabazine applied to the cortical surface increased the size of this component fivefold. The amplitude of the initial component was also sigmoidally related to stimulus strength suggesting that this component is predominantly controlled by the

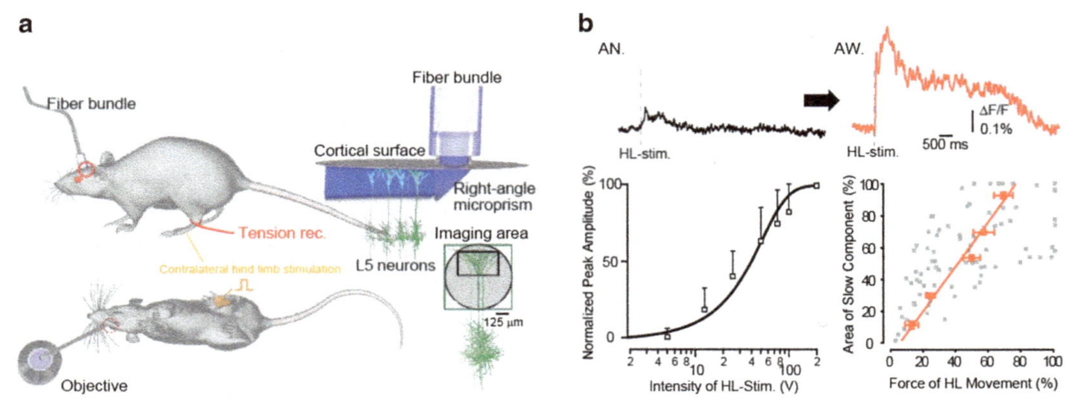

Fig. 4 Recording somatosensory evoked responses in pyramidal dendrites in vivo. (**a**) Schematic diagram of experiment. (**b**) Biphasic responses recorded in anesthetized (*upper left*) and awake (*upper right*) rats. The amplitude of the first component of the response is an approximately sigmoidal function of stimulus strength (*lower left*). The slow component correlates to the strength of the subsequent movement of the hindlimb in awake animals (*lower right*).

feed-forward information from the sensory stimulus through thalamic inputs to the column.

The first component is also regulated by local Martinotti inhibitory neurons located in the deeper layers. Application of TTX to these layers increases the initial dendritic component threefold and a similar effect is achieved by application of CNQX [39]. The effect is explained by the release of dendritic inhibition specifically to the distal apical compartment that is particularly potent in blocking dendritic Ca^{2+} activity [41]. The pyramidal → Martinotti → pyramidal dendrite loop therefore functions as a gain modulation system for setting the operating range of stimuli to maximize the diversity of dendritic activity over the population of apical dendrites within the cortical column.

The second component of the dendritic calcium signal following hindlimb stimulation is increased by an order of magnitude when the animal was released from anesthesia [40]. This activity correlated to the movement of the hindlimb in a manner suggesting volition on the part of the animal (complex movement lasting seconds). This suggests that top-down influences may well be mediating huge increases in dendritic activity in L5 pyramidal neurons although final confirmation remains to be tested. We could however show that no similar increase is detectable in local neuronal activity in the upper cortical layers upon release of anesthesia as has been also shown in several other studies [42–44]. This finding therefore showed that the periscope method could reliably distinguish dendritic activity in the apical dendrites of L5 pyramidal neurons coursing through the upper layers from other forms of activity in the surrounding neurons.

4 Discussion and Outlook

The main advantages of the periscope method are ease of implementation, low cost, and the fact that it can be used in awake, freely moving animals. These advantages make this approach attractive for deriving hypotheses about dendritic activity in L5 pyramidal neurons in vivo. It should also be readily adaptable to different species. The method as presented gives a very low spatial resolution recording of dendritic activity. The resolution is limited because of the dispersion of the fluorescence through several hundred microns of tissue; however, it is possible distinguish differences between the upper and lower halves of the ~600 μm wide field-of-view [32]. For more precise measurements of local changes in calcium in dendrites [10], we suggest the use of two-photon microscopy. The prospect of using two-photon, fiber-optic approaches [17–19] may also be appropriate in the future, but as yet, there have been no convincing reports of dendritic recordings with these systems.

With the periscope system, it is possible to investigate dendritic activity in the apical dendrites of L5 pyramidal neurons, however because of the various compromises of the method, caution should be used in interpreting the data. Calcium signals are regularly used as a proxy for activity but this comes with some caveats. When recording from dendrites, the interpretation of calcium signals is further challenged because of the multiple possible sources of calcium. Somatic (or axonal) activity produces action potentials that propagate back into the dendritic tree [45]. This tends to have relatively little impact on calcium in the dendrite except during high-frequency bursts (>80 Hz) [46, 47] or when coupled with simultaneous synaptic input to the dendrite [29, 47]. Synaptic input in the form of EPSPs was estimated to have very little influence on the signal recorded with the periscope system [32] whereas dendritic electrogenesis in the form of calcium spikes and/or NMDA spikes have a tremendous influence [32, 48, 49]. Therefore, although calcium changes recorded with the periscope system in all likelihood represent activity in apical dendrites, they do not translate simply to AP activity at the cell body.

This technique could be further developed in a number of ways that may make it interesting for various further projects. In principle, it should be possible to develop a wireless system that would allow recordings of dendritic activity in untethered animals. The system could also be readily adapted for providing illumination for excitation of light-sensitive protein such as channelrhodopsin [50]. Furthermore, it is also possible to insert the prism further into the cortex or to use a larger prism that spans more layers [51]. This could be used for excitation (or inhibition) of particular cell types in a subset of cortical lamina. The prism approach has also been used in conjunction with a two-photon microscope for high-resolution imaging in the vertical plane (equivalent to the perspective offered in vitro with brain slices) [51].

The periscope method has so far only been used in the somatosensory cortex of rats but it would be quite easy to probe different areas of the cortex and to use in different species. The prerequisite is that apical dendrites can be filled with a calcium indicator so that they are the only fluorescent elements within the view of the fiber-optic system. Cortical folding is therefore a potential hazard for the technique. However, it might also prove advantageous since - in theory - it could be used without the need for a prism in cortical areas where the pia runs vertically (e.g., on the side of a cortical fold, e.g., the olfactory or cingulate cortex in rats).

In summary, the field of dendritic research continues to burgeon with many new discoveries coming from in vivo preparations that lead directly to functional interpretations [10–14, 39]. The periscope method adds a simple, yet effective tool in pursuing these kinds of experiments that should be considered in designing the experimental program.

References

1. Svoboda K, Denk W, Kleinfeld D, Tank DW (1997) In vivo dendritic calcium dynamics in neocortical pyramidal neurons. Nature 385:161–165
2. Svoboda K, Helmchen F, Denk W, Tank DW (1999) Spread of dendritic excitation in layer 2/3 pyramidal neurons in rat barrel cortex in vivo. Nat Neurosci 2:65–73
3. Stosiek C, Garaschuk O, Holthoff K, Konnerth A (2003) In vivo two-photon calcium imaging of neuronal networks. Proc Natl Acad Sci USA 100:7319–7324
4. Kerr JN, Greenberg D, Helmchen F (2005) Imaging input and output of neocortical networks in vivo. Proc Natl Acad Sci USA 102:14063–14068
5. Ohki K, Chung S, Ch'ng YH, Kara P, Reid RC (2005) Functional imaging with cellular resolution reveals precise micro-architecture in visual cortex. Nature 433:597–603
6. Oka Y, Katada S, Omura M, Suwa M, Yoshihara Y, Touhara K (2006) Odorant receptor map in the mouse olfactory bulb: in vivo sensitivity and specificity of receptor-defined glomeruli. Neuron 52:857–869
7. Kerr JN, de Kock CP, Greenberg DS, Bruno RM, Sakmann B, Helmchen F (2007) Spatial organization of neuronal population responses in layer 2/3 of rat barrel cortex. J Neurosci 27:13316–13328
8. Rochefort NL, Grienberger CM, Konnerth A (2011) In vivo two-photon calcium imaging using multicell bolus loading of fluorescent indicators. In: Helmchen F, Konnerth A (eds) Imaging in neuroscience: a laboratory manual. Cold Spring Harbor Laboratory Press, Cold Spring Harbor, pp 491–500
9. Waters J, Schaefer A, Sakmann B (2005) Backpropagating action potentials in neurones: measurement, mechanisms and potential functions. Prog Biophys Mol Biol 87:145–170
10. Jia H, Rochefort NL, Chen X, Konnerth A (2010) Dendritic organization of sensory input to cortical neurons in vivo. Nature 464:1307–1312
11. Varga Z, Jia H, Sakmann B, Konnerth A (2011) Dendritic coding of multiple sensory inputs in single cortical neurons in vivo. Proc Natl Acad Sci USA 108:15420–15425
12. Chen X, Leischner U, Rochefort NL, Nelken I, Konnerth A (2011) Functional mapping of single spines in cortical neurons in vivo. Nature 475:501–505
13. Palmer LM, Schulz JM, Murphy SC, Ledergerber D, Murayama M, Larkum ME (2012) The cellular basis of GABAB-mediated interhemispheric inhibition. Science 335:989–993
14. Lavzin M, Rapoport S, Polsky A, Garion L, Schiller J (2012) Nonlinear dendritic processing determines angular tuning of barrel cortex neurons in vivo. Nature 490(7420):397–401
15. Dombeck DA, Khabbaz AN, Collman F, Adelman TL, Tank DW (2007) Imaging large-scale neural activity with cellular resolution in awake, mobile mice. Neuron 56:43–57
16. Dombeck D, Tank DW (2011) Two-photon imaging of neural activity in awake mobile mice. In: Helmchen F, Konnerth A (eds) Imaging in neuroscience: a laboratory manual. Cold Spring Harbor Laboratory Press, Cold Spring Harbor, pp 827–838
17. Helmchen F, Fee MS, Tank DW, Denk W (2001) A miniature head-mounted two-photon microscope. high-resolution brain imaging in freely moving animals. Neuron 31:903–912
18. Flusberg BA, Nimmerjahn A, Cocker ED, Mukamel EA, Barretto RP, Ko TH, Burns LD, Jung JC, Schnitzer MJ (2008) High-speed, miniaturized fluorescence microscopy in freely moving mice. Nat Methods 5:935–938
19. Sawinski J, Wallace DJ, Greenberg DS, Grossmann S, Denk W, Kerr JN (2009) Visually evoked activity in cortical cells imaged in freely moving animals. Proc Natl Acad Sci USA 106:19557–19562
20. London M, Häusser M (2005) Dendritic computation. Annu Rev Neurosci 28:503–532
21. Magee JC, Johnston D (2005) Plasticity of dendritic function. Curr Opin Neurobiol 15:334–342
22. Spruston N (2008) Pyramidal neurons: dendritic structure and synaptic integration. Nat Rev Neurosci 9:206–221
23. Losonczy A, Makara JK, Magee JC (2008) Compartmentalized dendritic plasticity and input feature storage in neurons. Nature 452:436–441
24. Larkum ME, Nevian T, Sandler M, Polsky A, Schiller J (2009) Synaptic integration in tuft dendrites of layer 5 pyramidal neurons: a new unifying principle. Science 325:756–760
25. Archie KA, Mel BW (2000) A model for intradendritic computation of binocular disparity. Nat Neurosci 3:54–63
26. Spratling MW (2002) Cortical region interactions and the functional role of apical dendrites. Behav Cogn Neurosci Rev 1:219–228
27. Wang J, Chen S, Nolan MF, Siegelbaum SA (2002) Activity-dependent regulation of HCN

pacemaker channels by cyclic AMP: signaling through dynamic allosteric coupling. Neuron 36:451–461

28. Larkum ME, Nevian T (2008) Synaptic clustering by dendritic signalling mechanisms. Curr Opin Neurobiol 18:321–331
29. Larkum ME, Senn W, Lüscher HR (2004) Top-down dendritic input increases the gain of layer 5 pyramidal neurons. Cereb Cortex 14: 1059–1070
30. Felleman DJ, Van Essen DC (1991) Distributed hierarchical processing in the primate cerebral cortex. Cereb Cortex 1:1–47
31. Boly M, Garrido MI, Gosseries O, Bruno MA, Boveroux P, Schnakers C, Massimini M, Litvak V, Laureys S, Friston K (2011) Preserved feedforward but impaired top-down processes in the vegetative state. Science 332:858–862
32. Murayama M, Perez-Garci E, Lüscher HR, Larkum ME (2007) Fiberoptic system for recording dendritic calcium signals in layer 5 neocortical pyramidal cells in freely moving rats. J Neurophysiol 98:1791–1805
33. Nagayama S, Zeng S, Xiong W, Fletcher ML, Masurkar AV, Davis DJ, Pieribone VA, Chen WR (2007) In vivo simultaneous tracing and Ca^{2+} imaging of local neuronal circuits. Neuron 53:789–803
34. Göbel W, Helmchen F (2007) New angles on neuronal dendrites in vivo. J Neurophysiol 98:3770–3779
35. Mank M, Griesbeck O (2008) Genetically encoded calcium indicators. Chem Rev 108: 1550–1564
36. Mittmann W, Wallace DJ, Czubayko U, Herb JT, Schaefer AT, Looger LL, Denk W, Kerr JN (2011) Two-photon calcium imaging of evoked activity from L5 somatosensory neurons in vivo. Nat Neurosci 14:1089–1093
37. Murayama M, Larkum ME (2009) In vivo dendritic calcium imaging with a fiberoptic periscope system. Nat Protoc 4:1551–1559
38. Lütcke H, Murayama M, Hahn T, Margolis DJ, Astori S, Zum Alten Borgloh SM, Göbel W, Yang Y, Tang W, Kügler S, Sprengel R, Nagai T, Miyawaki A, Larkum ME, Helmchen F, Hasan MT (2010) Optical recording of neuronal activity with a genetically-encoded calcium indicator in anesthetized and freely moving mice. Front Neural Circ 4:9
39. Murayama M, Pérez-Garci E, Nevian T, Bock T, Senn W, Larkum ME (2009) Dendritic encoding of sensory stimuli controlled by deep cortical interneurons. Nature 457:1137–1141
40. Murayama M, Larkum ME (2009) Enhanced dendritic activity in awake rats. Proc Natl Acad Sci USA 106:20482–20486
41. Pérez-Garci E, Gassmann M, Bettler B, Larkum ME (2006) The $GABA_{B1b}$ isoform mediates long-lasting inhibition of dendritic Ca^{2+} spikes in layer 5 somatosensory pyramidal neurons. Neuron 50:603–616
42. Greenberg DS, Houweling AR, Kerr JN (2008) Population imaging of ongoing neuronal activity in the visual cortex of awake rats. Nat Neurosci 11:749–751
43. de Kock CP, Sakmann B (2008) High frequency action potential bursts (>or = 100 Hz) in L2/3 and L5B thick tufted neurons in anaesthetized and awake rat primary somatosensory cortex. J Physiol 586:3353–3364
44. Lee AK, Manns ID, Sakmann B, Brecht M (2006) Whole-cell recordings in freely moving rats. Neuron 51:399–407
45. Stuart G, Schiller J, Sakmann B (1997) Action potential initiation and propagation in rat neocortical pyramidal neurons. J Physiol 505:617–632
46. Helmchen F, Svoboda K, Denk W, Tank DW (1999) In vivo dendritic calcium dynamics in deep-layer cortical pyramidal neurons. Nat Neurosci 2:989–996
47. Larkum ME, Kaiser KM, Sakmann B (1999) Calcium electrogenesis in distal apical dendrites of layer 5 pyramidal cells at a critical frequency of back-propagating action potentials. Proc Natl Acad Sci USA 96: 14600–14604
48. Schiller J, Schiller Y, Stuart G, Sakmann B (1997) Calcium action potentials restricted to distal apical dendrites of rat neocortical pyramidal neurons. J Physiol 505:605–616
49. Schiller J, Major G, Koester HJ, Schiller Y (2000) NMDA spikes in basal dendrites of cortical pyramidal neurons. Nature 404:285–289
50. Deisseroth K (2011) Optogenetics. Nat Methods 8:26–29
51. Chia TH, Levene MJ (2009) Microprisms for in vivo multilayer cortical imaging. J Neurophysiol 102:1310–1314

Chapter 8

Voltage-Sensitive Dye Imaging of Cortical Dynamics

Carl C.H. Petersen

Abstract

Voltage-sensitive dyes can be topically applied to the surface of the brain in order to label membranes in the supragranular layers of the neocortex. Fluorescence signals from voltage-sensitive dyes in vivo correlate linearly with membrane potential of excitatory layer 2/3 neurons with millisecond temporal resolution and with a spatial resolution resolving single cortical columns. Voltage-sensitive dye imaging across large cortical areas in awake behaving mice reveals interactions between somatosensory and motor cortex that depend strongly on behavioral state. The recent development of genetically encoded voltage-sensitive fluorescent proteins will likely help towards long-term measurements of the electrical activity of specific cell types.

Key words Voltage-sensitive dye, Membrane potential, Cortical spatiotemporal dynamics

1 Introduction

The neocortex can be divided into separate areas specialized for different functional roles, e.g., motor cortex for movement control, somatosensory cortex for sensation of touch, and visual cortex for vision. These broader cortical areas can often be further subdivided into primary and secondary areas, each of which usually contains well-organized maps. Spatial specificity of information processing in the neocortex is therefore a well-accepted concept. Nonetheless, it is also clear that information needs to be integrated across brain areas to generate unified and useful precepts for governing behavior. Very little is currently known about how information is transferred from one brain area to another. Voltage-sensitive dye imaging offers the possibility to monitor such interactions amongst different cortical areas with millisecond temporal resolution and subcolumnar spatial resolution. Voltage indicators are available as synthetic organic dyes, as well as fluorescent protein-based sensors. In this chapter, experiments with synthetic voltage indicators are described, whereas genetically encoded voltage-sensors are described in Chapter 4 of this volume.

Bruno Weber and Fritjof Helmchen (eds.), *Optical Imaging of Neocortical Dynamics*, Neuromethods, vol. 85,
DOI 10.1007/978-1-62703-785-3_8, © Springer Science+Business Media New York 2014

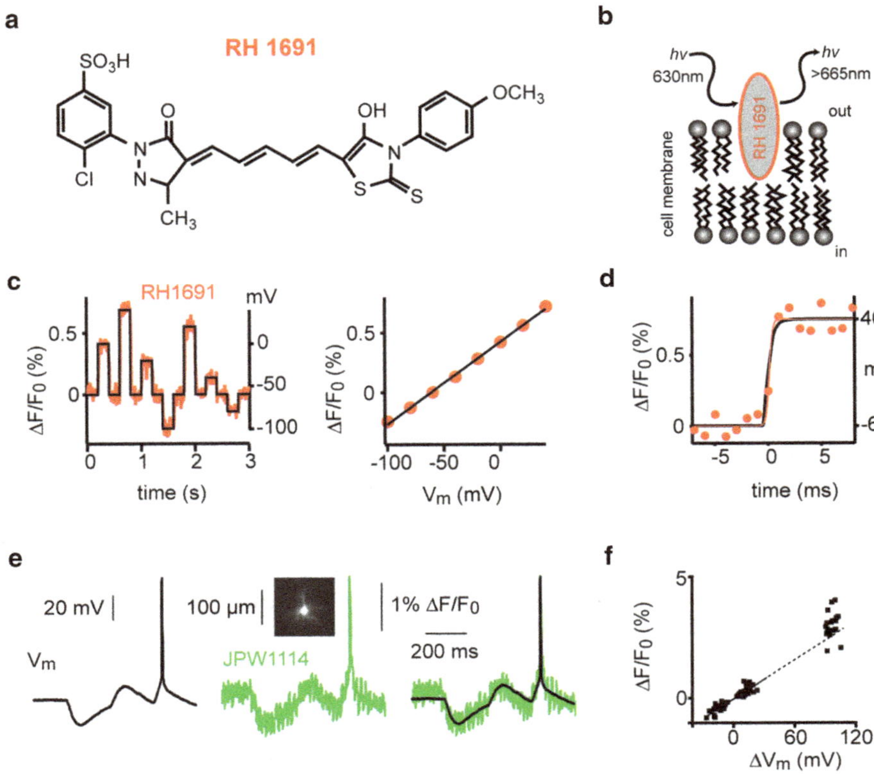

Fig. 1 Fast and linear fluorescence response of voltage-sensitive dyes. (**a**) Blue oxonal dyes, such as RH1691 shown here, are useful voltage-sensitive dyes for staining the neocortex for in vivo imaging. (**b**) The voltage-sensitive dye RH1691 is lipophilic and preferentially partitions into lipid membranes. When applied from the extracellular space, it preferentially inserts into the outer leaflet of the plasma membrane, where it senses the transmembrane potential. The electrical field alters the fluorescence of the dye molecule, which is excited by red light (~630 nm) and longer wavelength fluorescent photons (>665 nm) are typically measured. (**c**) RH1691 was applied to the large surface area of a *Xenopus* oocyte under two-electrode voltage-clamp. Changes in membrane potential result in linearly correlated changes in fluorescence. (**d**) Viewed at higher time resolution, the voltage-step induced in the *Xenopus* oocyte drives a rapid change in RH1691 fluorescence on the millisecond timescale. (**e**) Introduction of the voltage-sensitive dye JPW1114 into an excitatory pyramidal neuron in the somatosensory neocortex in a mouse brain slice in vitro. Hyperpolarizing or depolarizing current injections evoked membrane potential changes, which were followed optically at millisecond temporal resolution through voltage-sensitive dye imaging. Importantly both subthreshold membrane potential changes and action potentials are well-resolved. (**f**) Voltage-sensitive dye JPW1114 fluorescence from single pyramidal neurons in brain slices correlates linearly with membrane potential (Modified and reproduced from [4]).

A large variety of organic dyes showing voltage-dependent changes in spectral properties have been synthesized [1]. The key feature of any voltage-sensitive dye is that it must partition preferentially into the plasma membrane, so that it can sense the transmembrane potential. Different tissues appear to be stained better by some dyes than others, requiring extensive testing in the search for useful voltage-sensitive dyes for specific purposes. For in vivo voltage-sensitive dye imaging of neocortex, the blue dyes RH1691, RH1692, and RH1831 currently appear to be the best-suited dyes (Fig. 1a) [2, 3].

The long-wavelength excitation (630 nm) and emission (>665 nm) of these blue dyes (Fig. 1b) is helpful for in vivo voltage-sensitive dye imaging since hemoglobin absorbs much more strongly at shorter wavelengths. Minimizing hemoglobin absorption reduces heartbeat-related artifacts, allowing high signal-to-noise ratio (SNR) measurements on single trials without the need for averaging. Testing RH1691 on voltage-clamped *Xenopus* oocytes revealed fluorescence changes that were linearly related to membrane potential (Fig. 1c) [4]. In addition, the RH1691 fluorescence changes observed in voltage-clamped *Xenopus* oocytes had fast kinetics, following membrane potential changes on the millisecond timescale (Fig. 1d).

Voltage-sensitive dyes can also be loaded into single neurons in brain slices in vitro through whole-cell recording pipettes [4–8]. In general, different voltage-sensitive dyes appear well-suited for staining the plasma membrane from the inside compared to the outside. Voltage-sensitive dye JPW1114 can be loaded into individual neocortical pyramidal neurons through patch-pipettes and fluorescence transients were found to correlate linearly on the millisecond timescale with membrane potential changes evoked by current injections [4]. Subthreshold hyperpolarizations, subthreshold depolarizations, and action potentials were all closely matched between fluorescence signal and membrane potential recorded in the whole-cell configuration (Fig. 1e, f) [4].

Voltage-sensitive dyes therefore provide optical tools for imaging membrane potential in real-time with millisecond resolution. The fluorescence signals offer straightforward interpretation since they are linearly related to changes in membrane potential. The major difficulties result from the small changes in voltage-sensitive dye fluorescence associated with the physiological range of membrane potentials, typically much less than 1 % fractional change in fluorescence per 10 mV change in membrane potential. In order to resolve these small and rapid changes in voltage-sensitive dye signals from the background shot noise, it is critical to collect many emitted photons using fast cameras with large well-depths (Box 1).

Box 1 Camera Requirements

Voltage-sensitive dye signals are small (fractional changes in fluorescence are typically ~0.1 %) and fast (membrane potential changes occur on the millisecond timescale). Specialized cameras are therefore needed to detect these signals. Over the last decade, a large number of commercially available high-speed cameras have been developed providing frame rates up to ~10 kHz. However, only few commercially available cameras have sufficient SNRs to be capable of resolving the small fractional changes in light intensity from voltage-sensitive dyes. For in vivo voltage-sensitive dye imaging, the major source of measurement noise is the so-called shot noise. Dark noise, which typically dominates at low light levels, is not

(continued)

Box 1 (continued)

important for in vivo voltage-sensitive imaging, where there is typically a high level of fluorescence from the large area of the cortex stained with voltage-sensitive dye.

Shot noise results from the discrete nature of light in the form of photons. For any given stable light level, the actual number of photons emitted by the source varies when measured over a fixed period of time. For example, for the lowest light levels, either one photon or no photons would be counted. For higher light levels, the variability decreases. The standard deviation of the number of emitted photons per unit time from a stable light source varies as the square root of the number of emitted photons. The SNR can therefore be defined as:

$$\mathrm{SNR} = \mathrm{N}/\mathrm{sqrt}(\mathrm{N}) = \mathrm{sqrt}(\mathrm{N})$$

In order to resolve a 0.1 % signal change, we require an SNR of at least 1,000. From the above formula, we can see that we need to collect $1{,}000^2 = 1{,}000{,}000$ photons in order to resolve a 0.1 % signal change. The number of photons that can be usefully collected by a pixel on a camera and transformed into an electrical signal for further image processing is determined by the so-called well-depth. Cameras suitable for in vivo voltage-sensitive dye imaging should therefore have well-depths on the order of a million electrons, as well as having high frame rates.

2 Methods for In Vivo Voltage-Sensitive Dye Imaging of Large-Scale Network Activity

2.1 Staining the Cortex with Voltage-Sensitive Dye

The first technical aspect to consider is how to introduce voltage-sensitive dyes into the brain region of interest. In order to image large cortical areas, the dye obviously also needs to stain these regions. Whereas injections of dye are possible and work well, it is difficult to evenly label large brain areas with multiple injections. Fortunately, some voltage-sensitive dyes can simply be applied to the brain surface and over time diffuse into the neocortex. This is the case with the blue oxonal dye RH1691 [2], which is water-soluble (i.e., does not require DMSO) at the concentrations necessary for staining neocortex (0.1–1 mg/ml). The bone overlying the cortical region of interest must be removed and usually it is best to also remove the dura, although the dye can also penetrate the intact dura after extensive drying [9]. The dye diffuses into the cortex over a staining period of approximately 1 h and afterwards it is important to wash away unbound dye. The surgery must be performed under anesthesia, but the staining can be performed either under anesthesia or in the awake freely moving animal in its home cage, if a suitable sealed chamber is used. If the brain is subsequently fixed or rapidly frozen and cut into coronal sections, then the voltage-sensitive dye RH1691 staining of mouse somatosensory cortex is observed primarily in the supragranular layers, with less intense staining of deeper cortical layers (Fig. 2a, c) [10].

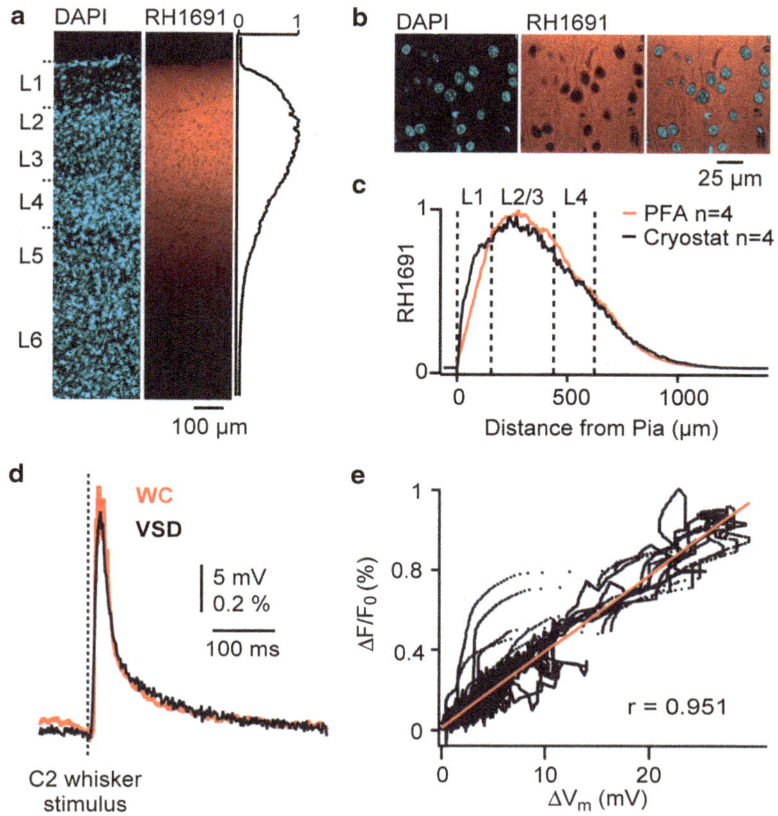

Fig. 2 Voltage-sensitive dye signals in vivo correlate with membrane potential changes of layer 2/3 excitatory pyramidal neurons. (**a**) RH1691 was applied to the exposed surface of mouse barrel cortex for a staining period of over 1 h followed by extensive rinsing to remove unbound dye. After a brief imaging session to confirm functional staining, the mouse was sacrificed and the brain sliced into coronal sections. The dye was found to have diffused into the neocortex, primarily staining the superficial neocortical layers 1–3. Lower dye concentration in layer 1 might reflect washout of the dye during the rinsing period. (**b**) Higher resolution confocal image of the same slice as shown in (**a**). The voltage-sensitive dye RH1691 is found in the neuropil and not inside the cell body. This staining pattern is consistent with the localization of the dye to plasma membranes. (**c**) Similar depth profile of RH1691 fluorescence was observed in different mice and did not depend upon whether the mouse was perfusion-fixed with paraformaldehye (PFA) or whether the brain was frozen and sectioned on a cryostat. (**d**) Whole-cell membrane potential recording was targeted to layer 2/3 of the C2 barrel column during simultaneous voltage-sensitive dye imaging. A brief deflection of the C2 whisker evoked a brief depolarization in the membrane potential of the neuron recorded in the whole-cell configuration. The voltage-sensitive dye signal increased fluorescence transiently in response to the whisker deflection with an almost identical time course to the membrane potential. (**e**) Fluorescence signals from RH1691 in vivo are highly correlated with membrane potential changes evoked by whisker stimulation (Modified and reproduced from [10], with permission from Elsevier).

Viewed at higher resolution with confocal microscopy, it is clear that the RH1691 stains neuropil and is excluded from intracellular compartments such as the cell bodies (Fig. 2b) [10]. This is consistent with localization of the voltage-sensitive dye in plasma membranes, which is where the dye needs to be in order to be sensitive to membrane potential. It is of the utmost importance not to damage the cortex during surgery and any blood on the cortical surface prevents successful imaging. Once a good preparation has been well-stained, then the dye remains in the cortex for several hours, during which functional imaging can reliably be performed. In order to stabilize the brain, it is important to cover the exposed and stained cortex with agarose (1–2 %) and a glass cover slip. Making a tight seal with the cover slip over the entire craniotomy helps considerably with stabilizing the brain from physical movement relative to the skull.

2.2 Imaging Cortex of Anesthetized Animals

Anesthesia offers a high level of stimulus control allowing detailed investigation of sensory responses in the neocortex. In addition, the lack of body movements in anesthetized animals means that the brain is also more stable. Head-posts should be directly attached to the skull to minimize head-movement relative to the voltage-sensitive dye imaging apparatus. In rats and mice with proper head-fixation, breathing does not cause any detectable artifacts during RH1691 voltage-sensitive dye imaging [10]. Despite the long-wavelength red light used for imaging RH1691 (which strongly reduces hemoglobin absorption compared to other dyes excited by green or blue light), a heartbeat-related artifact is evident during voltage-sensitive imaging experiments, especially in regions immediately surrounding large surface blood vessels. The heartbeat-related artifact is stable with respect to the electrocardiogram and if the electrocardiogram is recorded, the heartbeat-related artifacts can therefore be subtracted in post hoc image processing. In regions without large surface blood vessels, there is almost no heartbeat-related artifact. Depending upon the experimental goal, it may therefore not be necessary to process images to remove heartbeat-related artifacts. Dye bleaching is another important aspect to consider. Both spontaneous and evoked voltage-sensitive dye signals are small in amplitude (often around 0.1 %) and dye bleaching over a few seconds of imaging is typically of around the same magnitude. It is therefore of critical importance to subtract the bleaching component of the signal before any further data analyses. The best approach is often to interleave trials with and without sensory stimulation. Averaging the trials without stimuli should lead to a smooth bleaching curve for each pixel in the image, which can then be subtracted from individual stimulated trials. Alternatively, if trials without stimuli are not recorded, then a double exponential curve can be fitted to the averaged trials and subtracted from the single-trial data. However, it is important to note that stimuli can evoke long-lasting reverberations in cortical activity and it is obviously important that the fit of the exponential bleaching curve is not affected by the sensory-evoked response.

Comparison of membrane potential recorded from a single layer 2/3 neuron using the whole-cell patch-clamp technique in vivo with simultaneous voltage-sensitive dye imaging reveals that the fluorescence signal is highly correlated to subthreshold membrane potential changes (Fig. 2d, e) [4, 10–12]. The high correlation might seem surprising given that the membrane potential is measured in only one neuron whereas the voltage-sensitive dye signal results from neuronal membrane belonging to thousands of neurons. A likely explanation comes from dual whole-cell recordings, which reveal that subthreshold membrane potential fluctuations are in fact highly correlated in nearby neurons under many experimental conditions [11–15]. Importantly, action potential firing is uncorrelated in dual recordings from excitatory neurons [14, 15]. Therefore voltage-sensitive dye signals from a population of cortical neurons mainly reflect their correlated subthreshold membrane potential dynamics with only a minor impact of action potential firing.

The spatiotemporal dynamics of cortical processing in many different sensory systems of anesthetized animals has been successfully investigated using similar techniques to those described above. Voltage-sensitive dye imaging of visual cortex has been carried out in anesthetized cats [16–19], rats [20–22], and ferrets [23, 24]. Odor representations in the olfactory bulb of anesthetized rats and mice have also been imaged with voltage-sensitive dyes [25]. Tactile sensory responses have also been investigated in the somatosensory cortex of anesthetized rats [11, 12, 26–28] and mice [10, 29–31].

2.3 Imaging Cortex of Awake Head-Restrained Animals

Studying awake animals allows cortical processing to be investigated in more physiological states. Importantly, in experiments with awake animals it is possible to establish direct correlations between behavior and cortical activity. Many animals can be trained to accept head-restraint (including mice, rats, and monkeys), therefore in many respects offering similar experimental conditions to those of anesthetized animals. Although head-movements are minimized under these experimental conditions, other body movements can still be made and therefore stimulus control is usually somewhat less precise in the awake head-restrained preparation compared to anesthesia. Brain movement artifacts within the skull can be significant in awake animals despite head-restraint, especially during rapid movements often associated with the execution of learned goal-directed motor programs. Carefully sealed craniotomies and stable head-fixation are critical for imaging in awake head-restrained animals.

Imaging in awake animals is clearly more challenging than equivalent measurements under anesthesia and fewer studies have so far been published relating to voltage-sensitive dye imaging in awake animals. Sensory processing in early visual cortex of awake monkeys including during saccadic eye movements has been imaged using voltage-sensitive dye [32–35]. In awake head-restrained mice, spontaneous activity [30, 36] and whisker-deflection-evoked sensory responses in somatosensory cortex [30, 37] have been imaged using voltage-sensitive dyes.

2.4 Fiber-Optic Imaging of Cortex in Freely Moving Animals

For some experimental paradigms, it is essential that the animals are freely moving. In addition, freely moving animals are presumably in a more physiological state compared to during head-restraint and it is therefore important to be able to investigate spatiotemporal dynamics of cortical processing in freely moving animals. The simplest approach for imaging moving animals is to use fiber-optic image bundles [10], in which each individual fiber of the well-ordered array transmits a pixel. The diameter of each individual fiber can be as small as a few microns and the whole image bundle can be composed of many tens of thousands of fibers. The fiber array is then used as a 1:1 image relay. One end of the fiber is placed in the normal focal plane of the voltage-sensitive dye imaging setup, where the cortex would normally be located in a head-restrained experiment. The other end of the fiber can be placed in direct contact with the cortex or with a glass cover slip placed in between [10]. The spatial resolution of imaging through a fiber-optic image bundle is surprisingly similar to direct imaging of the cortex, presumably reflecting the relatively low (~50–100 μm) resolution of the in vivo voltage-sensitive dye imaging technique in general. There are however considerable light losses associated with imaging through the optical fibers. Collecting large numbers of emitted photons is critical for improving SNRs, and therefore imaging via fiber bundles has so far resulted in more noisy images. Movement artifacts in the voltage-sensitive dye imaging signals were surprisingly small in freely moving animals, presumably because of the direct contact of the optical fibers to the brain, which likely hinders their relative motion [10]. Nonetheless, it is important to note that large movement artifacts did occur when the animals abruptly ran into objects and walls. Future advances might include adding lenses to the fiber bundle to improve light collection and to alter field of view. Fiber optics might also be useful for imaging deep brain areas stained with voltage-sensitive dyes, but this has not yet been tested.

3 Example Application: Voltage-Sensitive Dye Imaging of Mouse Sensorimotor Cortex During Behavior

The most basic operation of the brain is to utilize sensory input to guide motor output. Somatosensation plays an important role in guiding most movements and therefore some of the simplest behaviors that can be studied in mammals are based on tactile sensorimotor loops. Reflexive sensorimotor loops can be executed via neural circuits in the spinal cord [38], but more complex goal-directed sensorimotor loops require participation of higher brain areas such as the cerebral cortex.

Mice extensively use mystacial vibrissae to obtain tactile information about their local environment [39–41]. The array of whisker hairs are highly organized in a stereotypical layout, allowing the same individual whiskers to be identified in different mice. The whiskers are actively moved forwards and backwards at high frequencies (5–20 Hz), when the mouse is actively exploring its environment. In order for mice to localize objects within the whisker trajectory, it is important to combine motor control signals with sensory input. Equally, sensory input is likely to alter motor control, so that optimal sensory information can be actively acquired. In this tactile sensory system, one would therefore expect strong interactions between sensory and motor areas of the brain. Voltage-sensitive dye imaging can be used to explore interactions between sensory and motor cortices.

3.1 Imaging a Cortical Sensorimotor Loop

By opening large craniotomies encompassing as much as possible of the dorsal neocortex, it is possible to stain almost the entire sensorimotor cortex of the mouse with voltage-sensitive dye [30, 36]. Voltage-sensitive dye imaging in awake head-restrained mice can then be used to examine the correlated membrane potential dynamics of different parts of the sensory and motor cortices in response to well-defined tactile stimuli during quantified whisker-related behavior. The primary somatosensory cortex of the mouse contains an exquisite map of the mystacial vibrissae, such that each whisker is individually represented in an anatomically defined "barrel" laid out in the same pattern as the whiskers on the snout.

Whisker deflections evoke activity in sensory neurons of the trigeminal nerve, which innervate the whisker follicle (Fig. 3a) [42–44]. The sensory information is signalled across glutamatergic synapses in the brainstem trigeminal complex, which project, in part, to the somatosensory thalamus. The thalamus in turn sends the tactile information to the somatosensory cortex [39–41]. At each station (brainstem, thalamus, and cortex) of the whisker signalling pathway there is a high degree of somatotopic order with segregated anatomical units ("barrelettes" in the brainstem, "barreloids" in the ventral posterior medial thalamus, and "barrels" in the cortex). Given the high degree of anatomical order, it is not surprising that a brief mechanical deflection applied to a single whisker evokes cortical activity that is initially highly localized in the primary somatosensory barrel cortex. The earliest whisker-evoked response that can be measured in the cortex (~7 ms after whisker deflection) is spatially confined to the somatotopically aligned cortical barrel column (Fig. 3b) [10, 11]. Depending upon stimulus amplitude [4, 11],

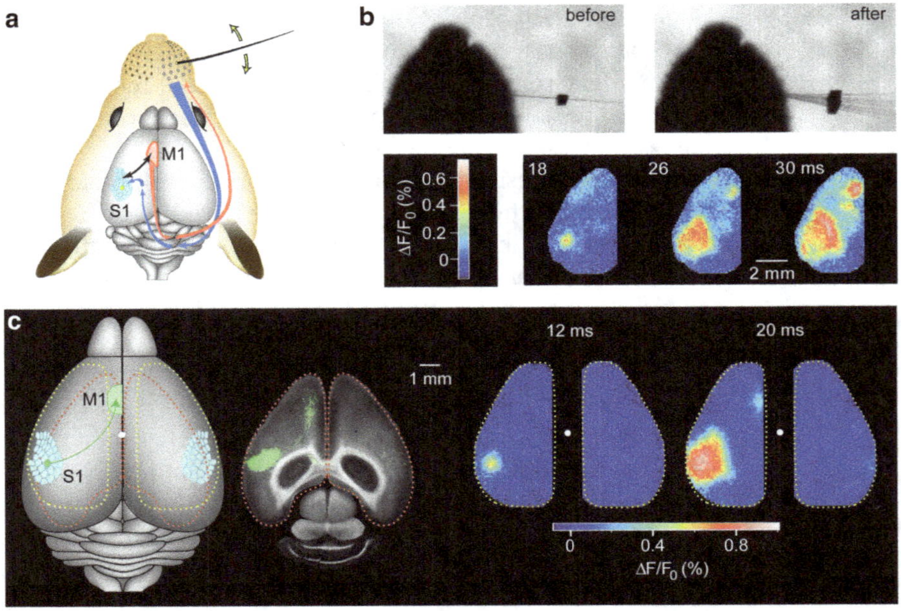

Fig. 3 Voltage-sensitive dye imaging of a cortical sensorimotor loop in awake head-restrained mice. (**a**) The mystacial vibrissae signal sensory information (*blue arrows*) to the primary somatosensory barrel cortex (S1) relayed via synapses in trigeminal brainstem and somatosensory thalamus. Primary whisker motor cortex (M1) drives exploratory whisker movements via brainstem central pattern generators and the motor neurons located in the facial nucleus (*red arrows*). Sensory cortex and motor cortex are highly interconnected. (**b**) A metal particle was attached to the C2 whisker of an awake head-restrained mouse (*black* particle visible on the whisker in the upper images of the silhouette of the mouse snout viewed from above). The mouse was located above an electromagnetic coil, through which brief 1–2 ms currents could be driven to evoke a brief and well-controlled whisker deflection. The behavioral images (above) are superimposed images over a 100 ms period before (*left*) and after (*right*) whisker stimulation. The whisker deflection drives a brief movement of the whisker, which is not visible in the behavioral filming, because it is in the same direction as the camera axis. The movements which are evoked by the sensory stimulus occur after a latency of many milliseconds and are driven by neural activity. Single-trial voltage-sensitive dye images (below) were simultaneously acquired with the behavioral filming. The early sensory response is localized to the primary somatosensory barrel cortex (18 ms). Subsequently, the signal spreads within S1 and a secondary localized hot spot is observed in M1 (26 ms). Both S1 and M1 hot spots evoke propagating waves of cortical activity. The depolarization of M1 may contribute to driving the sensory-evoked whisker movements. (**c**) Schematic drawing of the mouse brain viewed from above (*far left*) indicating the location of S1 barrel field and M1 vibrissae cortex. Viewed in a horizontal brain slice, anatomical labelling of the axonal projection from S1 to M1 using lentivirus expressing GFP (*center left*). A dense axonal termination field is found in M1. The whisker-deflection-evoked responses imaged with voltage-sensitive dye in S1 and M1 map closely to the anatomical locations of the projection from S1 to M1 (*right*). A monosynaptic connection from S1 to M1 might therefore drive the sensory responses in motor cortex evoked by whisker deflection (Modified and reproduced from [30], with permission from Elsevier).

age of animal [45], and brain state [10, 12, 30], the sensory response can spread horizontally across the somatotopic barrel map within the following milliseconds. The concept of a cortical column is therefore useful, but only for a short period of time of a few milliseconds, and subsequently it is important to consider processing of sensory information that is distributed over larger cortical areas. The spread of sensory information within the barrel cortex is likely to be mediated by the long horizontal axons of the neocortical excitatory pyramidal neurons [11], but longer loops might also contribute, for example, via thalamus or brainstem. The spread of sensory information beyond the aligned cortical column might be of functional importance. During normal whisker sensation, several whiskers are likely to contact any given object with a precise spatiotemporal sequence. The shape of an object might therefore be defined through comparing the relative timing of contacts of different whiskers with an object. For such a computation, it is obviously critical to be able to integrate the sensory information from different whiskers. The spreading sensory responses within the barrel cortex evoked by stimulation of a single whisker might underlie the integration of multi-whisker signals.

Approximately 7 ms after the initial localized response evoked by single whisker stimulation in primary somatosensory cortex, a second localized response is observed in the primary vibrissae motor cortex (Fig. 3b) [30, 37]. Over the next milliseconds this depolarization also begins to spread. The net result of a single whisker deflection is the initiation of two propagating waves of activity, the first starting from the somatosensory barrel cortex and the second starting from the motor cortex. A large part of the dorsal cortex is therefore informed about a single whisker deflection. The depolarization of motor cortex driven by the sensory stimulus correlates with initiation of whisker movements [30]. One might interpret such a behavioral response as the mouse perceiving the unexpected whisker stimulus and then wanting to find out more about the nature of the tactile input by moving its whiskers around in search of nearby objects that might have caused the initial sensation. The initial sensory input therefore triggers an exploratory motor output.

Mechanistically, we can explore the signalling pathways leading to depolarization of the motor cortex. Inactivation of barrel cortex blocks the sensory response in motor cortex and conversely stimulation of barrel cortex evokes a response in motor cortex [30]. Interestingly, neurons with soma located in the barrel cortex form prominent long-range axonal projections directly to the primary whisker motor cortex (Fig. 3c) [30, 46, 47]. These neurons appear likely to mediate the sensory-evoked responses in motor cortex, although more complex loops involving thalamus or other brain areas might also contribute.

3.2 State-Dependent Sensory Processing

Repeated presentations of the same whisker stimulus evoke highly variable responses in the neocortex, although the stimulus is reliably encoded by sensory neurons of the trigeminal ganglion [12, 43, 44, 48]. The cortex might therefore process sensory information in a context-dependent manner.

In analyzing whisker-related behavior, it is clear that there are two very different states. During quiet wakefulness, when the mouse is inactive, the whiskers do not move. Whisker deflections during this behavioral state must therefore reflect external stimuli coming from objects or animals moving into the whisker space. In other periods, mice actively explore their environment and in this active behavioral state the whiskers are moved back and forth scanning the local environment. During active exploration, it is likely that a great deal of the incoming sensory information is generated by the self-driven movements of whiskers against objects. The same stimulus delivered to the whiskers is processed very differently in the neocortex comparing these two behavioral states.

A brief whisker deflection delivered during a period of quiet wakefulness evokes the stereotypical response imaged with voltage-sensitive dye, involving sequential depolarization of somatosensory cortex followed by depolarization of motor cortex, as described above. However, if the same stimulus is delivered when the mouse is actively exploring with its whiskers moving, then the response is strongly reduced in both amplitude and spatial extent. During such active behavior the whisker-deflection-evoked response in barrel cortex is strongly suppressed in voltage-sensitive dye measurements from awake head-restrained mice (Fig. 4a) [30], in voltage-sensitive dye measurements from freely moving mice (Fig. 4b) [10], and in whole-cell membrane potential recordings from awake head-restrained mice [48]. Whereas the response during quiet wakefulness spreads over a large part of the barrel field and initiates activity in motor cortex, during active whisking the sensory response remains confined to a much smaller area of the barrel cortex and makes little impact upon the voltage-sensitive dye signal measured in the motor cortex (Fig. 4a). Sensory processing in the neocortex is therefore under strong behavioral control. The large amplitude spreading sensory responses evoked during quiet wakefulness might be viewed as a wide-spread wake-up call for the neocortex in response to an unexpected sensory stimulus. The small localized responses evoked during active whisking might be better suited to distinguish fine scale details of the sensory information related to single whiskers; for example, the frequency of whisker vibrations driven by sweeping a whisker across a rough texture [44, 49, 50].

4 Discussion and Outlook

In vivo voltage-sensitive dye imaging with the blue oxonal dyes (RH1691, RH1692, and RH 1838) [2, 3] has provided and continues to provide high temporal (~1 ms) and spatial (~100 μm) resolution views of single-trial subthreshold membrane potential dynamics in large-scale neocortical networks. Over the next years, we anticipate that voltage-sensitive dye imaging will provide an

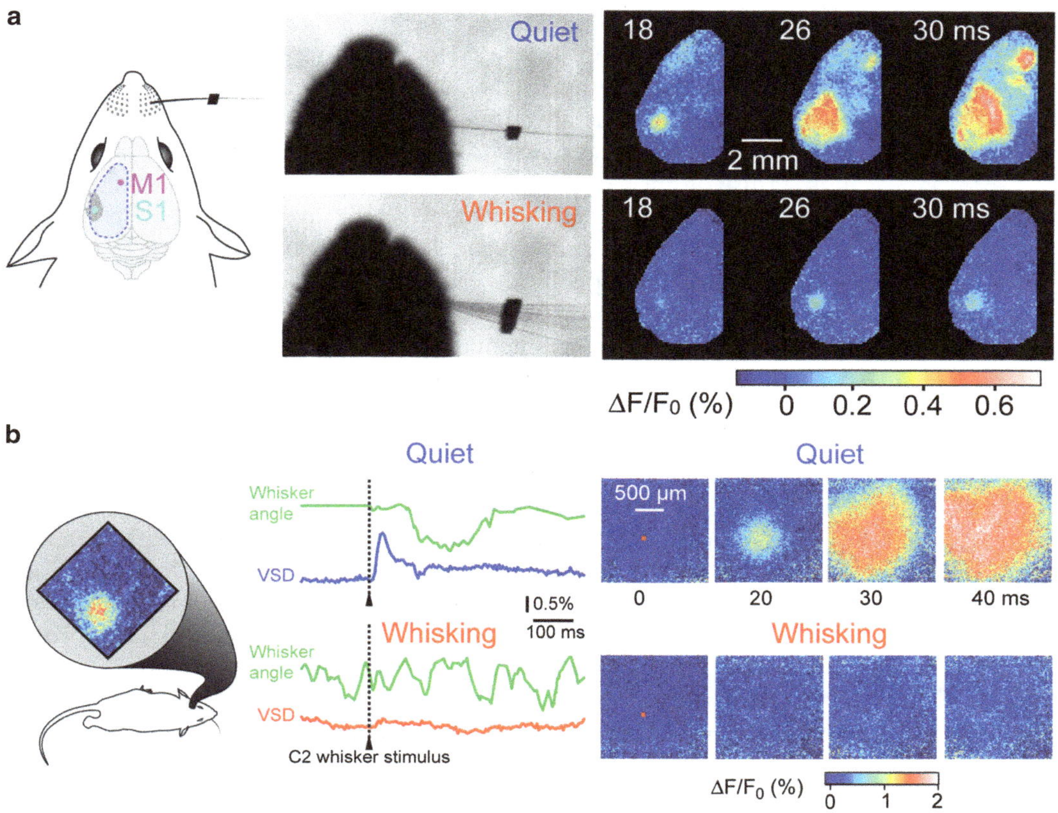

Fig. 4 State-dependent sensory processing imaged with voltage-sensitive dye in awake head-restrained and freely moving mice. (**a**) Voltage-sensitive dye imaging from an awake head-restrained mouse. A magnetic whisker deflection applied during a period of quiet wakefulness (when the whisker is not moving, pre-stimulus behavioral image, *upper left*) evokes a propagating sensory response in both S1 and M1. However, the same stimulus applied during a period of spontaneous pre-stimulus whisking (pre-stimulus behavioral image, *lower left*) evoked only a small and localized sensory response in the barrel field and no response in motor cortex. (**b**) Fiber-optic imaging of primary somatosensory barrel cortex stained with voltage-sensitive dye in a freely moving mouse. When the mouse is quiet and not moving its whisker, the whisker stimulus evokes a large amplitude spreading sensory response in S1. When the mouse is actively exploring and moving its whisker, the same stimulus evokes a sensory response that is strongly reduced in amplitude and barely resolved (**a**, Modified and reproduced from [30], with permission from Elsevier; **b**, Modified and reproduced from [10] with permission from Elsevier).

increasingly detailed view of the functional connectivity of the dorsal neocortex and how different brain areas interact during execution of simple learned behaviors. Mapping which brain areas are depolarized and which brain areas are hyperpolarized during specific periods of a behavioral task is an important first step towards obtaining mechanistic insight into brain function.

There are a large number of important limitations to the current methodology described in this chapter. The signals are small and it would be enormously helpful if SNRs could be further improved.

Further screening for better dyes, more stable light sources for excitation at optimal wavelengths and improved detectors will certainly make an impact over the coming years. However, the most important advances are likely to result from genetic engineering of voltage-sensitive fluorescent proteins (see Chapter 4 in this volume). Genetically encoded voltage-sensors would offer two major advantages over the synthetic dyes. Firstly, they could be targeted to specific neuronal cell types. In the current protocols, the dye non-specifically labels all membranes in the superficial layers of the cortex within the craniotomy, including both glia and neurons. Through genetic engineering one could specifically express voltage-sensitive fluorescent proteins in excitatory neurons or in inhibitory neurons, and it would be of great interest to learn about the membrane potential dynamics of specific classes of neurons in the neocortex. The second major advantage of a genetically encoded voltage-sensitive fluorescent protein over synthetic dyes is the possibility for longitudinal imaging over long periods of time. Whereas this has already been possible in monkey voltage-sensitive dye imaging studies [32–35] through repeated staining of a cortical area with an artificial dura, this has not yet been achieved in rodent experiments. Genetically encoded voltage-sensitive fluorescent proteins are likely to remain stably expressed over long periods of time, allowing changes in cortical dynamics to be monitored during learning or plasticity. After many years of frustrating efforts by many scientists [51, 52], there is now great hope that genetically encoded voltage-sensitive fluorescent proteins are close to becoming a useful reality. Thomas Knöpfel's laboratory recently developed VSFP2.3 [53], which is based on a voltage-sensitive phosphatase [54] to which a pair of two fluorescent proteins were attached to allow fluorescent resonance energy transfer (FRET). A more detailed description is provided in Chapter 4 of this volume. Importantly, expression of VSFP2.3 in the mammalian brain gave rise to signals in vivo with impressive SNRs, almost as good as those reported with the synthetic oxonal dyes. The current major drawbacks are (1) the kinetics of VSFP2.3, which are complex, voltage-dependent, and slower (~10 ms) than the synthetic voltage-sensitive dyes; and (2) the nonlinearity of VSFP2.3 with respect to membrane potential, which will complicate the quantitative interpretation of signals. Future improvements to VSFP2.3 or the development of genetically encoded voltage-sensors based on other voltage-sensitive proteins will likely resolve these issues. Interestingly, microbial opsins have recently been reported to give very large voltage-sensitive fluorescence signals, when expressed in mammalian cell lines allowing single action potentials to be imaged with sub-millisecond resolution and excellent SNR [55]. There is therefore currently enormous excitement in the field of voltage-sensitive dye imaging, with important technical developments opening new opportunities for visualizing neocortical network dynamics with unprecedented precision.

References

1. Loew LM (2010) Design and use of organic voltage sensitive dyes (Chapter 2). In: Canepari M, Zecevic D (eds) Membrane potential imaging in the nervous system: methods and applications. Springer, New York
2. Shoham D, Glaser DE, Arieli A, Kenet T, Wijnbergen C, Toledo Y, Hildesheim R, Grinvald A (1999) Imaging cortical dynamics at high spatial and temporal resolution with novel blue voltage-sensitive dyes. Neuron 24:791–802
3. Grinvald A, Hildesheim R (2004) VSDI: a new era in functional imaging of cortical dynamics. Nat Rev Neurosci 5:874–885
4. Berger T, Borgdorff A, Crochet S, Neubauer FB, Lefort S, Fauvet B, Ferezou I, Carleton A, Lüscher HR, Petersen CCH (2007) Combined voltage and calcium epifluorescence imaging in vitro and in vivo reveals subthreshold and suprathreshold dynamics of mouse barrel cortex. J Neurophysiol 97:3751–3762
5. Djurisic M, Antic S, Chen WR, Zecevic D (2004) Voltage imaging from dendrites of mitral cells: EPSP attenuation and spike trigger zones. J Neurosci 24:6703–6714
6. Palmer LM, Stuart GJ (2009) Membrane potential changes in dendritic spines during action potentials and synaptic input. J Neurosci 29:6897–6903
7. Canepari M, Willadt S, Zecevic D, Vogt KE (2010) Imaging inhibitory synaptic potentials using voltage sensitive dyes. Biophys J 98:2032–2040
8. Holthoff K, Zecevic D, Konnerth A (2010) Rapid time course of action potentials in spines and remote dendrites of mouse visual cortex neurons. J Physiol 588:1085–1096
9. Lippert MT, Takagaki K, Xu W, Huang X, Wu JY (2007) Methods for voltage-sensitive dye imaging of rat cortical activity with high signal-to-noise ratio. J Neurophysiol 98:502–512
10. Ferezou I, Bolea S, Petersen CCH (2006) Visualizing the cortical representation of whisker touch: voltage-sensitive dye imaging in freely moving mice. Neuron 50:617–629
11. Petersen CCH, Grinvald A, Sakmann B (2003) Spatiotemporal dynamics of sensory responses in layer 2/3 of rat barrel cortex measured in vivo by voltage-sensitive dye imaging combined with whole-cell voltage recordings and neuron reconstructions. J Neurosci 23:1298–1309
12. Petersen CCH, Hahn TTG, Mehta M, Grinvald A, Sakmann B (2003) Interaction of sensory responses with spontaneous depolarization in layer 2/3 barrel cortex. Proc Natl Acad Sci USA 100:13638–13643
13. Lampl I, Reichova I, Ferster D (1999) Synchronous membrane potential fluctuations in neurons of the cat visual cortex. Neuron 22:361–374
14. Poulet JFA, Petersen CCH (2008) Internal brain state regulates membrane potential synchrony in barrel cortex of behaving mice. Nature 454:881–885
15. Gentet LJ, Avermann M, Matyas F, Staiger JF, Petersen CCH (2010) Membrane potential dynamics of GABAergic neurons in the barrel cortex of behaving mice. Neuron 65:422–435
16. Tsodyks M, Kenet T, Grinvald A, Arieli A (1999) Linking spontaneous activity of single cortical neurons and the underlying functional architecture. Science 286:1943–1946
17. Sharon D, Grinvald A (2002) Dynamics and constancy in cortical spatiotemporal patterns of orientation processing. Science 295:512–515
18. Kenet T, Bibitchkov D, Tsodyks M, Grinvald A, Arieli A (2003) Spontaneously emerging cortical representations of visual attributes. Nature 425:954–956
19. Jancke D, Chavane F, Naaman S, Grinvald A (2004) Imaging cortical correlates of illusion in early visual cortex. Nature 428:423–426
20. Xu W, Huang X, Takagaki K, Wu JY (2007) Compression and reflection of visually evoked cortical waves. Neuron 55:119–129
21. Han F, Caporale N, Dan Y (2008) Reverberation of recent visual experience in spontaneous cortical waves. Neuron 60:321–327
22. Palagina G, Eysel UT, Jancke D (2009) Strengthening of lateral activation in adult rat visual cortex after retinal lesions captured with voltage-sensitive dye imaging in vivo. Proc Natl Acad Sci USA 106:8743–8747
23. Roland PE, Hanazawa A, Undeman C, Eriksson D, Tompa T, Nakamura H, Valentiniene S, Ahmed B (2006) Cortical feedback depolarization waves: a mechanism of top-down influence on early visual areas. Proc Natl Acad Sci USA 103:12586–12591
24. Ahmed B, Hanazawa A, Undeman C, Eriksson D, Valentiniene S, Roland PE (2008) Cortical dynamics subserving visual apparent motion. Cereb Cortex 18:2796–2810
25. Spors H, Grinvald A (2002) Spatio-temporal dynamics of odor representations in the mammalian olfactory bulb. Neuron 34:301–315
26. Derdikman D, Hildesheim R, Ahissar E, Arieli A, Grinvald A (2003) Imaging spatiotemporal dynamics of surround inhibition in the barrels somatosensory cortex. J Neurosci 23:3100–3105
27. Wallace DJ, Sakmann B (2008) Plasticity of representational maps in somatosensory cortex observed by in vivo voltage-sensitive dye imaging. Cereb Cortex 18:1361–1373

28. Ghosh A, Haiss F, Sydekum E, Schneider R, Gullo M, Wyss MT, Mueggler T, Baltes C, Rudin M, Weber B, Schwab ME (2010) Rewiring of hindlimb corticospinal neurons after spinal cord injury. Nat Neurosci 13:97–104
29. Civillico EF, Contreras D (2006) Integration of evoked responses in supragranular cortex studied with optical recordings in vivo. J Neurophysiol 96:336–351
30. Ferezou I, Haiss F, Gentet LJ, Aronoff R, Weber B, Petersen CCH (2007) Spatiotemporal dynamics of cortical sensorimotor integration in behaving mice. Neuron 56:907–923
31. Sigler A, Mohajerani MH, Murphy TH (2009) Imaging rapid redistribution of sensory-evoked depolarization through existing cortical pathways after targeted stroke in mice. Proc Natl Acad Sci USA 106:11759–11764
32. Slovin H, Arieli A, Hildesheim R, Grinvald A (2002) Long-term voltage-sensitive dye imaging reveals cortical dynamics in behaving monkeys. J Neurophysiol 88:3421–3438
33. Seidemann E, Arieli A, Grinvald A, Slovin H (2002) Dynamics of depolarization and hyperpolarization in the frontal cortex and saccade goal. Science 295:862–865
34. Ayzenshtat I, Meirovithz E, Edelman H, Werner-Reiss U, Bienenstock E, Abeles M, Slovin H (2010) Precise spatiotemporal patterns among visual cortical areas and their relation to visual stimulus processing. J Neurosci 30: 11232–11245
35. Meirovithz E, Ayzenshtat I, Bonneh YS, Itzhack R, Werner-Reiss U, Slovin H (2010) Population response to contextual influences in the primary visual cortex. Cereb Cortex 20: 1293–1304
36. Mohajerani MH, McVea DA, Fingas M, Murphy TH (2010) Mirrored bilateral slow-wave cortical activity within local circuits revealed by fast bihemispheric voltage-sensitive dye imaging in anesthetized and awake mice. J Neurosci 30:3745–3751
37. Matyas F, Sreenivasan V, Marbach F, Wacongne C, Barsy B, Mateo C, Aronoff R, Petersen CCH (2010) Motor control by sensory cortex. Science 330:1240–1243
38. Sherrington CS (1906) The integrative action of the nervous system. Yale University Press, New Haven, CT
39. Brecht M (2007) Barrel cortex and whisker-mediated behaviors. Curr Opin Neurobiol 17:408–416
40. Petersen CCH (2007) The functional organization of the barrel cortex. Neuron 56:339–355
41. Diamond ME, von Heimendahl M, Knutsen PM, Kleinfeld D, Ahissar E (2008) 'Where' and 'what' in the whisker sensorimotor system. Nat Rev Neurosci 9:601–612
42. Ebara S, Kumamoto K, Matsuura T, Mazurkiewicz JE, Rice FL (2002) Similarities and differences in the innervation of mystacial vibrissal follicle-sinus complexes in the rat and cat: a confocal microscopic study. J Comp Neurol 449:103–119
43. Jones LM, Depireux DA, Simons DJ, Keller A (2004) Robust temporal coding in the trigeminal system. Science 304:1986–1989
44. Arabzadeh E, Zorzin E, Diamond ME (2005) Neuronal encoding of texture in the whisker sensory pathway. PLoS Biol 3:e17
45. Borgdorff AJ, Poulet JFA, Petersen CCH (2007) Facilitating sensory responses in developing mouse somatosensory barrel cortex. J Neurophysiol 97:2992–3003
46. Chakrabarti S, Alloway KD (2006) Differential origin of projections from SI barrel cortex to the whisker representations in SII and MI. J Comp Neurol 498:624–636
47. Aronoff R, Matyas F, Mateo C, Ciron C, Schneider B, Petersen CCH (2010) Long-range connectivity of mouse primary somatosensory barrel cortex. Eur J Neurosci 31:2221–2233
48. Crochet S, Petersen CCH (2006) Correlating whisker behavior with membrane potential in barrel cortex of awake mice. Nat Neurosci 9:608–610
49. Ritt JT, Andermann ML, Moore CI (2008) Embodied information processing: vibrissa mechanics and texture features shape micromotions in actively sensing rats. Neuron 57:599–613
50. Jadhav SP, Wolfe J, Feldman DE (2009) Sparse temporal coding of elementary tactile features during active whisker sensation. Nat Neurosci 12:792–800
51. Baker BJ, Lee H, Pieribone VA, Cohen LB, Isacoff EY, Knöpfel T, Kosmidis EK (2007) Three fluorescent protein voltage sensors exhibit low plasma membrane expression in mammalian cells. J Neurosci Methods 161:32–38
52. Baker BJ, Mutoh H, Dimitrov D, Akemann W, Perron A, Iwamoto Y, Jin L, Cohen LB, Isacoff EY, Pieribone VA, Hughes T, Knöpfel T (2008) Genetically encoded fluorescent sensors of membrane potential. Brain Cell Biol 36:53–67
53. Akemann W, Mutoh H, Perron A, Rossier J, Knöpfel T (2010) Imaging brain electric signals with genetically targeted voltage-sensitive fluorescent proteins. Nat Methods 7:643–649
54. Murata Y, Iwasaki H, Sasaki M, Inaba K, Okamura Y (2005) Phosphoinositide phosphatase activity coupled to an intrinsic voltage sensor. Nature 435:1239–1243
55. Kralj JM, Douglass AD, Hochbaum DR, Maclaurin D, Cohen AE (2011) Optical recording of action potentials in mammalian neurons using a microbial rhodopsin. Nat Methods 9:90–95

Chapter 9

Two-Photon Imaging of Neuronal Network Dynamics in Neocortex

Benjamin F. Grewe, Fritjof Helmchen, and Björn M. Kampa

Abstract

Optical investigation of fast neuronal network dynamics in the intact neocortex—using appropriate activity-dependent indicators—requires single-cell resolution at large imaging depths and sufficient acquisition speed. These requirements are met by two-photon laser scanning microscopy, which has become one of the key methods for functional measurements of neuronal population activity in vivo, primarily in combination with calcium indicators. In this chapter we focus on various advanced two-photon imaging techniques that were recently developed to improve scanning speed, to enable 3D sampling from large numbers of neurons, or to extend imaging towards measurements in freely behaving animals. In general, sampling speed and population size trade off against each other. Currently, about 1,000 neurons can be measured with good signal-to-noise ratio at 1–10 Hz or a few tens of neurons can be optically recorded at 1 kHz. Measurements of local network activity have been used to either characterize spatial distributions of functional properties, such as orientation tuning in visual cortex, or reveal neuronal activation patterns on a fast time scale. We illustrate these new opportunities with examples from in vivo two-photon calcium imaging in mouse visual cortex. The chapter concludes with a discussion of the advantages and limitations of the various techniques and of future perspectives. The direct observation of neuronal network dynamics in living animals no doubt will help to elucidate principles of operation of neocortical microcircuits.

Key words Calcium indicator, Laser scanning, Neurons, Two-photon microscopy, In vivo imaging, Acousto-optic deflectors

1 Introduction

One of the most powerful advantages of two-photon microscopy is its intrinsic 3D resolution, which enables imaging with micrometer resolution several hundred micrometer deep inside light-scattering tissue [1] (see Chapters 1 and 2 in this volume). In combination with functional calcium indicators, two-photon microscopy therefore has made it possible to characterize the spatiotemporal dynamics of neuronal populations in the neocortex of living animals (reviewed in [2–4]). A standard two-photon microscope setup for in vivo imaging includes a Titanuim:Sapphire (Ti:Sa) laser source, a Pockels cell for laser intensity adjustment, *x*–*y*-scanners, scan optics

Bruno Weber and Fritjof Helmchen (eds.), *Optical Imaging of Neocortical Dynamics*, Neuromethods, vol. 85,
DOI 10.1007/978-1-62703-785-3_9, © Springer Science+Business Media New York 2014

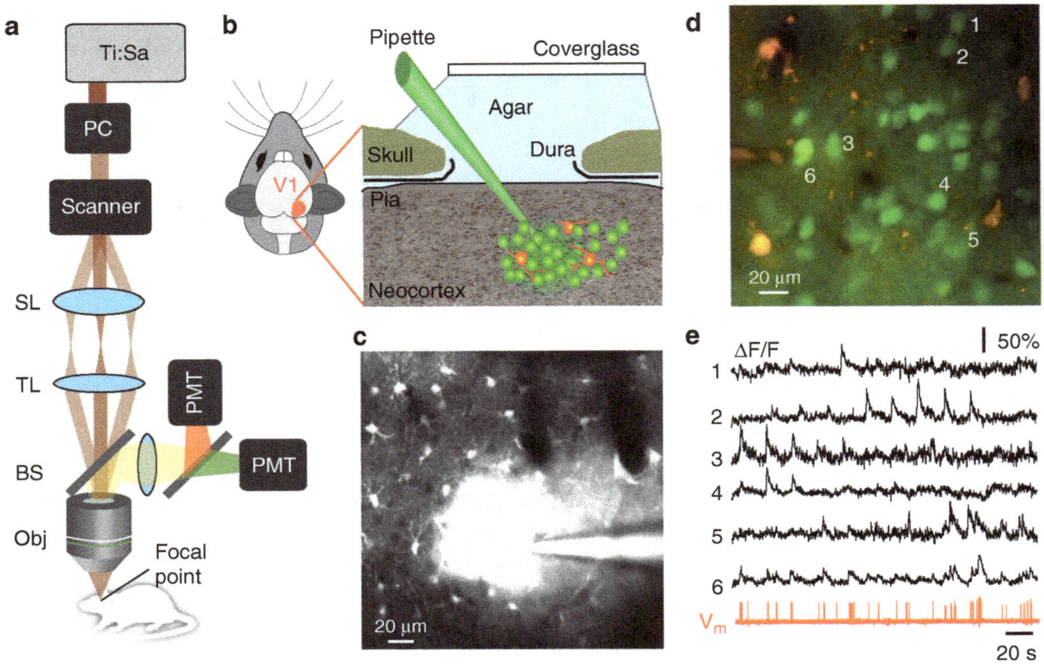

Fig. 1 Two-photon calcium imaging of in vivo neuronal population activity. (**a**) Schematized setup of a generic two-photon microscope system, comprising the Ti:Sa laser source, a Pockels Cell (PC), a laser scanning unit (LS), scan optics (scan lens, SL, and tube lens, TL), a dichroic beam splitter (BS), photomultipliers (PMTs) as fluorescence detectors, and the microscope objective (Obj). (**b**) Schematic drawing of the rodent visual cortex and how it is targeted with a glass pipette for MCBL of local neuronal cell populations. (**c**) Fluorescence image taken during MCBL with the calcium-sensitive dye OGB-1. The *red* fluorescence channel is shown, in which astrocytes are visible, previously labeled by brief application of dissolved SR101 to the pial surface. For better visualization of the dye injection the red fluorescent dye Alexa-594 (50 µM) was added to the injection pipette. (**d**) Two-photon fluorescence image of a layer 2/3 cell population in the intact mouse visual cortex following MCBL with OGB-1 and counterstaining of astrocytes with SR101. The patch pipette on the *left* (filled with red fluorescent dye) was used to obtain a simultaneous juxtacellular recording from one particular labeled neuron. (**e**) Example calcium transients from six selected neurons as indicated in (**d**), recorded with conventional 2D frame scanning at 5 Hz frame scanning rate. The simultaneous electrical recording of the spike train in neuron 6 is shown at the *bottom*.

(scan lens and tube lens), a dichroic beam splitter to reflect the fluorescence light into the detection pathway, photomultipliers (PMTs) for fluorescence detection, and the microscope objective (Fig. 1a). In addition, a broad palette of calcium-sensitive fluorescent indicator dyes is available as reporters of neural activity including synthetic organic dyes as well as genetically encoded indicators (see Chapter 4 in this volume). In this chapter, we highlight the use of synthetic calcium indicators for network imaging and in particular the recent advances in microscopy techniques that have been made for improving readout of network dynamics. Single-cell dye loading for the investigation of localized dendritic calcium signals and the use of genetically encoded calcium indicators for long-term imaging are covered in accompanying Chapters 6 and 10, respectively.

Calcium indicators offer a number of advantages for optically recording neuronal activity. First, because of the low intracellular resting calcium concentration (50–100 nM) action potential-evoked calcium influx causes substantial changes in intracellular calcium concentration that are detectable with high-affinity calcium indicators. Second, calcium transients are several orders of magnitude slower than the underlying action potentials. Rather than being a drawback, slow dynamics are actually of great advantage for laser scanning techniques because these typically have relatively slow acquisition rates. Whereas direct imaging of membrane potential changes with voltage indicators is possible (see also Chapter 4 in this volume), the signal-to-noise ratio (SNR) is still rather low and imaging rates in the kilohertz range are required.

For staining of large cell populations, synthetic calcium indicators are applied in the acetoxymethyl-esther (AM) form, in which dye molecules are membrane-permeable so that they can enter cells but then become trapped inside after ester-cleavage by endogenous esterases [5]. Typically, the AM-ester form of indicators such as Oregon Green BAPTA-1 (OGB-1), Fluo-4, or Rhod-2 is injected as a bolus directly into the neural tissue of interest using a glass micropipette (Fig. 1b, c). This technique was introduced for in vivo application in 2003 and is referred to as "multi-cell bolus loading" (MCBL) [6, 7]. Meanwhile MCBL has found widespread applications, in particular for in vivo population imaging as it offers the opportunity to label large numbers of neurons and glial cells (Fig. 1d, e). Furthermore, due to the high sensitivity of calcium indicators such as OGB-1 and Fluo-4 this method permits recording of calcium signals with high SNR and in the best cases enables the reconstruction of action potential pattern with single-spike resolution [8–10] (see Box 1). During recent years MCBL has been used extensively in rodents and other species for in vivo population imaging in various cortical regions, including visual cortex [11–20, 47], auditory cortex [21–23], somatosensory cortex [9, 24, 25], and motor cortex [26].

With now available in vivo population staining methods, microscopy technology rather than dye loading has become the bottleneck for optical measurements of fast neural network activity. Two key aspects are the size of the sampled area (or volume), which directly relates to the number of recorded neurons, and the imaging speed, which determines the effective sampling rate per cell. Recent developments of novel laser scanning modes have addressed these issues, aiming at recording from larger networks and/or with higher speed. Unfortunately, network size and acquisition speed relate inversely to each other because a minimum dwell time per cell is necessary in order to generate sufficient numbers of fluorescence photons for an adequate SNR. For this reason, most of the developments have aimed at improving either one or the other aspect. While standard laser scanning microscopes are suitable for 1D or 2D measurements at relatively high speed

Box 1 Measuring Action Potential Activity with Fluorescent Calcium Indicators

Inferring neuronal spiking activity from calcium imaging data is based on the brief Ca^{2+} influx that is evoked during each action potential through activation of voltage-gated Ca^{2+} channels. Ca^{2+} influx is evoked at the soma and, in most neuronal cell types, also in the dendrites (at least the proximal ones where backpropagating action potentials are effective) [27]. Ca^{2+} influx may also occur from other sources, e.g., through local activation of receptor channels or by release from intracellular stores. However, in pyramidal neurons, for which action potential-evoked calcium dynamics has been best characterized, a one-to-one relationship between somatic calcium signals and spikes can be assumed in most cases. Each action potential evokes a transient increase in intracellular free Ca^{2+} concentration ($[Ca^{2+}]_i$) with a rather stereotype amplitude and time course, showing an initial fast rise and a slower decay of single- or double-exponential shape. The amplitude and decay time constant of these elementary "calcium transients" are determined by multiple factors including surface-to-volume ratio, Ca^{2+}-buffering, and extrusion mechanisms, and not least by the affinity and concentration of the calcium indicator itself (see [27]). For typical concentrations of the high-affinity indicator OGB-1 (as reached, for example, by MCBL), the decay time lasts several hundred milliseconds. While some calcium indicators allow quantification of absolute $[Ca^{2+}]_i$ levels, calcium signals are often simply expressed as relative percentage changes from baseline ($\Delta F/F$). During spike trains the calcium transients summate and the compound somatic calcium signal represents a convolution of the spike train with the elementary calcium transient. Various "deconvolution" methods have been devised to reconstruct best estimates of spike trains from the fluorescence traces, given the limited SNR [28, 29]. For example, we recently introduced a "peeling algorithm" that searches for the next calcium transient, subtracts a presumed elementary calcium transient from the fluorescence trace, jumps back, and then iteratively repeats this procedure until an entire spike train is uncovered and a noisy residual trace remains [10]. High-speed methods of population imaging in addition make it possible to determine the time of occurrence of spikes with near-millisecond precision [10].

(line scans and movies of 2D frames, respectively), they typically require seconds for imaging even a small volume. On the other hand, it is obvious that cortical networks are 3D structures with intricate spatial arrangements of the neurons. One specific goal therefore has been to extend fast imaging methods to the third spatial dimension—the dimension along the optical axis of the microscope—in order to capture network dynamics on the subsecond time scale in 3D space. A second major goal has been to push the temporal resolution of network measurements towards the millisecond regime, aiming at enabling accurate reconstructions of action potential trains from the fluorescence recordings. Major steps towards this goal were achieved using nonmechanical, acousto-optic scanning devices and special scanning modes as will be discussed below. Finally, another important goal is to enable imaging of neuronal network dynamics in awake, behaving animals. Such measurements can now be achieved either by training animals to tolerate head-fixation during imaging sessions ([13, 26, 30, 31]; see also Chapters 8 and 10 in this volume) or through the development of portable, miniaturized two-photon microscopes that allow imaging in freely moving animals [32, 33]. In this chapter we summarize recent progress with regard to these three goals, providing some technical details as well as examples of applications.

2 Novel Two-Photon Microscopy Techniques for Neuronal Network Imaging

Optical readout of neuronal network dynamics in the intact brain remains challenging. Even though two-photon microscopy provides single-cell resolution down to nearly a millimeter of cortical depth, the major task is to measure the neuronal calcium signals—typically lasting only a few hundred milliseconds—from hundreds of cells distributed in 3D space and with a sufficient sensitivity and temporal resolution to faithfully reconstruct action potential dynamics in the local population. Here we summarize recent advances in laser scanning techniques that led to improvements in signal acquisition from cell populations, with regard to either network size or speed of acquisition. In the standard two-photon imaging mode the laser focus is scanned in the focal plane in a raster-like fashion (Fig. 2), using either two galvanometric scan mirrors or—for faster frame rates—one fast resonant scan mirror in combination with a second galvanometric mirror [19, 34, 35]. 3D imaging is realized by acquiring image stacks in a stepwise fashion with a standard focusing device, but this is typically too slow to follow sub-second dynamics. Relatively fast volume scanning can be achieved by combining rapid 2D raster-scanning with continuous movement of the objective, e.g., using a piezoelectric focusing element [36]. Even scanning of tilted and curved planes in 3D space in principle is feasible (Fig. 2) [37].

Raster-scanning techniques face the problem of wasting scan time on background regions that are of no real interest. A useful strategy to increase acquisition speed therefore is to restrict fluorescence excitation to the structures of interest. As a first step, standard laser scanning can be adapted to scan arbitrary free line scans across preselected subsets of cells [22, 23, 37–39]. The scanning speed can be adjusted on the fly to optimize dwell time on the cell bodies [38]; however, this approach is eventually limited by the inertia of the mechanical scan mirrors. For volume imaging a closed spiral scan trajectory in 3D space can be generated by synchronizing appropriate drive signals to the x–y-scan mirrors with a continuous movement of the microscope objective along the opti-

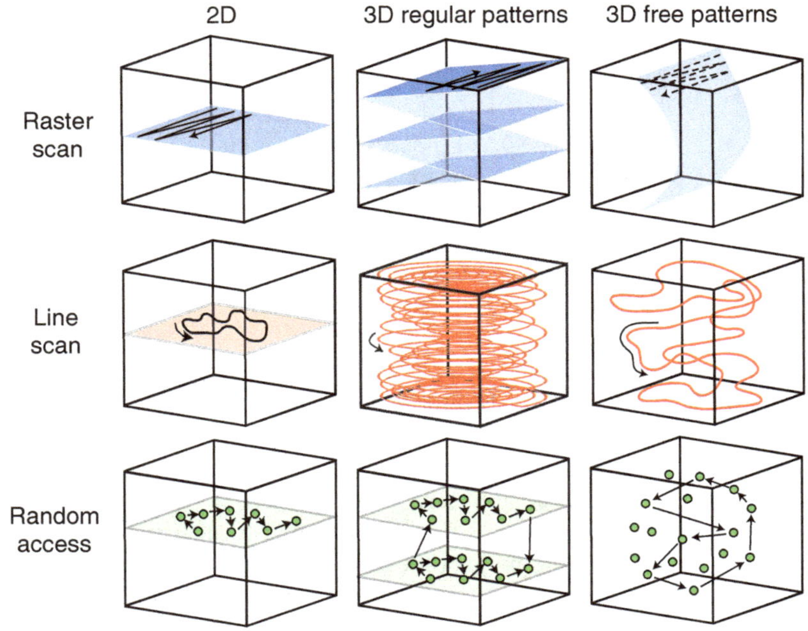

Fig. 2 Laser scanning modes in two-photon microscopy. Overview of the principal types of laser scanning, including raster-like scanning (*top row*), free line scans (*middle row*), and random-access scanning (*bottom row*). For each type the possible variations for 2D scanning (*left*), 3D scanning with a regular scan pattern (*middle*), and 3D scanning with a free-form pattern (*right*) are shown. While raster-scanning and line-scanning can be implemented either with galvanometric scan mirrors or with acousto-optic scanners (combined with a piezoelectric focusing element for 3D scanning), the random-access modes require the use of two or more acousto-optic scanners. Nearly all of the modes depicted here have been applied for in vivo calcium imaging, except for the curved plane imaging.

cal axis (Fig. 2) [40]. Moreover, user-defined 3D scan trajectories based on preselected target positions can be created for effective scanning of large populations (Fig. 2) [40, 41]. Finally, full and deliberate restriction of scan positions to the cells of interest is possible by employing acousto-optic deflectors (AODs) for laser scanning (see Box 2). With AODs the laser focus can be shifted arbitrarily between any two positions in the field-of-view (FOV) within a few microseconds, enabling random-access scanning at high speed (e.g., 83,333 addressable points per second in our implementation; [10]). In combination with fast axial scanning devices, e.g., a novel electrically tunable lens, AOD scanning allows imaging of neuronal cell populations in multiple focal planes with almost video rate [41]. Finally, novel setups with four AODs arranged in series have demonstrated high-speed random-access scanning in three dimensions [43, 44], literally enabling "jumping" of the laser focus from cell to cell, which should be beneficial for acquiring functional data from entire local networks. This overview shows the high degree of flexibility for choosing an

Box 2 Acousto-optic Deflectors

An AOD is a crystal (usually TeO_2) with a piezoelectric transducer attached to one side generating acoustic sound waves in the crystal (excitation in the MHz range). Based on the Bragg diffraction law AODs permit spatial control of the angular deflection of a laser beam, with the change in angle given by $\Delta\theta = (\lambda/v)\cdot\Delta f$, where λ is the laser wavelength, Δf the change in acoustic wave frequency, and v the velocity of the acoustic wave. AODs are characterized by the following parameters: (1) the range of the deflection angle, which is linked to the frequency range of the device; (2) the resolution of the deflector, given as the number of resolvable spots that can be addressed; (3) the transition time, which corresponds to the time that is needed until the acoustic wave has traveled across the laser beam width (typically in the microsecond-range and below); and (4) the deflection efficiency into the first-order beam, which usually is around 75 % per AOD (for TeO_2). To realize 2D scanning two crossed AODs are used in series. The main advantage of using AOD scanners is the ability to rapidly (within a few microseconds) address arbitrary positions within the scan field allowing for fluorescence excitation in a random-access manner. Disadvantages are the relatively poor transmission efficiency (~55 % for a pair of AODs) and the large optical dispersion, both temporal and spatial, which has to be dealt with in AOD-based two-photon microscopes. Temporal dispersion, also known as group delay dispersion (GDD), causes broadening of the ultrashort Ti:Sa laser pulses, which would lead to a strong reduction of nonlinear fluorescence excitation. In addition, the spatial dispersion of laser wavelengths in AODs destroys the sharp laser focus if uncompensated. To restore a high-quality laser focus under the microscope objective, both effects need to be minimized, which can be effectively achieved with additional compensatory dispersive elements, e.g., a single prism mounted at appropriate orientation and distance from the AOD scanner in the beam path [10, 45].

appropriate laser scanning technique. In the following we describe the implementations of a 3D imaging approach and of an AOD-based high-speed two-photon microscope in more detail.

2.1 Imaging Neural Network Dynamics in 3D

In principle any custom-built or commercially available two-photon microscope can be upgraded for mechanical 3D scanning. While typically two standard galvanometric scan mirrors are used for *x–y*-scanning, fast *z*-scanning is achieved with a piezoelectric focusing element that allows rapid movement of the microscope objective along the optical axis (Fig. 3a) [40]. Almost any standard microscope objective can be mounted onto such piezoelectric focusing elements, which support travel ranges of 400 μm and more. Typically, however, mirror scanning and movement of the piezo-element have to be synchronized by custom-written software. The principle idea of fast 3D laser scanning is to set the objective in continuous "swinging" motion and to create a 3D line scan that samples relatively homogenously from a particular observation volume. Although, the spatial resolution of this approach is relatively

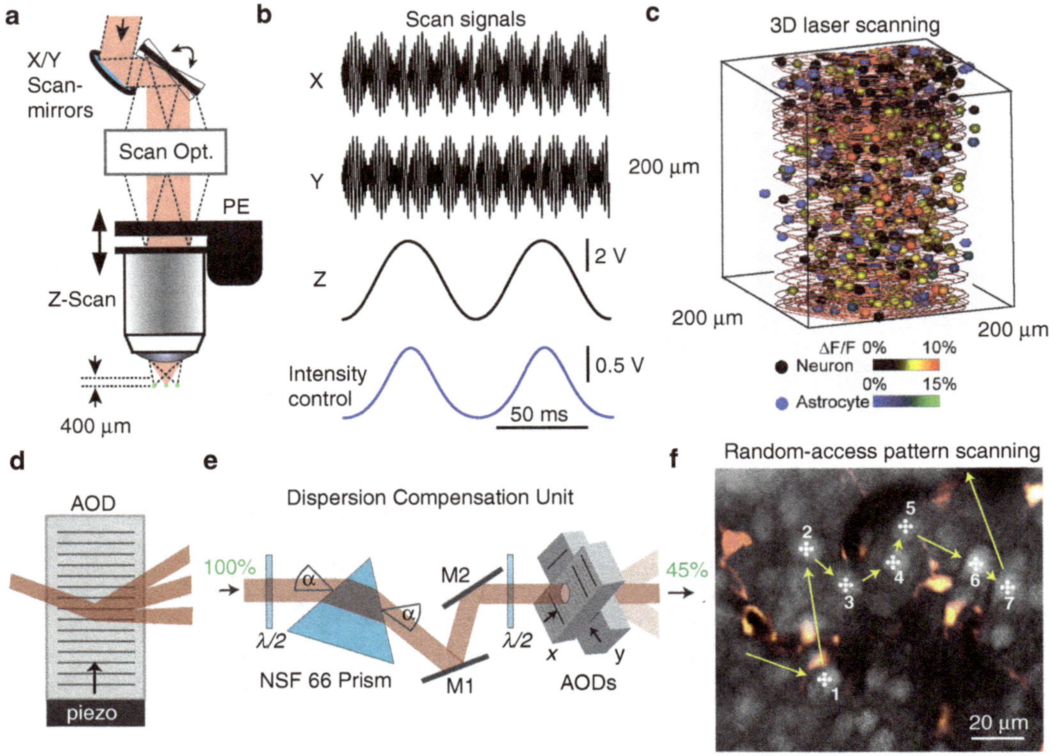

Fig. 3 Laser scanning technology for 3D and high-speed imaging. (**a**) Scheme of mechanical 3D laser scanning. Lateral *x*–*y*-scanning is achieved with a pair of standard, galvanometric scan mirrors while fast *z*-scanning is performed with a piezoelectric focusing element that continuously displaces the microscope objective along the optical axis. (**b**) A 3D scan trajectory is generated by synchronization of appropriate drive signals to the *x*–*y*-scanner with the objective's *z*-movement. Here, drive signals to produce a series of opening and closing spirals are shown. For obtaining the intended 3D trajectory, amplitude reduction and phase shifts caused by the inertia of the moving objective need to be taken into account (not shown). For deep tissue in vivo imaging the laser intensity needs to be automatically adjusted with imaging depth, e.g., with a Pockels cell. (**c**) Example of a 3D scan trajectory based on a series of opening spirals, overlaid over the positions of cell bodies. (**d**) Schematic drawing of an acousto-optic deflector. The piezoelectric transducer attached to the crystal generates sound waves of different spatial frequencies, so that the laser beam is diffracted at different angles. (**e**) For imaging with AOD scanners, their dispersive effects need to be compensated. The schematic drawing shows a single-prism dispersion compensation unit (DCU), consisting of two λ/2-waveplates and a custom prism oriented at 45°-angle to the AOD axes. (**f**) L2/3 neuronal population in mouse barrel cortex labeled with OGB-1 (depth 210 μm). High-speed random-access pattern scanning was performed using 5-point patterns targeted to 53 cells in total (7 example cells are shown). On each cell the small cross-like pattern was scanned. Cell positions were manually preselected from a reference image. (**e**, **f** Adapted from [10]).

low (only a line is scanned) the 3D scanning method can be tuned to effectively "hit" a large fraction of the cell bodies within a volume of 150–300 μm side length [40]. For example, a volume-covering 3D line scan can be created by spiral scanning in the *x*–*y*-dimension while simultaneously driving the piezo-element with a sinusoidal voltage so that it mechanically oscillates along the optical axis (Fig. 3b). As a result the focal plane is continually moved up and down through the scanning volume, producing a 3D spiral trajectory that is repeatedly scanned (Fig. 3c). With piezo-actuators, repeat frequencies of 5–20 Hz are easily achieved. Higher repeat frequencies are possible, in particular with the new large-aperture electrically tunable lens as an alternative focusing element [42]. A critical issue is, however, to correct for frequency-dependent phase shift and amplitude reduction caused by the inertia of the objective (or the resonance properties of a tunable lens, respectively). To correct these effects, the inertia properties have to be measured beforehand for a particular objective by comparing the actual position of the objective with the intended position change and adjusting the drive signal to the piezo-element accordingly.

The simple mechanical 3D imaging approach is suitable for in vivo calcium imaging from large neuronal populations in neocortical layer 2/3. Within volumes of 150–250 μm side length, sampling of several hundred cells with up to 10 Hz temporal resolution has been demonstrated [40, 46, 47]. A critical step is to assign the collected fluorescence signals correctly to individual cells. This assignment can be done based on a standard reference image stack of the sampled volume, from which positions of cell somata are selected manually or semiautomatically using image segmentation procedures. Then fluorescence signals from all scan line segments that pass through a particular neuron are assigned to this cell and averaged (often cells are hit multiple times by the scan line during one repeat cycle). In this way, a set of fluorescence traces with rather high SNR is built up for an entire local neuronal network. Many variations of the 3D scanning approach are possible. For example, the same setup can be used to obtain vertical or tilted views by rotating the focal plane or to create ribbon-like scanning, which is useful for scanning neuronal dendrites [37]. In brain slices, calcium imaging from dendritic branches localized in 3D has been demonstrated as yet another application of mechanical 3D scanning [41, 48, 49].

2.2 High-Speed Calcium Imaging of Neuronal Population Activity

A laser beam can be rapidly deflected at controllable angles by employing acoustic waves in two crossed AODs, enabling fast random-access scanning in 2D (Fig. 3d). As mentioned in Box 2, dispersion compensation is a critical issue when using AODs for two-photon microscopy because spatial and temporal dispersive effects cause laser focus deformation and temporal pulse broadening, respectively. These effects can be compensated, for example, with a single, appropriately arranged prism so that the original pulse duration is nearly restored and the elliptic deformation of the laser focus in the FOV center is abolished (Fig. 3e) [10, 45]. With the single-prism approach laser beam transmission of the complete scanning and dispersion compensation apparatus can be optimized to reach 45 %. With nowadays available Ti:Sa lasers (average output power >2.5 W) such throughput is sufficient to perform deep in vivo calcium imaging in the neocortex down to several hundred microns [10, 21]. The key advantage of AODs is that the laser focus can be switched between any two positions in the scanning field within few microseconds. The transition time is determined by the laser beam size at the AOD aperture and the sound velocity in the crystal, i.e., the time duration it takes to fill the AOD aperture with an optical grating of different frequency. In our setup we could reach a transition time of 10 μs. Adding an additional minimum on-position dwell time of 2 μs, this means that in principle 83,333 points can be addressed per second in an arbitrary, fully flexible way. A limitation of the AOD approach is that the number of resolvable spots is limited, depending on the beam size and the maximum deflection angle. Therefore, even with optimal dispersion compensation, a compromise needs to be found between FOV size and effective spatial resolution [10]. For fast imaging of neuronal cell bodies, highest possible (diffraction-limited) resolution is, however, not absolutely necessary.

AOD-based random-access scanning can be optimized for high-speed calcium imaging from neuronal populations. For example, the signal integration time on each neuron can be extended somewhat to increase fluorescence yield per neuron and improve SNR. For this purpose we have devised a new random-access pattern scanning (RAPS) mode, in which a predefined fixed, small spatial pattern of a few points is scanned on each neuron, increasing effective dwell time while avoiding stationary parking of the laser focus on the cell, which may cause photo-bleaching and cell damage [10] (see also [50]). After completing a pattern scan on each neuron, the laser focus jumps to the next neuron where the next pattern scan is started anew (Fig. 3f). The small pattern, for example, may consist of a cross of 5 points, which in our setup leads to a fluorescence integration time of 50 μs per neuron (integration is interrupted for the additional 10 μs it takes to "jump" to the next cell). With the 5-point RAPS "averaging" approach 100 neurons can be accessed at high speed (167 Hz) and 33 neurons can be sampled even at 500 Hz. The dwell time per neuron results in a high SNR, which we could demonstrate enables not only the detection of single action potential-evoked calcium transients but in addition near-millisecond precise reconstruction of spike times from the onsets of the calcium transients [10] (5–15 ms accuracy for 200–500 Hz cell sampling rate). The AOD-based high-speed scanning method thus allows monitoring neuronal populations containing ten to hundred cells at much higher speed but with similar SNR as compared to galvanometric approaches. This technique therefore opens new opportunities to measure spatiotemporal spiking activity in local cortical circuits in vivo on the fast (down to a few ten milliseconds) time scale. We will present an example of its application further below.

2.3 Alternative Special Laser Scanning Modes

Several other approaches have been explored during recent years to advance functional imaging of neuronal populations. One strategy to improve laser scanning speed is to simultaneously excite fluorescence at multiple focal spots. These multifocal methods usually necessitate the use of a CCD camera to detect the fluorescence of each spot separately. Multi-spot excitation can be implemented by splitting the laser beam in multiple beamlets [51, 52] or even by creating arbitrary excitation patterns with a "scanless" approach using a diffractive spatial light modulator [53]. Multifocal approaches can in principle efficiently collect fluorescence from multiple structures in parallel. Disadvantages of multi-spot excitation are, however, the reduced available laser intensity per focal spot and the problem of cross talk when imaging deep (>150 μm) in scattering tissue where correct assignment of fluorescence photons to the individual excitation spots becomes difficult. As a result multifocal imaging applications have so far been mainly applied to brain slice preparations.

The assignment problem of multifocal imaging in scattering tissue can be circumvented by using a time-division multiplexing approach. Multiplexing is based on the temporal separation of individual laser pulses in different optical laser beam pathways (for example, using beam splitters and delay lines). Laser pulse separation in both space and time then allows to separate fluorescence signals originating from different focal spots even under scattering conditions. This approach can be used to either perform simultaneous imaging in multiple, neighboring areas or, as has recently been used for in vivo two-photon calcium imaging [54], perform high-speed imaging of multiple focal planes, extending high-speed acquisition of network signals to 3D. With these advances it has now become possible to sample the spatiotemporal distribution of action potential patterns in cortical networks of more than 100 neurons at more than 100 Hz temporal resolution in the living brain.

2.4 Towards Neuronal Population Imaging in Freely Moving Animals

Finally, great progress has also been made in recent years to adapt two-photon imaging approaches so that they can be applied in awake, behaving animals. This research direction is of utmost importance because neuronal population activity clearly is altered in anesthetized animals [15] and—even more importantly—neuronal correlates of complex behavior can only be investigated in awake behaving animals. Two different avenues have emerged that now make it possible for researchers to measure neuronal network dynamics in awake animals during behavior. One approach employs rather standard two-photon microscopes (adapted to the specific requirements of the behavioral experiment) to perform imaging in head-immobilized animals [26, 30, 31, 55]. This approach is described in more detail in Chapter 10 in this volume. The alternative approach is technically more challenging as it aims to develop miniaturized, portable microscopes that can be carried by a rat (or even mouse) and thus should be suitable for cellular imaging in freely moving animals. Ten years ago this approach was pioneered in a study presenting a head-mounted miniaturized microscope that allowed high-resolution imaging of blood capillaries in freely moving rats [32]. Key to such an approach is the use of fiber optics at least for fluorescence excitation, which poses several nontrivial technical challenges regarding fluorescence excitation efficiency and laser scanning (for reviews, see [56–58]). Only recently, after a several-years exploration period of various technical designs [59–61], a milestone was achieved in 2009 by measuring for the first time calcium transients in the visual cortex of freely moving rats using a light weight two-photon microscope [33]. As yet another alternative, miniaturized fluorescence microscopes using single-photon excitation have been successfully applied in freely moving mice to resolve neural activity with single-cell resolution, e.g., in the molecular layer of cerebellar cortex, where dendritic trees of Purkinje neurons are nicely stacked, and even in hippocampus through an endoscopic lens [62, 63]. Although microscope miniaturization still is in its infancy and clearly suffers from limitations in resolution, SNR, and penetration depth, further technical improvements should fully establish this method in the coming years and promote new opportunities to optically probe neuronal ensemble activity during natural animal behavior.

3 Examples of Application from Rodent Visual Cortex

In the following we provide two examples of how the methodological advances described above can be used to collect information about functional properties of local cortical circuits. Specifically, we discuss the application of 3D network imaging using mechanical laser scanning and of high-speed AOD-based two-photon microscopy to investigate rodent visual cortex.

3.1 Orientation Preference in 3D Neural Networks of Visual Cortex

In visual cortex, neurons respond best to moving grating stimuli presented to the animal's eyes. The neuronal responses are sustained throughout the duration of the stimulus and can reach high firing rates. This strong activation results in large calcium signals that can be detected with in vivo two-photon imaging. Whereas neurons in cat visual cortex have been shown to be highly organized in cortical columns [14], the orientation selectivity of neurons in rat visual cortex is more randomly dispersed [13]. To assess the 3D organization of visual responses we used the fast 3D scanning technique to identify a complete set of neurons in a supragranular cube of 200 μm side length. Neurons in layer 2/3 of the rat visual cortex were bolus-loaded with OGB-1 [6, 7] and astrocytes were counterstained with sulforhodamine 101 (SR101) [64]. Animals were placed under a custom-built two-photon microscope and neuronal calcium transients were measured using 3D spiral

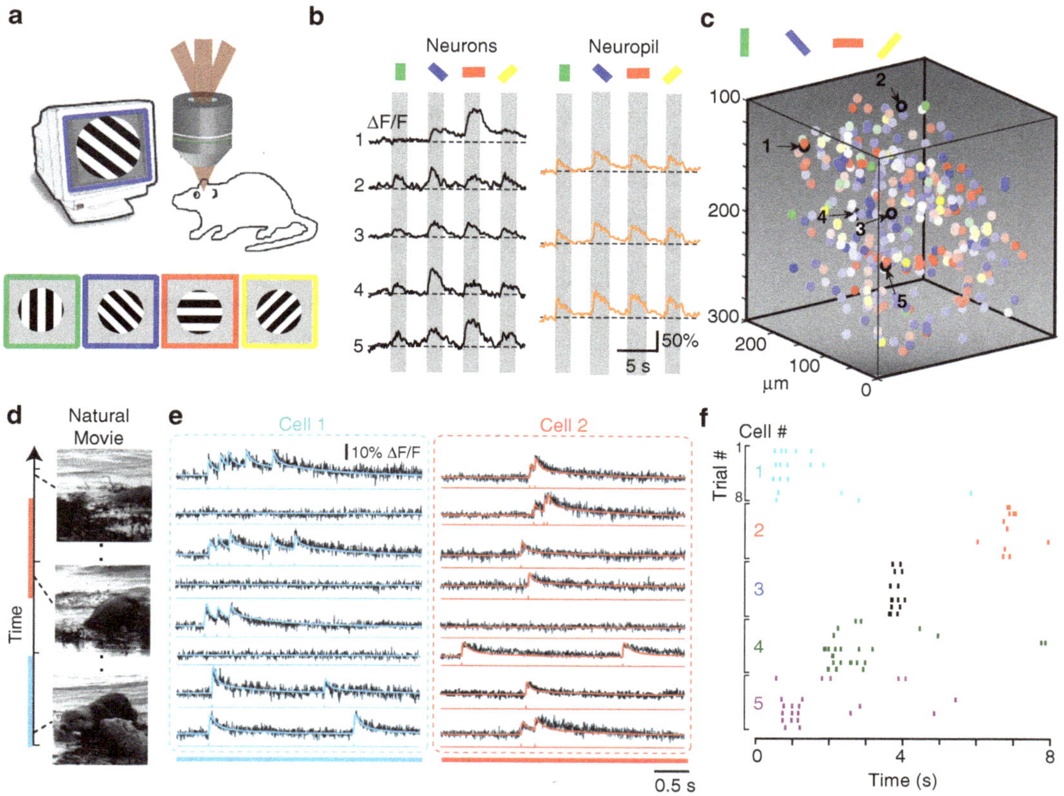

Fig. 4 Imaging neuronal network activity in visual cortex of rodents. (**a**) Calcium transients in OGB-1-labeled neurons of visual cortex were measured in response to visual stimuli presented on a monitor in front of the contralateral eye. (**b**) Example calcium transients from a 3D population imaging experiment performed in the visual cortex of a rat while presenting high-contrast gratings at four different orientations. Acquisition rate was 5 Hz. Example traces from five neurons (*left*) and neuropil activity (*right*) obtained at different depths are shown. *Grey* vertical bars indicate stimulation periods. Orientation of the moving grating is shown at the *top*. (**c**) 3D visualization of orientation preference for all 328 recorded neurons. Neurons are shown at their respective 3D positions with their orientation tuning color-coded as indicated at the *top*. Increasing color saturation indicates stronger orientation selectivity. White cells possess no orientation preference. (**d**) In a different type of experiment visual stimulation consisted of a 10-s long movie showing a natural scene that was repeatedly presented to the contralateral eye. High-speed 2D imaging was performed using an AOD-based microscope. (**e**) Example fluorescence transients in two neurons evoked by the eight consecutive presentations of the movie (traces are from two different 5-point RAPS experiments at 278 Hz and 309 Hz, respectively). The respective time periods during the movie, when cells 1 and 2 were active, are indicated in (**d**). Lower traces with ticks indicate the spike times reconstructed by the automated peeling algorithm. (**f**) Spike raster plots of reconstructed spiking responses in eight trials of movie presentation for five neurons, including cells 1 and 2 that showed reliable activity during specific epochs of the natural movie. (**e**, **f** Adapted from [10]).

scanning in response to visual stimulation with moving gratings of different orientations (Fig. 4a). The location of the neurons in the imaged volume was manually determined from a reference stack and small volumes of interest were placed around the cell bodies for assigning signals to individual neurons. Neurons with a clear orientation preference could be recognized by a peak response during the preferred stimulus (Fig. 4b). Neuropil signals that were analyzed from segments of the 3D scan trajectory between cells at various depths showed little orientation preference (Fig. 4b). For visualization of a 3D map of orientation tuning, the imaged neuronal population was reconstructed according to the neurons' individual spatial position and neurons were schematically color-coded in green, blue, red, or yellow according to their orientation preference (Fig. 4c). Color saturation in addition indicated the orientation selectivity index (OSI), which was calculated as $OSI = (A_{max} - A_{max,\ 90°})/(A_{max} + A_{max,\ 90°})$, where A_{max} refers to the maximal $\Delta F/F$ response amplitude and $A_{max,90°}$ to the $\Delta F/F$ amplitude for the perpendicular orientation. These results and similar findings in recent mouse studies [21, 47] show a non-columnar organization of orientation-tuned neurons in rodent visual cortex and demonstrate the fidelity of fast 3D scanning techniques to detect neuronal tuning properties in populations of several hundred neurons simultaneously.

3.2 High-Speed In Vivo Calcium Imaging in Mouse Visual Cortex

As an example of high-speed in vivo calcium imaging we measured spiking activity in the primary visual cortex of mice in response to the presentation of a natural contrast movie of 10 s duration (Fig. 4d). Here, we applied the AOD-based two-photon microscope for 2D scanning of neuronal populations of about 50–60 neurons using the 5-point RAPS mode resulting in an effective temporal resolution of >300 Hz. At this acquisition rate and given the SNR in our measurements, we estimate that spike times can be revealed with a confidence interval of about 10–15 ms [10]. In several experiments we found individual neurons that responded consistently during brief epochs (around 1–2 s duration) at particular time periods during the movie (Fig. 4e). For automated extraction of visually evoked spiking activity from fluorescent calcium traces we used the "peeling algorithm" [10], which results in a model of the calcium traces that can be overlaid on the raw fluorescence traces (Fig. 4e, colored overlay). The peeling algorithm thus allowed us to extract spike raster plots fairly accurately, with some remaining jitter, revealing the within-epoch variability of spike numbers and precise spike timing for the responsive neurons (Fig. 4f). This example demonstrates that high-speed AOD-RAPS can resolve spatiotemporal dynamics of sensory-evoked spiking activity in neocortical neural networks on a near-millisecond time scale.

4 Discussion and Outlook

The combination of fast laser scanning techniques in two-photon microscopy and methods for bolus-loading of neuronal populations with fluorescent calcium indicators has opened a new door for the investigation of brain activity on the network level. While standard multi-electrode recordings usually offer much higher temporal resolution [65], the identity of the recorded neurons often remains unknown. Furthermore, the spacing of the recording electrodes prevents sampling of neurons at high spatial resolution and therefore in principle will not allow us to obtain a comprehensive measure of local network activity. In this chapter we have highlighted the recent advances that now enable in vivo network imaging of either a few tens of neurons at very high speed (>500 Hz) or several hundreds of neurons, even distributed in 3D, at moderate speed (about 10 Hz). Population imaging using two-photon microscopy thus offers the advantage to record activity patterns from complete samples within a local neuronal population. Further expansion of imaging volumes can be expected and in vivo recordings of several thousands of neurons at moderate speeds seem possible in the very near future.

In addition, the recorded neurons can be visualized and further identified using genetic markers or post hoc immunohistochemistry [66, 67]. This has recently been demonstrated in the mouse visual cortex, by investigating the orientation tuning properties of different types of interneurons [66, 68]. The cell type-specific expression of genetically encoded calcium indicators is a further emerging possibility to target individual neuronal subtypes for investigating their function in the intact brain. Thus, a multitude of options is arising to collect functional data sets from neuronal networks with identified neuronal subtypes. Most recently, in vivo two-photon imaging has been combined with post hoc electron microscopic reconstruction of anatomical connectivity [69] or physiological probing of synaptic connectivity [18], opening fascinating opportunities to understand brain microcircuits in terms of both network structure and dynamics.

We also have seen significant progress in overcoming the speed limitation of laser scanning approaches compared to electrophysiological recordings by using high-speed AOD scanning methods. While two-photon population imaging originally was mainly used for mapping neuronal tuning properties [13, 21, 22], fast scanning approaches now allow the study of ongoing network dynamics during naturalistic visual stimulation (see Fig. 4). Thus, not only the distributions of static tuning properties but also dynamic state space trajectories of local networks can now be analyzed [70]. Moreover, bringing together novel scanning techniques with chronic imaging of cell type-specific GECIs (perhaps also voltage indicators; see Chapter 4 in this volume) promises an exciting future for studying the dynamics of cortical circuits in great detail, particularly in head-fixed animals in realistic behavioral scenarios [31, 55]. The introduction of GECIs further enables chronic studies of neuronal activity over several months (see Chapter 10 in this volume), which has been difficult to achieve with implanted microelectrodes. Targeting specific cell types or pathways for expression of GECIs will allow following changes in activation patterns of specific individual neurons during development and learning. In conclusion, a new door to an exciting field in neuroscience—the study of neural circuit dynamics on the mesoscopic scale—is now wide open.

References

1. Helmchen F, Denk W (2005) Deep tissue two-photon microscopy. Nat Methods 2:932–940
2. Garaschuk O, Milos RI, Grienberger C, Marandi N, Adelsberger H, Konnerth A (2006) Optical monitoring of brain function in vivo: from neurons to networks. Pflugers Arch 453:385–396
3. Grewe BF, Helmchen F (2009) Optical probing of neuronal ensemble activity. Curr Opin Neurobiol 19:520–529
4. Grienberger C, Konnerth A (2012) Imaging calcium in neurons. Neuron 73:862–885
5. Tsien RY (1981) A non-disruptive technique for loading calcium buffers and indicators into cells. Nature 290:527–528
6. Stosiek C, Garaschuk O, Holthoff K, Konnerth A (2003) In vivo two-photon calcium imaging of neuronal networks. Proc Natl Acad Sci USA 100:7319–7324
7. Garaschuk O, Milos RI, Konnerth A (2006) Targeted bulk-loading of fluorescent indicators for two-photon brain imaging in vivo. Nat Protoc 1:380–386
8. Kerr JND, Greenberg D, Helmchen F (2005) Imaging input and output of neocortical networks in vivo. Proc Natl Acad Sci USA 102:14063–14068
9. Sato TR, Gray NW, Mainen ZF, Svoboda K (2007) The functional microarchitecture of the mouse barrel cortex. PLoS Biol 5:e189
10. Grewe BF, Langer D, Kasper H, Kampa BM, Helmchen F (2010) High-speed in vivo calcium imaging reveals neuronal network activity with near-millisecond precision. Nat Methods 7:399–405
11. Ohki K, Chung S, Ch'ng YH, Kara P, Reid RC (2005) Functional imaging with cellular resolution reveals precise micro-architecture in visual cortex. Nature 433:597–603
12. Ohki K, Chung S, Kara P, Hübener M, Bonhoeffer T, Reid RC (2006) Highly ordered arrangement of single neurons in orientation pinwheels. Nature 442:925–928
13. Greenberg DS, Houweling AR, Kerr JN (2008) Population imaging of ongoing neuronal activity in the visual cortex of awake rats. Nat Neurosci 11:749–751
14. Mrsic-Flogel TD, Hofer SB, Ohki K, Reid RC, Bonhoeffer T, Hübener M (2007) Homeostatic regulation of eye-specific responses in visual cortex during ocular dominance plasticity. Neuron 54:961–972
15. Li Y, Van Hooser SD, Mazurek M, White LE, Fitzpatrick D (2008) Experience with moving visual stimuli drives the early development of cortical direction selectivity. Nature 456:952–956
16. Kara P, Boyd JD (2009) A micro-architecture for binocular disparity and ocular dominance in visual cortex. Nature 458:627–631
17. Rochefort NL, Garaschuk O, Milos RI, Narushima M, Marandi N, Pichler B, Kovalchuk Y, Konnerth A (2009) Sparsification of neuronal activity in the visual cortex at eye-opening. Proc Natl Acad Sci USA 106:15049–15054
18. Ko H, Hofer SB, Pichler B, Buchanan KA, Sjöström PJ, Mrsic-Flogel TD (2011) Functional specificity of local synaptic connections in neocortical networks. Nature 473:87–91
19. Katona G, Szalay G, Maak P, Kaszas A, Veress M, Hillier D, Chiovini B, Vizi ES, Roska B, Rozsa B (2012) Fast two-photon in vivo imaging with three-dimensional random-access scanning in large tissue volumes. Nat Methods 9:201–208
20. Roth MM, Helmchen F, Kampa BM (2012) Distinct functional properties of primary and posteromedial visual area of mouse neocortex. J Neurosci 32(28):9716–9726
21. Bandyopadhyay S, Shamma SA, Kanold PO (2010) Dichotomy of functional organization in the mouse auditory cortex. Nat Neurosci 13:361–368
22. Rothschild G, Nelken I, Mizrahi A (2010) Functional organization and population dynamics in the mouse primary auditory cortex. Nat Neurosci 13:353–360
23. Bathellier B, Ushakova L, Rumpel S (2012) Discrete neocortical dynamics predict behavioral categorization of sounds. Neuron 76: 435–449
24. Kerr JN, de Kock CP, Greenberg DS, Bruno RM, Sakmann B, Helmchen F (2007) Spatial organization of neuronal population responses in layer 2/3 of rat barrel cortex. J Neurosci 27:13316–13328

25. Winship IR, Murphy TH (2008) In vivo calcium imaging reveals functional rewiring of single somatosensory neurons after stroke. J Neurosci 28:6592–6606
26. Dombeck DA, Graziano MS, Tank DW (2009) Functional clustering of neurons in motor cortex determined by cellular resolution imaging in awake behaving mice. J Neurosci 29:13751–13760
27. Helmchen F, Imoto K, Sakmann B (1996) Ca^{2+} buffering and action potential-evoked Ca^{2+} signaling in dendrites of pyramidal neurons. Biophys J 70:1069–1081
28. Yaksi E, Friedrich RW (2006) Reconstruction of firing rate changes across neuronal populations by temporally deconvolved Ca^{2+} imaging. Nat Methods 3:377–383
29. Vogelstein JT, Watson BO, Packer AM, Yuste R, Jedynak B, Paninski L (2009) Spike inference from calcium imaging using sequential Monte Carlo methods. Biophys J 97:636–655
30. Dombeck DA, Khabbaz AN, Collman F, Adelman TL, Tank DW (2007) Imaging large-scale neural activity with cellular resolution in awake, mobile mice. Neuron 56:43–57
31. Huber D, Gutnisky DA, Peron S, O'Connor DH, Wiegert JS, Tian L, Oertner TG, Looger LL, Svoboda K (2012) Multiple dynamic representations in the motor cortex during sensorimotor learning. Nature 484:473–478
32. Helmchen F, Fee MS, Tank DW, Denk W (2001) A miniature head-mounted two-photon microscope. High-resolution brain imaging in freely moving animals. Neuron 31:903–912
33. Sawinski J, Wallace DJ, Greenberg DS, Grossmann S, Denk W, Kerr JN (2009) Visually evoked activity in cortical cells imaged in freely moving animals. Proc Natl Acad Sci USA 106:19557–19562
34. Bonin V, Histed MH, Yurgenson S, Reid RC (2011) Local diversity and fine-scale organization of receptive fields in mouse visual cortex. J Neurosci 31:18506–18521
35. Keller GB, Bonhoeffer T, Hübener M (2012) Sensorimotor mismatch signals in primary visual cortex of the behaving mouse. Neuron 74:809–815
36. Andermann ML, Kerlin AM, Reid RC (2010) Chronic cellular imaging of mouse visual cortex during operant behavior and passive viewing. Front Cell Neurosci 4:3
37. Göbel W, Helmchen F (2007) New angles on neuronal dendrites in vivo. J Neurophysiol 98:3770–3779
38. Lillis KP, Eng A, White JA, Mertz J (2008) Two-photon imaging of spatially extended neuronal network dynamics with high temporal resolution. J Neurosci Methods 172:178–184
39. Wilson NR, Runyan CA, Wang FL, Sur M (2012) Division and subtraction by distinct cortical inhibitory networks in vivo. Nature 488:343–348
40. Göbel W, Kampa BM, Helmchen F (2007) Imaging cellular network dynamics in three dimensions using fast 3D laser scanning. Nat Methods 4:73–79
41. Katona G, Kaszas A, Turi GF, Hajos N, Tamas G, Vizi ES, Rozsa B (2011) Roller Coaster Scanning reveals spontaneous triggering of dendritic spikes in CA1 interneurons. Proc Natl Acad Sci USA 108:2148–2153
42. Grewe BF, Voigt FF, van't Hoff M, Helmchen F (2011) Fast two-layer two-photon imaging of neuronal cell populations using an electrically tunable lens. Biomed Opt Express 2:2035–2046
43. Kirkby PA, Srinivas Nadella KM, Silver RA (2010) A compact Acousto-Optic Lens for 2D and 3D femtosecond based 2-photon microscopy. Opt Express 18:13721–13745
44. Reddy GD, Kelleher K, Fink R, Saggau P (2008) Three-dimensional random access multiphoton microscopy for functional imaging of neuronal activity. Nat Neurosci 11:713–720
45. Zeng S, Lv X, Zhan C, Chen WR, Xiong W, Jacques SL, Luo Q (2006) Simultaneous compensation for spatial and temporal dispersion of acousto-optical deflectors for two-dimensional scanning with a single prism. Opt Lett 31:1091–1093
46. Kampa BM, Göbel W, Helmchen F (2011) Measuring neuronal population activity using 3D laser scanning. Cold Spring Harb Protoc 2011:1340–1349
47. Kampa BM, Roth MM, Göbel W, Helmchen F (2011) Representation of visual scenes by local neuronal populations in layer 2/3 of mouse visual cortex. Front Neural Circuits 5:18
48. Botcherby E, Smith C, Booth M, Juskaitis R, Wilson T (2010) Arbitrary-scan imaging for two-photon microscopy. Proc SPIE 7569:756917–756918
49. Botcherby EJ, Smith CW, Kohl MM, Debarre D, Booth MJ, Juskaitis R, Paulsen O, Wilson T (2012) Aberration-free three-dimensional multiphoton imaging of neuronal activity at kHz rates. Proc Natl Acad Sci USA 109: 2919–2924
50. Ranganathan GN, Koester HJ (2010) Optical recording of neuronal spiking activity from unbiased populations of neurons with high spike detection efficiency and high temporal precision. J Neurophysiol 104:1812–1824
51. Kurtz R, Fricke M, Kalb J, Tinnefeld P, Sauer M (2006) Application of multiline two-photon microscopy to functional in vivo imaging. J Neurosci Methods 151:276–286

52. Niesner R, Andresen V, Neumann J, Spiecker H, Gunzer M (2007) The power of single and multibeam two-photon microscopy for high-resolution and high-speed deep tissue and intravital imaging. Biophys J 93:2519–2529
53. Nikolenko V, Watson BO, Araya R, Woodruff A, Peterka DS, Yuste R (2008) SLM microscopy: scanless two-photon imaging and photostimulation with spatial light modulators. Front Neural Circuits 2:5
54. Cheng A, Gonçalves J, Golshani P, Arisaka K, Portera-Cailliau C (2011) Simultaneous two-photon calcium imaging at different depths with spatiotemporal multiplexing. Nat Methods 8:139–142
55. Harvey CD, Coen P, Tank DW (2012) Choice-specific sequences in parietal cortex during a virtual-navigation decision task. Nature 484:62–68
56. Helmchen F (2002) Miniaturization of fluorescence microscopes using fibre optics. Exp Physiol 87:737–745
57. Flusberg BA, Cocker ED, Piyawattanametha W, Jung JC, Cheung ELM, Schnitzer MJ (2005) Fiber-optic fluorescence imaging. Nat Methods 2:941–950
58. Lee CM, Engelbrecht CJ, Soper TD, Helmchen F, Seibel EJ (2010) Scanning fiber endoscopy with highly flexible, 1 mm catheterscopes for wide-field, full-color imaging. J Biophotonics 3:385–407
59. Engelbrecht CJ, Johnston RS, Seibel EJ, Helmchen F (2008) Ultra-compact fiber-optic two-photon microscope for functional fluorescence imaging in vivo. Opt Express 16:5556–5564
60. Piyawattanametha W, Cocker ED, Burns LD, Barretto RP, Jung JC, Ra H, Solgaard O, Schnitzer MJ (2009) In vivo brain imaging using a portable 2.9 g two-photon microscope based on a microelectromechanical systems scanning mirror. Opt Lett 34:2309–2311
61. Flusberg BA, Jung JC, Cocker ED, Anderson EP, Schnitzer MJ (2005) In vivo brain imaging using a portable 3.9 gram two-photon fluorescence microendoscope. Opt Lett 30:2272–2274
62. Flusberg BA, Nimmerjahn A, Cocker ED, Mukamel EA, Barretto RP, Ko TH, Burns LD, Jung JC, Schnitzer MJ (2008) High-speed, miniaturized fluorescence microscopy in freely moving mice. Nat Methods 5:935–938
63. Ghosh KK, Burns LD, Cocker ED, Nimmerjahn A, Ziv Y, Gamal AE, Schnitzer MJ (2011) Miniaturized integration of a fluorescence microscope. Nat Methods 8:871–878
64. Nimmerjahn A, Kirchhoff F, Kerr JN, Helmchen F (2004) Sulforhodamine 101 as a specific marker of astroglia in the neocortex in vivo. Nat Methods 1:31–37
65. Buzsaki G (2004) Large-scale recording of neuronal ensembles. Nat Neurosci 7:446–451
66. Kerlin AM, Andermann ML, Berezovskii VK, Reid RC (2010) Broadly tuned response properties of diverse inhibitory neuron subtypes in mouse visual cortex. Neuron 67:858–871
67. Langer D, Helmchen F (2012) Post hoc immunostaining of GABAergic neuronal subtypes following in vivo two-photon calcium imaging in mouse neocortex. Pflugers Arch 463:339–354
68. Runyan CA, Schummers J, Van Wart A, Kuhlman SJ, Wilson NR, Huang ZJ, Sur M (2010) Response features of parvalbumin-expressing interneurons suggest precise roles for subtypes of inhibition in visual cortex. Neuron 67(5):847–57
69. Bock DD, Lee WC, Kerlin AM, Andermann ML, Hood G, Wetzel AW, Yurgenson S, Soucy ER, Kim HS, Reid RC (2011) Network anatomy and in vivo physiology of visual cortical neurons. Nature 471:177–182
70. Lütcke H, Helmchen F (2011) Two-photon imaging and analysis of neural network dynamics. Rep Prog Phys 74(8):086602

Chapter 10

Chronic Two-Photon Imaging of Neural Activity in the Anesthetized and Awake Behaving Rodent

David J. Margolis, Henry Lütcke, Fritjof Helmchen, Bruno Weber, and Florent Haiss

Abstract

Chronic in vivo two-photon imaging of genetically encoded sensors has recently enabled the measurement of activity from the same individual neurons repeatedly in different imaging sessions over months, opening new possibilities to investigate the function and plasticity of neuronal activity in both anesthetized and awake animals. A successful chronic imaging experiment involves the combination of several key techniques, including expression of genetically encoded indicators, optical measurement with cellular resolution through a cranial window, and, in awake animals, behavioral paradigms adapted to the two-photon microscope. Here we cover methods and advances in chronic imaging of cortical activity in the awake, behaving rodent for the investigation of learning and performance of sensory and cognitive tasks.

Key words Neuronal networks, Action potential, GECIs, Cortical activity, Sensory perception, Behavior

1 Introduction

Intricate sequences of action potentials (APs) within distributed populations of neurons constitute a major part of the brain's information processing. At the local population level, even neighboring cortical neurons show heterogeneity in their properties of spontaneous and evoked activity. Thus, sampling from many cells simultaneously and densely from a large fraction of the local population is important in order to capture an accurate picture of cortical network function. In vivo two-photon calcium imaging with cellular resolution has emerged as a powerful technique to reveal the nature of network coding within local cortical populations (reviewed in refs. [1–3]; see also Chapter 9). While many important questions related to the neuronal network basis of coding and transmission of information can be studied over relatively short time scales of milliseconds to hours (i.e., within single experimental sessions), other processes including learning, memory, plasticity, and disease progression can occur over days to weeks and are therefore not

Bruno Weber and Fritjof Helmchen (eds.), *Optical Imaging of Neocortical Dynamics*, Neuromethods, vol. 85, DOI 10.1007/978-1-62703-785-3_10,

amendable for investigation within single experimental sessions. Furthermore, from a practical standpoint, it would be desirable in almost any experiment to collect data from the same cells in multiple sessions in order to increase statistical power. This has not been feasible using standard calcium imaging techniques because synthetic dyes do not optimally re-label the same cells in more than one application.

Chronic recordings of cortical activity can be performed using other methods, including wide-field optical imaging and extracellular electrophysiology, but depending on the requirements of the experiment, these techniques have disadvantages compared to two-photon imaging. Chronic optical imaging of intrinsic signals and chronic voltage-sensitive dye imaging have been used to investigate map-level stability or plasticity of cortical representations in the mouse [1, 2], rat [3–5], and monkey [6]. A genetically encoded reporter of synaptic activity, and recently a genetically encoded calcium indicator (GECIs), have also been used [7, 8]. However, these wide-field methods lack cellular resolution. While electrophysiology has cellular resolution, better temporal resolution and signal-to-noise characteristics than two-photon population imaging, and also has the potential to track single-unit activity over time [9, 10], it still has major limitations. One issue is the correspondence between extracellular activity and neuron identity over different recording sessions. The number of stable units drops to less than 50 % within 5 days [9, 10]. Even if a unit appears stable over time in terms of its recorded spike waveform or inter-spike interval distribution, it is still difficult to unequivocally rule out the possibility that the original unit went silent and a nearby neuron became active. Additionally, there is no anatomical information about the recorded cells, so cell classification into anatomical types is not possible. Even if unit recordings can be maintained over days, extracellular techniques provide much sparser sampling of neural activity. Inserting large arrays of electrodes can increase cell number and spatial information [11], but the electrodes are usually spaced hundreds of microns apart and therefore reveal little about the network microcircuitry. Lastly, metal electrodes are not optimal for recording from smaller cells, and cannot be used to target genetically identified subpopulations as can two-photon imaging.

Chronic in vivo two-photon imaging has been used to measure changes in the fine structure of neuronal dendritic spines over weeks and months (reviewed in [12]), but long-term imaging of cell function has been challenging. Chronic recording of neuronal population activity has not been feasible using standard calcium imaging techniques because synthetic dyes cannot easily be used to re-label the same cells at later time points. Improvements in expression, design, and imaging of GECIs have recently allowed chronic measurement of activity from the same cells over long time periods [13–17], with sensitivity in some cases beginning to approach that of synthetic dyes. Genetically encoded reporters of other physiological signals including membrane voltage ([18]; Chapter 4, this volume) and intracellular signaling molecules (e.g., [19]) appear promising in vivo but have not yet been tested in chronic imaging experiments. We propose that future studies using chronic imaging of genetically encoded indicators will provide new fundamental information about the long-term function and plasticity of individual cells and circuits. In this chapter, we cover methods and advances related to chronic in vivo imaging of activity in cortical neurons of the behaving rodent.

2 Methods for Chronic Two-Photon Imaging

2.1 In Vivo Measurement of Neuronal Activity Using GECIs

In vivo two-photon imaging of cellular activity using synthetic calcium indicators has emerged as a powerful technique to survey spontaneous and sensory- or motor-evoked activity of dense local neuronal populations in the cortex (reviewed in [20–22]). This topic is covered further in detail in Chapter 9. Synthetic calcium indicators, until recently [23] exhibited better signal-to-noise properties than GECIs for reporting AP activity [24, 25], but a critical disadvantage is that they have proven difficult to apply repeatedly for chronic imaging experiments [26, 27]. Andermann et al. [13] recorded stimulus-evoked signals in the visual cortex of mice from the same cells re-labeled with Oregon Green BAPTA-1-AM 2 days after the initial experiment, and managed a third session on the subsequent day. However, re-injecting dye required reopening of the craniotomy and continued health of the tissue may be a significant factor in this type of experiment [26]. Furthermore, it is not clear that it would be possible to re-label cells after more than two injections and therefore experiments would be limited to a few days.

GECIs are fluorescent calcium indicator proteins that can be expressed in neurons and have some crucial advantages compared to synthetic dyes for chronic imaging. GECIs achieve stable expression levels after one-time introduction into tissue and, therefore, do not require repeated application. Unlike synthetic dyes, which bulk-label all cells near the injection site including neurons and glial cells, expression of fluorescent proteins including GECIs can be targeted to specific cell types using cell-specific promoters [28], developmentally timed in utero electroporation, or inherent tropism of viral vector serotypes [29]. Subcellular targeted expression can be achieved by using targeting sequences [19] or fusion proteins [30].

Specific expression can be an advantage or a disadvantage depending on the needs of the experiment. Cell-type specificity can mean that labeling is sparse, which is a potential drawback for experiments that aim to record from a large fraction of cells within a population. For large-scale in vivo population imaging of GECIs,

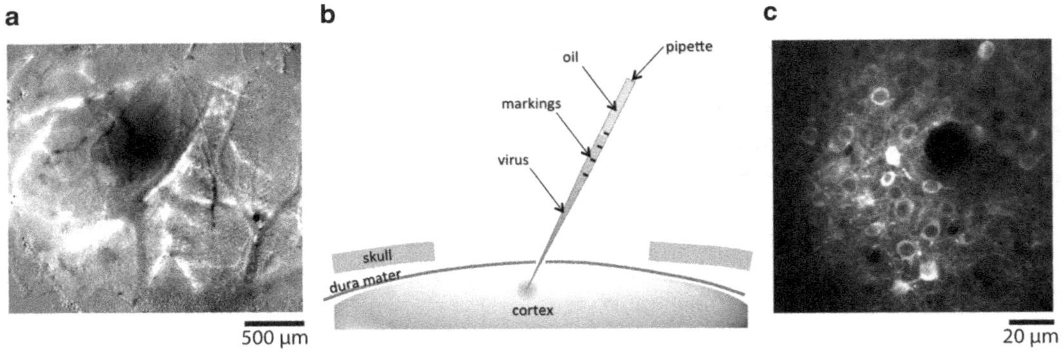

Fig. 1 Identification of the area for the injection, injection of the virus, and the resulting expression pattern of the GECI YC3.60 in neocortex. (**a**) Intrinsic optical imaging map of a single whisker stimulation. *Dark area* indicates reduced reflected light at 630 nm which corresponds to the stimulus-induced activated cortical area. (**b**) Schematic of the virus injection procedure. (**c**) Neocortical neurons expressing YC3.60 in layer 2/3 of the mouse somatosensory cortex.

viral vectors have been the expression method of choice. Typically, at least in primary sensory areas, mapping techniques such as intrinsic optical imaging are used to target the location of the virus injection. Volumes in the nanoliter range are injected with glass micropipettes into the desired cortical area (Fig. 1). While different combinations of promoter, serotype, and type of virus have been used, lentiviral and adeno-associated virus (AAV) vectors lead to high transfection rates with low levels of immune responses. The AAV hybrid serotype 1/2 with the human synapsin promoter (AAV1/2-hSyn) has features amenable to long-term population imaging, including neuron-specific expression, high rates of transduction (up to approximately 80 % of neurons near the injection site), a large spread (~1 mm for a 200 nl injection), and stable expression over time (see Fig. 1c for an example). This vector has been used in recent studies to express the GECIs GCaMP3, d3cpv, and YC3.60 [14, 16, 31–36]. Dense expression means that many neighboring cells can be imaged at one time, allowing comprehensive sampling of local population activity. In addition, individual neurons are labeled brightly throughout the dendrites and axon; for single-cell or dendritic imaging this can be a potential disadvantage because neuronal processes from individual cells are lost in the thicket of processes from neighboring cells. To visualize dendrites it is better to have sparse expression, such as that achieved by different expression methods, including in utero electroporation (e.g., [15]; where labeling density can vary from ~5 to 50 % of neurons) or AAVs with different promoters ([13]; CBA promoter). Methods to achieve sparse expression are potentially useful for correlating anatomy on the level of spines and single-cell physiology [37]. There is evidence from [17] that in utero electroporation may be associated with altered cell physiology.

Other viruses including semiliki forest virus also result in sparse expression, but are lethal to cells and therefore not suitable for long-term experiments [15]. Optimizing in vivo expression of GECIs is an area of active interest and new transgenic mouse lines might offer significant advantages in some of these respects [38, 39].

A crucial feature of GECIs that has so far limited their widespread use is the sensitivity to reporting AP activity in neurons in vivo. AP sensitivity is a critical concern for measuring activity in sparsely firing neurons, which dominate population activity in many brain regions [40]. Recent improvements in design and expression of three indicators in particular have boosted GECIs toward the long-awaited goal of in vivo single AP sensitivity. Wallace et al. [35] showed that AAV-mediated expression of d3cpv, from the Cameleon family of fluorescence resonance energy transfer (FRET; [41]) indicators, allowed optical detection of most APs (>70 %) at low false positive rates (1–2 APs per 1 min recording time; Fig. 2a, b). The widely used GCaMP3 indicator from the lab of Loren Looger was shown to report AP-evoked activity with similar sensitivity at low numbers of APs, then gigantic relative changes with more than three APs ([17]; Fig. 2c, d). Other indicators such as TN-XXL, which has been shown to be useful for chronic imaging ([15], Fig. 2e), may also not be sensitive enough to report single APs in vivo. A recent study compared YC variants with GCaMP3 and showed that YCs are better at detecting low numbers of APs in spite of the large signal change of GCaMP3 with large bursts [42]. Further development of GCaMPs have dramatically improved their kinetics and sensitivity [23, 43].

How stable are the properties of GECIs over time? If GECIs are used to investigate stability and plasticity of neuronal activity, it is important to calibrate the properties of the indicator itself after long-term expression. Margolis et al. [16] targeted YC3.60 expressing neurons using in vivo juxtacellular patch clamp recordings after months of expression and imaging through a cranial window. Comparison with previous work using acute experiments revealed no difference in terms of sensitivity to AP activity (Fig. 2f–h). Sensory map dynamics measured with wide-field imaging also displayed stability of over multiple weeks [8]. It is not clear so far whether the functional properties of other GECIs remain stable over long time periods. This is a critical point that could potentially limit the utility of some GECIs for long-term experiments.

Sensitivity to AP-evoked calcium influx depends on a sufficiently high concentration of indicator protein, which is in turn a factor of the expression method. In the case of the YC3.60, for example, expression level may be as important as the properties of the sensor itself. Nagai et al. [44] created a transgenic mouse to express membrane-bound YC3.60 but expression was not specific to neurons and activity was observed only in response to strong stimulation. When the same protein was expressed using an

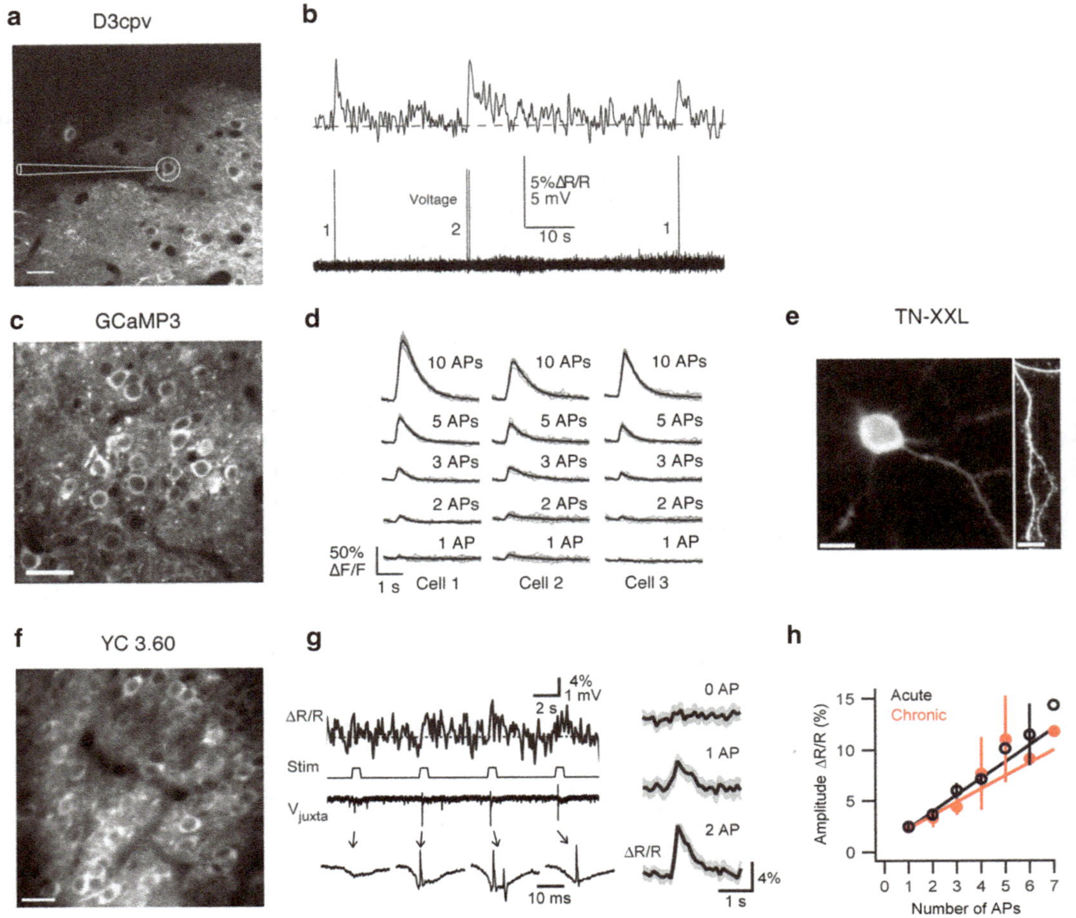

Fig. 2 In vivo calcium imaging using genetically encoded calcium indicators. (**a**) L2/3 neurons in the somatosensory cortex expressing d3cpv (150 μm deep; scale bar, 20 μm; 6 weeks after infection) together with the tip of the patch pipette (*arrow*). (**b**) Simultaneous fluorescence and voltage recordings. Shown are the ratios of signal from within the soma. (**c**) GCaMP3 expression in L2/3 neurons of the primary motor cortex at 72 days after injection (*top*; scale bar, 30 μm). (**d**) Examples of single-trial responses (*gray*) and average across ten trials (*black*) from three neurons to evoked APs at 50 Hz with mouse under anesthesia. (**e**) *Left*, in vivo two-photon image of a pyramidal neuron expressing TN-XXL (average of 200 frames), 275 μm below the cortical surface. *Right*, dendrites and spines from this cell at a depth 120 μm. Scale bars, 10 μm. (**f**) Neocortical neurons expressing YC3.60 in L2/3 of the mouse somatosensory cortex. (**g**) Juxtacellular voltage recording from an L2/3 neuron in barrel cortex and simultaneous two-photon Ca^{2+} measurement from the soma. Mean Ca^{2+} transients are shown at right for 1, 2, and 3 APs. (**h**) Relationship between Ca^{2+} transient amplitude and number of APs. In chronic experiments, neurons were targeted for juxtacellular recordings after removal of the cranial window. See Margolis et al. [16] for more details. (**a**, **b**) adapted from [35]; (**c**, **d**) adapted from [17]; (**e**) adapted from [15]; (**f-h**) adapted from [16].

AAV1/2-hSyn vector, expression levels were higher (likely in the range of tens of μM; [35]) and neuron-specific, and sensitivity was dramatically improved [32]. New generation transgenic mice show improved expression and functional properties [38, 39], and this is likely to keep improving in the near future.

Dynamic range is another important feature, which is the increase of the calcium transient amplitude proportional to increasing numbers of APs; it appears to be reasonable for GCaMP3, YC3.60, and TN-XXL, but may be problematic with d3cpv. Efforts to improve many of these indicators have been made [45, 46] and recently came to fruition with the introduction of GCaMP6 [23]. At present there is no clear best indicator for detecting sparse firing in chronic imaging experiments; the choice must be motivated by a number of other considerations as well, as outlined below.

2.2 Chronic In Vivo Imaging

Expression of calcium binding proteins at concentrations of tens of μM is a concern for potential negative effects on cell physiology. Instead of calmodulin in the case of Cameleons, the Troponin C-based family of indicators (e.g., TN-XXL) has possible advantages in this respect because the calcium binding element is derived from nonneuronal sources and is therefore not engaged in neuronal signal transduction ([47]; Fig. 2g). However, GCaMP3 was found to be the only indicator whose expression did not perturb turning behavior when expressed in worms [17]. A specific concern is the potential effect on calcium-dependent processes such as synaptic transmission and plasticity. Tian et al. [17] performed control experiments that demonstrated GCaMP3 expressing neurons received the same amount of total synaptic input as non-expressing neighboring neurons. In addition, Huber et al. [14] provided further controls that indicated GCaMP3 expression did not perturb LTP in hippocampal slice cultures (see also Discussion, below).

The critical issue is that GECIs can be used for chronic applications, as long as expression and functional properties are stable over time. Repeated imaging of calcium signals in individual cells over periods of ~20–50 days has been demonstrated in the visual cortex of anesthetized [15] and awake mice [13] using in utero electroporation of TN-XXL and viral expression of YC3.60, respectively. These studies used sparse expression techniques and the total number of cells sampled was limited to only a few neurons. The small sample size thus provided limited information about population coding. Tian and coworkers achieved dense expression of GCaMP3 by intracortical injection of AAV2/1 under the synapsin promoter [17]. Expression in the same neurons appeared similar up to 120 days after infection, although data showing increased brightness after 4 weeks were presented more recently [38]. Calcium transients imaged in motor cortex while mice ran on a treadmill also appeared similar when imaged 48 days apart, but no further analysis was done in these studies to assess the similarity of calcium signals over time.

Two recent studies directly investigated the stability of neuronal function using long-term GECI imaging and measuring from the same neurons over extended time periods (Fig. 3). Margolis et al. [16] imaged whisker-evoked activity in primary somatosensory (barrel) cortex in anesthetized mice using YC3.60 repeatedly in the same

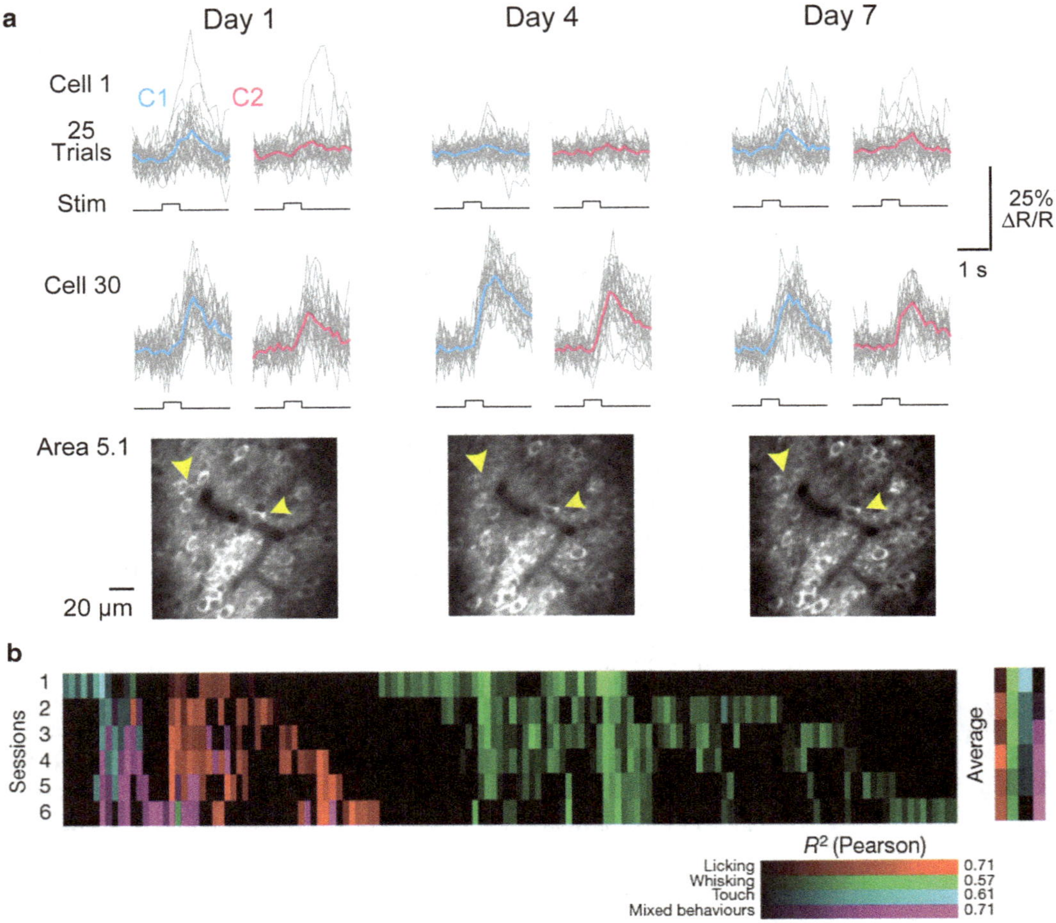

Fig. 3 Chronic imaging of neuronal activity in individual cells and local populations. (**a**) Whisker-evoked Ca^{2+} transients from two example neurons imaged over 1 week in repeated sessions. For each neuron, responses are shown for stimulation of the C1 and C2 whiskers. Gray lines are single trials and color lines are means. Cell 30 is a rare high-responsive cell that displayed reliable sensory-evoked Ca^{2+} transients to nearly every stimulus. Adapted from [16]. (**b**) Neuronal response types imaged over six sessions as mice learned a whisker-dependent discrimination task. Individual neurons are represented as color boxes along the horizontal axis and the response type is color-coded over sessions along the vertical axis. Neurons were classified as related to licking, whisking, touch, or mixed. Individual neurons sometimes appeared and disappeared from the representation, although major changes between response types were not prominent. Adapted from [14].

neurons for up to 4 months. This allowed measurement of both the responsiveness and tuning of individual neurons and populations over time in response to stimulation of the same two whiskers. The most active (high responsive) and least active (low responsive, or silent) neurons displayed stable responsiveness over time. While it is known that activity within cortical populations is heterogeneous and sparse [48], this provided direct evidence that sparse activity is stable. The whisker preference (tuning) of individual neurons was more labile and could change from session to session, even though the larger populations were relatively stable, similar to the results of [14] for sensorimotor representations.

Huber et al. [14] repeatedly imaged GCaMP3 expressing neurons in primary motor cortex as mice performed a whisker-dependent object localization task over a period of 2–3 weeks. This allowed for tracking of population and individual neuron activity throughout sensorimotor learning. The study found intermingled representations of behavioral parameters including whisking, touch, and licking. While populations represented these behavioral parameters faithfully over time, individual neurons were relatively more unstable, dropping in and out of the representation from session to session. However, because GCaMP3 does not report ≤2 APs, it is not clear if these neurons were truly silent or active at a level lower than the detection threshold in certain sessions.

These studies went on to follow the same neurons throughout changes in experience [16] and learning [14], opening up the field for investigation of how individual neurons and populations undergo experience-dependent and learning-related functional plasticity.

In summary, GECIs can be used successfully for chronic imaging of activity of individual neurons, local neuronal populations, and sensory maps over time periods of months. Choice of GECI should be motivated by its expression properties, response properties, and signaling relative to the particular needs of the experiment. In addition to sensitivity to neuronal activity, chronic experiments require that indicator expression and function is stable over long time periods. It will be important in future experiments to determine the functional properties of GECIs after long-term expression.

2.3 Gaining Chronic Optical Access to the Brain

Chronic in vivo imaging depends on not only expression and functional properties of indicators but also reliable methods to optically access the brain with cellular resolution over long time periods. To this end, recent studies have adopted cranial window methods from the pioneering in vivo investigations of dendritic spine dynamics (reviewed in [49]). The basic premise of the open skull cranial window technique is to permanently replace a piece of the skull (up to 5–16 mm^2 area) with transparent glass. This is done by performing a craniotomy over the region of interest, then fixing a cover glass in place using composite dental cement. If the window remains clear over the next weeks, the preparation is suitable for chronic in vivo imaging in repeated imaging sessions. In addition to the dendritic spine studies, this method has recently been used to repeatedly measure functional signals using GECIs [13, 15, 17]. It is easily adapted for imaging in different cortical areas including motor, somatosensory, and visual cortex, and has also been more laboriously modified for imaging deep brain structures including the hippocampus [50]. Variations of the cranial window for structural or functional cellular imaging have so far been applied in mouse and macaque monkey [51, 52] (Fig. B1, Box 1). We present a brief protocol in Box 1, including extension of the technique to the rat.

An alternative approach is to image through the near-transparent thin-skull [55]. An advantage of the thin-skull method is that it is less invasive, reducing possible perturbation of the underlying brain tissue as a result of the surgery. Xu et al. [56] suggested that such disturbance might be substantial, causing changes in dendritic spine dynamics and glial cell activation. Although this issue is still controversial, many investigators now use longer recovery times (2–3 weeks) before starting to image in order to avoid this potential problem. Another source of glial cell activation is the immune response to the injected virus and the expressed transgene. Astrocytosis and microgliosis can be observed right after virus injection but these inflammatory reactions are strongly reduced around the injection site after a few weeks [57, 58]. A disadvantage of the thin-skull window is that only a limited area (0.1–0.3 mm^2) is rendered optically accessible, precluding studies of large-scale measurements of neuronal populations. A further disadvantage is that the number of possible imaging days is limited to two to three because the skull does not tolerate repeated re-thinning. Choice of imaging window must ultimately be informed by the type of experiment desired, as each method for in vivo brain visualization has its own distinct advantages. However, for applications involving measurement of cellular activity using GECIs, the glass window has some important advantages. The larger optically accessible area allows imaging of larger populations of neurons across the cortical surface. Furthermore, neurons deeper below the surface are more easily imaged because depth penetration is not reduced by the remaining skull. Finally, measuring population activity, as opposed to dendritic spine structure, requires long periods of data collection during sensory-evoked or motor-related neuronal processing. Because cortical neurons show sparse and variable activity, it is important to collect as much data as possible over multiple repeated sessions to achieve an accurate picture of neuronal activity dynamics. Thus, the limitation on the number of imaging sessions is perhaps the most important factor undermining the utility of the thin-skull window for chronic measurement of cellular activity.

A possible compromise between the open- and thin-skull type windows was introduced recently by Drew et al. [53]. Their polished and reinforced thinned skull (PoRTS) window creates a kind of hybrid "thindow" that may resolve some of the problems associated with each of the alternatives described above. In particular, the PoRTS window is not limited to a small imaging area like the thin-skull window or even the mid-scale sizes of the glass window. However, because the skull is not removed, depth penetration will

Box 1 Chronic Protocols: Injection, Cranial Window, and Head-Post

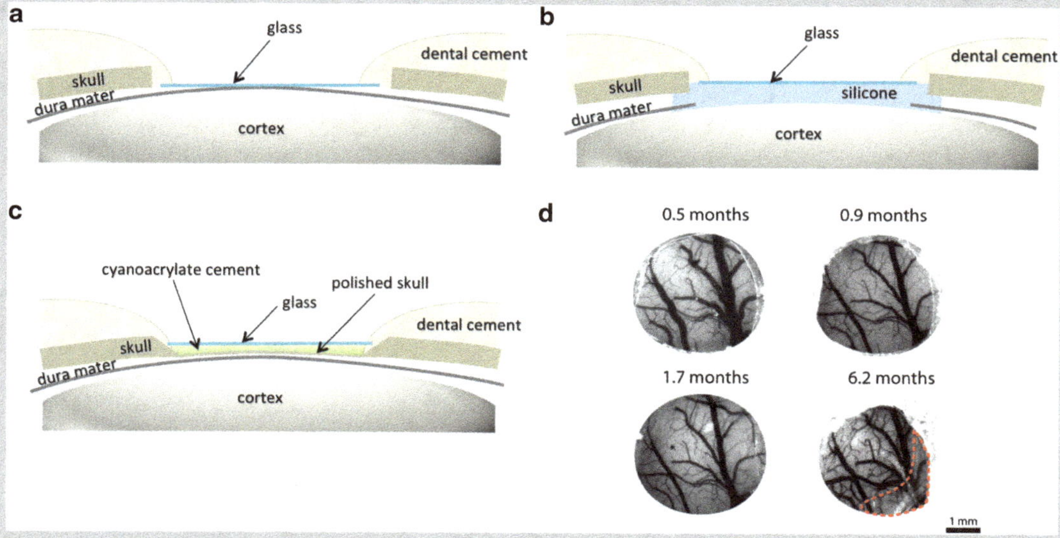

Fig. B1 Schematics of chronic cranial windows. (**a**) Cranial window with removed skull used in mice. (**b**) Cranial window with removed skull and dura used in rats. (**c**) Cranial window with polished and reinforced thinned skull. (**d**) Images taken through a cranial window as depicted in (**b**) at 570 nm illumination 0.5, 0.9, 1.7, and 6.2 months after implantation. *Red dashed line* indicates a region where dura regrowth was visible.

This box contains a step-by-step procedure for a chronic cranial window in rats or mice (Fig. B1a,b). The PoRTS is depicted in Fig. B1c and described in [53].

Step 1: On the day before surgery inject dexamethasone (2 μg/g bodyweight intraperitoneal) to reduce edema.

Step 2: Anesthetize the mouse or rat with isoflurane (4 % for the induction, 1.5–2 % for surgery with oxygen at 400 ml/min). Fix the animal's head in a stereotaxic frame and place it on a heating blanket in order to keep the body temperature at 37 °C.

Step 3: Shave the animal's head and disinfect the scalp with 70 % ethanol.

Step 4: Make an incision into the scalp to expose the underlying bone. The skull has to be cleaned from the connective tissue. The temporal muscles have to be disconnected from the skull on both sides.

Step 5 (optional): Intrinsic optical imaging at 630 nm can be performed through the thinned skull to locate the desired area for the virus injection. A surface vessel image taken with same camera and optics with green (570 nm) light allows the experimenter to identify the region where the trepanation/injection has to be performed in steps 8 and 9.

Step 6 (only for rats): Drill holes for the subsequent insertion of scull screws (Modus 1.5, 3 mm length; Medartis, Basle, Switzerland). The exact location of these screws has been described in Fig. 1 of [54].

Step 7: A dental bonding agent (e.g., Optibond FL; Kerr Corporation) is applied to the skull and polymerized with blue light (460 nm) for 20 s. The skull is then covered with a layer of light curing dental cement (1 or 3 mm thick for mice and rats, respectively) sparing the area where the cranial window will be located but covering the area where the temporal muscle was connected. This second layer has to be polymerized with blue light for 40 s.

Step 8: Remove the skull (<1 mm diameter) to expose the dura above the area where the virus will be injected.

Step 9: A sharp pulled glass pipette with a broken tip is used to inject the virus into the cortex. The pipette is inserted in a 10 μl syringe (Nanofil syringe, World Precision Instruments, Sarasota, USA) and filled with low viscosity oil. Using a micro-pump system (UltraMicroPump III, World Precision

Box 1 (continued)

Instruments, Sarasota) a 1:1 mixture of the virus with 20 % mannitol solution is filled from the tip into the pipette.

Step 10: Penetrate the cortex through the dura to the desired depth (300 μm from the surface of the brain for a layer 2/3 injection) and inject 20–200 nl of virus mixture at 20 nl/min. Wait for 10 min after the injection before retracting the pipette to prevent back-spilling of the injected fluid.

Step 11: Make a trepanation that is slightly larger than the round cover glass (3 mm diameter for a mouse, 4 mm diameter for a rat) centered on the injected area.

Step 12 (only for rats): Make a small incision into the dura and tear it apart with fine forceps. Alternatively, the dura can be cut with Vannas spring scissors (Fine Science Tools, Heidelberg, Germany). In some areas this may lead to rupture of large vessels of the dura mater, which leads to bleedings. These bleedings can be stopped with fine bipolar forceps connected to an electro-surgery unit (Erbe, Tubingen, Germany). In order to prevent the coagulation of the underlying brain it is recommended to lift up the dura flap with the bipolar forceps and to perform the whole procedure under constant irrigation. Once all bleedings are stopped a ring of silicone (Kwik-Sil, WPI, Sarasota, FL, USA) is applied to the dura edges to prevent future regrowth.

Step 13: Place the round cover glass onto the brain (with dura intact in mice) and apply a slight pressure to put the window on the same level as the surrounding bone.

Step 14: Connect the window edges with the rest of the head implant using light curing dental cement. Place a head-post onto the head-cap and connect it with cement. Polymerize the dental cement with blue light for at least 40 s.

Step 15: Suture the skin and glue it to the head-cap with cyanoacrylate glue (Histoacryl, Braun, Melsungen, Germany). As analgesic administer caprofen (5 mg/kg; Pfizer GmbH, Karlsruhe, Germany; recommended for rats) or buprenorphine (0.05 mg/kg; Essex Pharma GmbH, Munich, Germany; for mice) twice a day subcutaneously for 2 days after surgery.

still not be as high as with the glass window. The PoRTS window has yet to be tested for measurement of neuronal activity. If it turns out to be difficult to combine with intracortical viral injection, or if cellular motion is a problem in awake animals, the PoRTS window may not be optimal. These possibilities will have to be tested in future experiments.

2.4 Imaging in Awake Animals

In the last decade, the two-photon laser-scanning microscope has become a workhorse in neuroscience labs for a wide variety of in vitro and in vivo experimental applications requiring high-resolution fluorescence imaging of cellular structure and function [59, 60]. Adapting the microscope to the living animal is challenging, especially when the animal is awake. This can be achieved by mounting a miniaturized version of the microscope on the head of the freely moving animal, or by immobilizing the head of the animal under a traditional fixed microscope. While improving

Fig. 4 (continued) two-photon microscope (titanium:sapphire laser (Ti:S), long-pass filter (LP), galvanometers (X-Y), scan lens (SL), mirror (M), tube lens (TL), dichroic mirror (DM), collection lens (CL), biconcave lens (L), bandpass filter (BP), focusing lens (FL), photomultiplier tube (PMT), sliding stage (used to move microscope for treadmill access), *X-Y* translation (moves treadmill and mouse), *Z*-translation (objective focus control), and rubber tube (shown in cross section, for light shielding)). Adapted from [50].

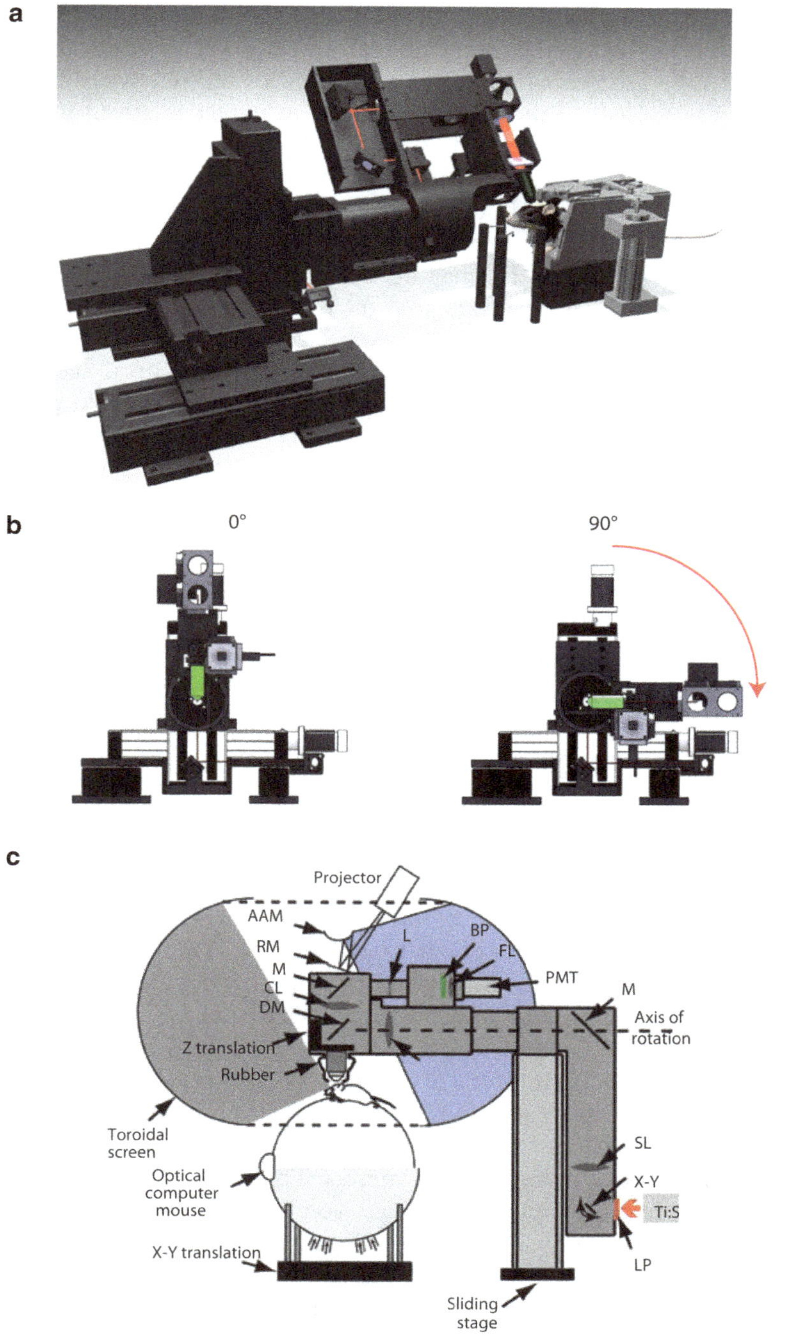

Fig. 4 Two-photon microscopes with behavioral apparatus. (**a**) CAD rendering of the two-photon microscope (lab of Bruno Weber, University of Zurich) with a head-posted rat in the behavioral apparatus (left perspective). The laser beam is colored in *red*. The microscope is able to move 13 cm in *x*, *y*, and *z* dimensions. (**b**) Front view of the microscope in the upright and 90° tilted position. Rotation is done around the focal point of the objective (*green*). (**c**) Microscope setup for imaging during virtual navigation [26]. The experimental apparatus, consisting of a spherical treadmill, a virtual reality apparatus (projector, reflecting mirror (RM), angular amplification mirror (AAM), toroidal screen, and optical computer mouse to record ball rotation), and a custom

experimental control, head fixation severely limits the behavioral repertoire of the animal. On the other hand, miniaturized microscopes suffer from inferior optics and relative difficulty of everyday use. Although there have been recent advances in miniaturized two-photon microscopy [61–63] we focus here on adaptations of the traditional two-photon microscope for the study of rodent head-fixed behavior.

Cellular and systems neuroscience has begun to see major advances in imaging during behavior in head-fixed animals. It is important that the microscope itself does not inhibit animal motion, and that it does not restrict access for electrophysiology and behavioral equipment. A moveable upright microscope allows freedom in positioning the subject as well as devices for behavioral experiments. For example, the whisker-related task in rats discussed below requires a number of piezo stimulators, sensors, and lick ports at the front and sides of the animal. To this end, a mobilized two-photon microscope was implemented by Winfried Denk and colleagues [51] originally for imaging axonal dynamics in visual cortex of macaque monkeys. The microscope was fitted with a moveable objective that allowed movement in three dimensions and also rotation around one axis with a stable beam position with respect to the scanning head [64]. Large rotation angles allow imaging of the most dorsal and lateral parts of the brain. This type of design, which inspired the commercially available Moveable Objective Microscope from Sutter Instruments, allows scanning over a large area while keeping the animal stationary. A similar design has recently been used in behaving monkeys [65].

A microscope design with even higher degrees of flexibility is depicted in Fig. 4a, b. This design is fully motorized and allows movements of up to 13 cm in x, y, and z dimensions, and ±90° rotation around the focal point of the objective. The use of motorized linear and rotational stages allows quick positioning of the microscope and therefore access to different locations on the cortex within tens of seconds. With the addition of a fast linear piezo stage, the microscope objective can be rapidly moved up or down to image at different cortical depths or to quickly retract the objective by several centimeters. This feature is especially useful to avoid a collision of the objective with the animal's implant in behavioral paradigms that allow slight head rotation during reward retrieval [66]. Alternatively, the microscope can be elevated and the head-fixed animal can be placed on a moveable platform, allowing limited ambulation. Dombeck and coworkers imaged head-immobilized mice walking and running upon a floating Styrofoam ball (Fig. 4c) [26]. This has allowed for the study of movement-related cellular activity in motor cortex [67, 68] and running-induced modulation of spiking activity in visual cortex [69]. Other strategies have been used to allow some degree of mobility during head-fixed imaging, including a planar treadmill [17, 70] and fixed-axis (not floating) ball [71]. David Tank's group went a step further and added a theater around the microscope, so the mouse could navigate in a virtual environment. For example, Harvey and coworkers showed that mice can be trained to run back and forth along a virtual linear track for rewards, virtually recreating classic spatial navigation studies to study hippocampal place cells [72] and mazes to study neural processes underlying decision making [31]. This approach has been adopted by many laboratories and has opened new avenues into mouse behavior and cognition.

For many lines of investigation it is not necessary or even desirable to allow walking or running during the imaging session. Sensorimotor integration has been successfully studied in the whisker system of awake head-fixed rodents using a fixed platform [33, 73–75]. Whisking does not depend on ambulation and might be easier to study without it.

Brain movement poses one technical problem for imaging cellular activity in awake animals. Because the cellular structures being imaged are on the order of 10 μm, small movements of the cells can cause large artifacts in the fluorescence signals. Precautions must be taken to reduce cell movement during the experiment, and corrections must be implemented offline to eliminate the possibility of motion-induced contamination of fluorescence transients. Techniques have been used to mechanically dampen brain movement, such as silicon elastomer "plugs" between the brain tissue and glass window, or fitting the window snugly against the dura matter (Fig. B1; Box 1). However, residual motion in awake animals of a few microns, especially during brief periods of sudden movement, is essentially unavoidable. Thus, post-processing of the image data to correct for this motion is an important step in the analysis of functional signals. See Box 2 for discussion of this point.

Some GECIs have advantages for avoiding motion artifacts. The two-color ratiometric indicators including the Troponin (e.g., TN-XXL) and Cameleon families (e.g., YC3.60, d3cpv) comprise two fluorescent proteins with spectrally distinct emissions. Upon calcium binding, FRET occurs between the shorter- and longer wavelength protein, causing a decrease in cyan and increase in yellow fluorescence emission (in the case of the above probes). Practically, this means that motion results in correlated changes in both fluorophores, while AP-related signals result in anticorrelated signals that can be distinguished by taking the ratio of the two emission channels. For example, using viral expression of YC3.60, Andermann et al. [13] observed that measuring calcium signals using the ratio of YFP/CFP fluorescence provided increased robustness to motion artifacts as compared with non-ratiometric indicators, even without offline motion correction, and even when the mouse was licking. However, without implementation of online motion correction, it is necessary to correct for motion artifacts offline regardless of the indicator used.

Box 2 Chronic Imaging in Awake Animals: Motion Correction

Brain movement poses a challenge for in vivo two-photon microscopy. During imaging of anesthetized animals, pulsatile tissue displacements due to heartbeat or breathing can be effectively eliminated by the use of agar and cover glass. In the awake rodent, however, brain motion may lead to large artifactual changes in fluorescence intensity, especially when the animal is actively engaged in a task. Movements of only a few micrometers may destroy the small functional signals observed in response to action potential firing. Current approaches focus on mechanical reduction of motion (see main text), and automatic detection of image frames corrupted by motion and the subsequent correction using offline algorithms. It is useful to distinguish between movements that occur within the imaging plane—and are therefore in principle correctable—and through-plane motion, which has to be discarded from further analysis. Because image pixels in laser-scanning microscopy are acquired line-by-line, with relatively low repetition rates, movements typically lead to complex, nonrigid distortions in the image, which are not correctable by image-based rigid-body transformations commonly employed for motion correction of functional magnetic resonance imaging data. Instead, existing approaches attempt to correct for motion on a line-by-line basis. For example, Dombeck et al. [26] used a Hidden Markov Model to compute the displacement probability for each line and subsequently correct the image line-by-line using the most likely offset. While computationally expensive, the Hidden Markov Model ensures correction with the optimal displacement trajectory given all possible paths. An alternative approach based on Lucas–Kanade image registration has recently been proposed which may be computationally faster and does not require uncorrupted reference frames [76]. Once imaging speed becomes fast enough (i.e., >10 Hz), in-frame distortions are less severe and standard correction techniques such as cross-correlation across frames become possible [50]. Because correction algorithms are sensitive to changes in pixel brightness, it is important to estimate and correct movements on a static fluorescence channel, such as SR101-labeled astrocytes, and subsequently apply the results to the functional channel. For chronic imaging experiments, a static channel may be obtained by additional expression of a red fluorescent protein or the use of ratiometric calcium indicators, which show common motion-related fluorescence changes in both channels in addition to the opposite-direction functional changes in either channel. Finally, the described approaches for movement correction largely apply to whole-frame scanning approaches only. ROI-based techniques, such as line or random-access scanning ([77]; see also Chapter 9, this volume), preclude the application of offline algorithms and will necessitate the development of prospective motion correction using, for example, a high-speed camera movie of the brain surface.

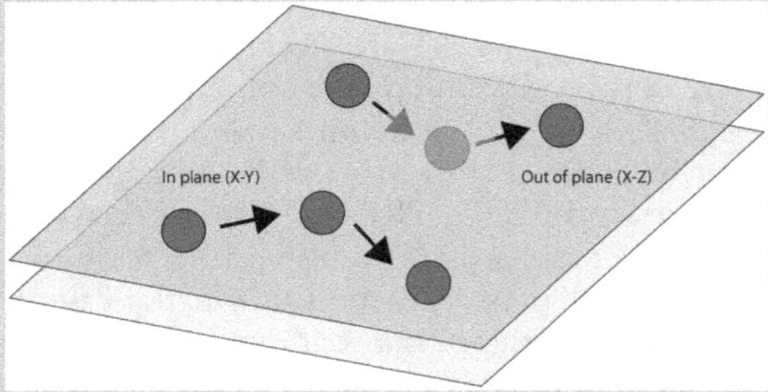

Fig. B2 Correcting for motion artifacts in awake rodents. Regimes of brain motion.

3 Using Chronic Two-Photon Imaging to Study Cortical Neural Networks

The studies to date that have used chronic imaging of GECIs have focused primarily on methodological advances [13, 15, 17, 50]. Thus, in terms of neuroscientific questions, the field has just begun to apply these methods to interesting experimental paradigms

[14, 16, 31, 33, 34, 36, 50, 78]. Because GECIs offer the promise of following cellular activity over time, one clear goal is to determine how neuronal activity changes during learning, plasticity, or disease. However, a more basic issue relates to the stability of cellular response properties over time. It has so far not been possible using imaging or electrophysiological techniques to record activity from the same unequivocally identified cells over days and weeks. It has therefore been unclear to what extent the functionally heterogeneous cells of the cortex maintain their characteristic responses over time [48]. Recent data has provided direct evidence for stability and plasticity of function, as recently reviewed [79] and discussed above in Section 2.1. This work has gone beyond the initial reports of functional stability in anesthetized [15] and awake [13] mouse visual cortex. The sparse expression methods in these studies restricted data collection to just a few cells, limiting the conclusions that could be drawn about populations. It will be interesting to extend these types of experiments to other brain areas. Dombeck et al. [50] found that the place fields of hippocampal CA1 cells during virtual navigation underwent dramatic remapping across 7 days. A similar phenomenon has not yet been reported in the neocortex, and the consistency of neuronal responses across days may critically depend on brain area and the behavioral and cognitive demands of the task.

These examples represent some of the first applications of GECIs for chronic imaging of cortical function with cellular resolution. Many ongoing and future studies will use chronic imaging of GECIs to address fundamental questions about stability and plasticity of neural network function.

4 Discussion and Outlook

We have presented methods and data for the chronic study of neuronal activity in the anesthetized and head-fixed behaving rodent. Some of the advantages and disadvantages associated with these methods are discussed below, as well as outlooks toward future applications.

4.1 Advantages of GECIs for Chronic Studies of Neuronal Function

Perhaps the greatest advantage of GECIs is that their expression can be restricted to certain types of cells. Specificity is achieved by genetic means: cell-specific promoters, inherent tropism of viruses, or by using Cre-dependent viruses in mouse lines that express Cre-recombinase in specific cell types. The power of these methods has been skillfully shown in the last 5 years by Deisseroth and colleagues to selectively express the light-sensitive cation channel, Channelrhodopsin-2 (ChR2), in cholinergic and dopaminergic cells of the brainstem/midbrain and in subpopulations of cortical excitatory or inhibitory cells, thereby allowing selective control of the action potential activity of these cells (see Chapters 5 and 11, this volume). Cell-specific promoters have been used to some extent (e.g., synapsin for neuronal-specific expression) but are generally more unpredictable and less flexible, in part because AAV vectors have limited capacity for the often large promoter sequences. These methods have not yet been used to their full potential for measuring neural activity with GECIs. The future should see many advances in measuring cortical activity enabled by the selective expression of GECIs, bringing together the complementary stimulation and recording components of the optogenetic toolbox.

Selective expression of GECIs is a potentially powerful technique because it allows cell-type discrimination beyond what a microelectrode can provide. This is also true for small structures, like neuronal dendrites, axons, and synaptic terminals. GECIs label these subcellular compartments in addition to the perinuclear region of the soma that is used for population imaging. Although data on dendritic and axonal calcium signaling using GECIs are currently limited, this is an active area of research. A new strategy of compartmentalized GECI expression was introduced for targeting of GCaMP2 to synaptic boutons by fusion of the protein to synaptophysin [30]. This approach could open avenues for further subcellular specificity and may even enhance the signaling properties of the indicator.

One interesting application of axonal labeling is the dual function as an anatomical tracer. For example, labeling input axons or synapses from distant connected sites allow measurement of activity from the projection site without direct imaging of its origin [34]. The strategy is similar to [80] but using expression of GECIs instead of ChR2. Photostability of the indicator may be important for axonal or synaptic measurements and GCaMP3 may have advantages over other indicators in this respect.

It is also now possible to bypass injection of viral vectors altogether and instead use transgenic mouse lines that express GECIs. Attempts using inverse pericam and camgaroo-2 [81] and the Troponin C-based sensor CerTN-L15 [82] achieved expression but labeled neurons still suffered from relatively low-sensitivity to action potential activity. Three new GCaMP3 expressing transgenic mouse lines were recently introduced, one Cre-inducible and two Thy1 lines [38, 39], which show improved signaling capabilities compared to earlier transgenic lines.

Mice are not the subject of choice for every experiment. Rats may be desirable in some types of complex learning tasks. Other experimental questions are only addressable with rats, cats, ferrets, or monkeys. For such lines of investigation, transgenic techniques are not practical and it will be important to transduce neurons by other means. Fortunately, lentivirus and AAV viral vectors have proven viable for expressing ChR2 and the GECIs mem-TNXXL in macaque monkeys [65, 83]. Furthermore, YC3.60 can be expressed and enable long-term imaging in behaving rats (Haiss et al., unpublished). Thus, viral vectors will likely continue to be an indispensable tool for applications across species.

What is the scientific benefit of long-term measurement of neuronal activity? The major advantage relates to the ability to follow activity in the same individual cells over time. Fundamentally, the extent to which cell function remains constant over time is not well understood. The functional properties of individual cells have been followed through learning- or experience-dependent plasticity in just a few recent studies [14, 16]. Previous studies of long-term learning, plasticity, and disease used a cross-sectional group approach to compare effects of a manipulation with a control/un-manipulated group. This classical approach provides valuable information about large effects but cannot answer more detailed and fundamental questions regarding how individual cells change over time. Chronic recording will further reveal fundamental information about the time course and direction of functional changes on the single-cell and population levels [37, 84].

4.2 Disadvantages of GECIs

Optical reporters seldom approach the temporal resolution and signal-to-noise afforded by electrophysiological measurements. Even though imaging increases the number and density of cells sampled within neuronal populations, the confidence that activity occurred is reduced. In addition, GECIs only report suprathreshold spiking activity and are largely blind to subthreshold membrane potential dynamics. Thus, specific questions regarding sparsely firing neurons or subthreshold activity may inevitably need to be addressed with intracellular recordings of membrane voltage or extracellular recording of spiking activity. In any case, electrophysiology will remain an indispensable tool for verifying and calibrating the signals of GECIs [85].

Genetically encoded voltage indicators represent a potential alternative to GECIs (see also Chapter 4, this volume). Unlike GECIs, they report mainly subthreshold membrane potential [18]. This could be a blessing and a curse: distinguishing which cells are spiking rather than undergoing a large postsynaptic potential may prove challenging. Importantly, new generation probes are able to report stimulation of single whiskers in vivo [18]. If expression is stable over time, and if expressing cells can be imaged with two-photon microscopy, these indicators are potentially suitable for chronic imaging, but this has so far not been tested.

As long as GECIs do not perturb the basic physiology of neurons, the advantages should outweigh potential disadvantages.

4.3 Outlook

In summary, this chapter describes chronic imaging techniques and reviews studies that benefit from the new possibilities enabled by long-term two-photon GECI imaging.

Recent technological improvements of two-photon microscopy aim to improve two critical issues: increasing the speed of functional signal acquisition and penetrating deeper into cortical tissue. Random-access microscopy using acousto-optic deflectors (see also Chapter 9 in this volume) and optical multiplexing [86] that enables simultaneous acquisition of multiple imaging planes are just two examples toward high-speed functional imaging. Longer wavelength lasers [87], regenerative amplifiers [88], or three-photon excitation [89] are examples of new methods to enhance depth penetration using multiphoton microscopy.

There is no doubt that these technological advances in multiphoton imaging will shed new light on the principles of neuronal computation within cortical microcircuits. Together with the improvement of GECIs and the chronic imaging strategies described in this chapter, we will possibly soon be able to image the functional properties of all neurons in a cortical layer or even the entire microcircuit on the scale of a cortical column.

The "perfect experiment," involving high-resolution imaging to dissect the functional microcircuitry of the cortex throughout the time course of plasticity, learning, or disease progression, is perhaps not too far away.

References

1. Hofer SB, Mrsic-Flogel TD, Bonhoeffer T, Hübener M (2006) Prior experience enhances plasticity in adult visual cortex. Nat Neurosci 9:127–132
2. Kaneko M, Hanover JL, England PM, Stryker MP (2008) TrkB kinase is required for recovery, but not loss, of cortical responses following monocular deprivation. Nat Neurosci 11: 497–504
3. Masino SA, Frostig RD (1996) Quantitative long-term imaging of the functional representation of a whisker in rat barrel cortex. Proc Natl Acad Sci USA 93:4942–4947
4. Polley DB, Chen-Bee CH, Frostig RD (1999) Two directions of plasticity in the sensory-deprived adult cortex. Neuron 24:623–637
5. Polley DB, Kvasnak E, Frostig RD (2004) Naturalistic experience transforms sensory maps in the adult cortex of caged animals. Nature 429:67–71
6. Slovin H, Arieli A, Hildesheim R, Grinvald A (2002) Long-term voltage-sensitive dye imaging reveals cortical dynamics in behaving monkeys. J Neurophysiol 88:3421–3438
7. Bozza T, McGann JP, Mombaerts P, Wachowiak M (2004) In vivo imaging of neuronal activity by targeted expression of a genetically encoded probe in the mouse. Neuron 42:9–21
8. Minderer M, Liu W, Sumanovski LT, Kügler S, Helmchen F, Margolis DJ (2012) Chronic imaging of cortical sensory map dynamics using a genetically encoded calcium indicator. J Physiol 590:99–107
9. Dickey AS, Suminski A, Amit Y, Hatsopoulos NG (2009) Single-unit stability using chronically implanted multielectrode arrays. J Neurophysiol 102:1331–1339
10. Tolias AS, Ecker AS, Siapas AG, Hoenselaar A, Keliris GA, Logothetis NK (2007) Recording chronically from the same neurons in awake, behaving primates. J Neurophysiol 98: 3780–3790
11. Buzsaki G (2004) Large-scale recording of neuronal ensembles. Nat Neurosci 7:446–451
12. Holtmaat A, Svoboda K (2009) Experience-dependent structural synaptic plasticity in the mammalian brain. Nat Rev Neurosci 10:647–658
13. Andermann ML, Kerlin AM, Reid RC (2010) Chronic cellular imaging of mouse visual cortex during operant behavior and passive viewing. Front Cell Neurosci 4:3
14. Huber D, Gutnisky DA, Peron S, O'Connor DH, Wiegert JS, Tian L, Oertner TG, Looger LL, Svoboda K (2012) Multiple dynamic representations in the motor cortex during sensorimotor learning. Nature 484:473–478
15. Mank M, Santos AF, Direnberger S, Mrsic-Flogel TD, Hofer SB, Stein V, Hendel T, Reiff DF, Levelt C, Borst A, Bonhoeffer T, Hübener M, Griesbeck O (2008) A genetically encoded calcium indicator for chronic in vivo two-photon imaging. Nat Methods 5:805–811
16. Margolis DJ, Lütcke H, Schulz K, Haiss F, Weber B, Kügler S, Hasan MT, Helmchen F (2012) Reorganization of cortical population activity imaged throughout long-term sensory deprivation. Nat Neurosci 15:1539–1546
17. Tian L, Hires SA, Mao T, Huber D, Chiappe ME, Chalasani SH, Petreanu L, Akerboom J,

McKinney SA, Schreiter ER, Bargmann CI, Jayaraman V, Svoboda K, Looger LL (2009) Imaging neural activity in worms, flies and mice with improved GCaMP calcium indicators. Nat Methods 6:875–881

18. Akemann W, Mutoh H, Perron A, Rossier J, Knopfel T (2010) Imaging brain electric signals with genetically targeted voltage-sensitive fluorescent proteins. Nat Methods 7:643–649
19. Zhong H, Sia GM, Sato TR, Gray NW, Mao T, Khuchua Z, Huganir RL, Svoboda K (2009) Subcellular dynamics of type II PKA in neurons. Neuron 62:363–374
20. Garaschuk O, Milos RI, Grienberger C, Marandi N, Adelsberger H, Konnerth A (2006) Optical monitoring of brain function in vivo: from neurons to networks. Pflugers Arch 453:385–396
21. Göbel W, Helmchen F (2007) In vivo calcium imaging of neural network function. Physiology (Bethesda) 22:358–365
22. Grewe BF, Helmchen F (2009) Optical probing of neuronal ensemble activity. Curr Opin Neurobiol 19:520–529
23. Chen TW, Wardill TJ, Sun Y, Pulver SR, Renninger SL, Baohan A, Schreiter ER, Kerr RA, Orger MB, Jayaraman V, Looger LL, Svoboda K, Kim DS (2013) Ultrasensitive fluorescent proteins for imaging neuronal activity, Nature 499, 295–300
24. Hendel T, Mank M, Schnell B, Griesbeck O, Borst A, Reiff DF (2008) Fluorescence changes of genetic calcium indicators and OGB-1 correlated with neural activity and calcium in vivo and in vitro. J Neurosci 28:7399–7411
25. Pologruto TA, Yasuda R, Svoboda K (2004) Monitoring neural activity and [Ca^{2+}] with genetically encoded Ca^{2+} indicators. J Neurosci 24:9572–9579
26. Dombeck DA, Khabbaz AN, Collman F, Adelman TL, Tank DW (2007) Imaging large-scale neural activity with cellular resolution in awake, mobile mice. Neuron 56:43–57
27. Mrsic-Flogel TD, Hofer SB, Ohki K, Reid RC, Bonhoeffer T, Hübener M (2007) Homeostatic regulation of eye-specific responses in visual cortex during ocular dominance plasticity. Neuron 54:961–972
28. Shevtsova Z, Malik JM, Michel U, Bahr M, Kügler S (2005) Promoters and serotypes: targeting of adeno-associated virus vectors for gene transfer in the rat central nervous system in vitro and in vivo. Exp Physiol 90:53–59
29. Nathanson JL, Yanagawa Y, Obata K, Callaway EM (2009) Preferential labeling of inhibitory and excitatory cortical neurons by endogenous tropism of adeno-associated virus and lentivirus vectors. Neuroscience 161:441–450
30. Dreosti E, Odermatt B, Dorostkar MM, Lagnado L (2009) A genetically encoded reporter of synaptic activity in vivo. Nat Methods 6:883–889
31. Harvey CD, Coen P, Tank DW (2012) Choice-specific sequences in parietal cortex during a virtual-navigation decision task. Nature 484:62–68
32. Lütcke H, Murayama M, Hahn T, Margolis DJ, Astori S, Zum Alten Borgloh SM, Gobel W, Yang Y, Tang W, Kügler S, Sprengel R, Nagai T, Miyawaki A, Larkum ME, Helmchen F, Hasan MT (2010) Optical recording of neuronal activity with a genetically-encoded calcium indicator in anesthetized and freely moving mice. Front Neural Circuits 4:9
33. O'Connor DH, Peron SP, Huber D, Svoboda K (2010) Neural activity in barrel cortex underlying vibrissa-based object localization in mice. Neuron 67:1048–1061
34. Petreanu L, Gutnisky DA, Huber D, Xu NL, O'Connor DH, Tian L, Looger L, Svoboda K (2012) Activity in motor-sensory projections reveals distributed coding in somatosensation. Nature 489:299–303
35. Wallace DJ, Meyer zum Alten Borgloh S, Astori S, Yang Y, Bausen M, Kügler S, Palmer AE, Tsien RY, Sprengel R, Kerr JN, Denk W, Hasan MT (2008) Single-spike detection in vitro and in vivo with a genetic Ca^{2+} sensor. Nat Methods 5:797–804
36. Xu NL, Harnett MT, Williams SR, Huber D, O'Connor DH, Svoboda K, Magee JC (2012) Nonlinear dendritic integration of sensory and motor input during an active sensing task. Nature 492(7428):247–251
37. Hübener M, Bonhoeffer T (2010) Searching for engrams. Neuron 67:363–371
38. Zariwala HA, Borghuis BG, Hoogland TM, Madisen L, Tian L, De Zeeuw CI, Zeng H, Looger LL, Svoboda K, Chen TW (2012) A Cre-dependent GCaMP3 reporter mouse for neuronal imaging in vivo. J Neurosci 32: 3131–3141
39. Chen Q, Cichon J, Wang W, Qiu L, Lee SJ, Campbell NR, Destefino N, Goard MJ, Fu Z, Yasuda R, Looger LL, Arenkiel BR, Gan WB, Feng G (2012) Imaging neural activity using Thy1-GCaMP transgenic mice. Neuron 76: 297–308
40. Wolfe J, Houweling AR, Brecht M (2010) Sparse and powerful cortical spikes. Curr Opin Neurobiol 20:306–312
41. Miyawaki A, Llopis J, Heim R, McCaffery JM, Adams JA, Ikura M, Tsien RY (1997)

Fluorescent indicators for Ca^{2+} based on green fluorescent proteins and calmodulin. Nature 388:882–887
42. Yamada Y, Michikawa T, Hashimoto M, Horikawa K, Nagai T, Miyawaki A, Hausser M, Mikoshiba K (2011) Quantitative comparison of genetically encoded Ca indicators in cortical pyramidal cells and cerebellar Purkinje cells. Front Cell Neurosci 5:18
43. Ohkura, M., Sasaki, T., Sadakari, J., Gengyo-Ando, K., Kagawa-Nagamura, Y., Kobayashi, C., Ikegaya, Y., and Nakai, J. (2012) Genetically encoded green fluorescent Ca^{2+} indicators with improved detectability for neuronal Ca^{2+} signals, PLoS One 7, e51286
44. Nagai T, Yamada S, Tominaga T, Ichikawa M, Miyawaki A (2004) Expanded dynamic range of fluorescent indicators for Ca^{2+} by circularly permuted yellow fluorescent proteins. Proc Natl Acad Sci USA 101:10554–10559
45. Akerboom J, Chen TW, Wardill TJ, Tian L, Marvin JS, Mutlu S, Calderon NC, Esposti F, Borghuis BG, Sun XR, Gordus A, Orger MB, Portugues R, Engert F, Macklin JJ, Filosa A, Aggarwal A, Kerr RA, Takagi R, Kracun S, Shigetomi E, Khakh BS, Baier H, Lagnado L, Wang SS, Bargmann CI, Kimmel BE, Jayaraman V, Svoboda K, Kim DS, Schreiter ER, Looger LL (2012) Optimization of a GCaMP calcium indicator for neural activity imaging. J Neurosci 32:13819–13840
46. Horikawa K, Yamada Y, Matsuda T, Kobayashi K, Hashimoto M, Matsu-ura T, Miyawaki A, Michikawa T, Mikoshiba K, Nagai T (2010) Spontaneous network activity visualized by ultrasensitive Ca^{2+} indicators, yellow Cameleon-Nano. Nat Methods 7:729–732
47. Mank M, Griesbeck O (2008) Genetically encoded calcium indicators. Chem Rev 108:1550–1564
48. Barth AL, Poulet JF (2012) Experimental evidence for sparse firing in the neocortex. Trends Neurosci 35:345–355
49. Holtmaat A, Bonhoeffer T, Chow DK, Chuckowree J, De Paola V, Hofer SB, Hubener M, Keck T, Knott G, Lee WC, Mostany R, Mrsic-Flogel TD, Nedivi E, Portera-Cailliau C, Svoboda K, Trachtenberg JT, Wilbrecht L (2009) Long-term, high-resolution imaging in the mouse neocortex through a chronic cranial window. Nat Protoc 4:1128–1144
50. Dombeck DA, Harvey CD, Tian L, Looger LL, Tank DW (2010) Functional imaging of hippocampal place cells at cellular resolution during virtual navigation. Nat Neurosci 13:1433–1440
51. Stettler DD, Yamahachi H, Li W, Denk W, Gilbert CD (2006) Axons and synaptic boutons are highly dynamic in adult visual cortex. Neuron 49:877–887
52. Yamahachi H, Marik SA, McManus JN, Denk W, Gilbert CD (2009) Rapid axonal sprouting and pruning accompany functional reorganization in primary visual cortex. Neuron 64:719–729
53. Drew PJ, Shih AY, Driscoll JD, Knutsen PM, Blinder P, Davalos D, Akassoglou K, Tsai PS, Kleinfeld D (2010) Chronic optical access through a polished and reinforced thinned skull. Nat Methods 7:981–984
54. Schwarz C, Hentschke H, Butovas S, Haiss F, Stuttgen MC, Gerdjikov TV, Bergner CG, Waiblinger C (2010) The head-fixed behaving rat—procedures and pitfalls. Somatosens Mot Res 27:131–148
55. Yang G, Pan F, Parkhurst CN, Grutzendler J, Gan WB (2010) Thinned-skull cranial window technique for long-term imaging of the cortex in live mice. Nat Protoc 5:201–208
56. Xu HT, Pan F, Yang G, Gan WB (2007) Choice of cranial window type for in vivo imaging affects dendritic spine turnover in the cortex. Nat Neurosci 10:549–551
57. Lowenstein PR, Mandel RJ, Xiong WD, Kroeger K, Castro MG (2007) Immune responses to adenovirus and adeno-associated vectors used for gene therapy of brain diseases: the role of immunological synapses in understanding the cell biology of neuroimmune interactions. Curr Gene Ther 7:347–360
58. Reimsnider S, Manfredsson FP, Muzyczka N, Mandel RJ (2007) Time course of transgene expression after intrastriatal pseudotyped rAAV2/1, rAAV2/2, rAAV2/5, and rAAV2/8 transduction in the rat. Mol Ther 15:1504–1511
59. Helmchen F, Denk W (2005) Deep tissue two-photon microscopy. Nat Methods 2:932–940
60. Kerr JN, Denk W (2008) Imaging in vivo: watching the brain in action. Nat Rev Neurosci 9:195–205
61. Engelbrecht CJ, Johnston RS, Seibel EJ, Helmchen F (2008) Ultra-compact fiber-optic two-photon microscope for functional fluorescence imaging in vivo. Opt Express 16: 5556–5564
62. Piyawattanametha W, Cocker ED, Burns LD, Barretto RP, Jung JC, Ra H, Solgaard O, Schnitzer MJ (2009) In vivo brain imaging using a portable 2.9 g two-photon microscope based on a microelectromechanical systems scanning mirror. Opt Lett 34:2309–2311
63. Sawinski J, Wallace DJ, Greenberg DS, Grossmann S, Denk W, Kerr JN (2009) Visually evoked activity in cortical cells imaged in freely moving animals. Proc Natl Acad Sci USA 106:19557–19562

64. Euler T, Hausselt SE, Margolis DJ, Breuninger T, Castell X, Detwiler PB, Denk W (2009) Eyecup scope—optical recordings of light stimulus-evoked fluorescence signals in the retina. Pflugers Arch 457:1393–1414
65. Heider B, Nathanson JL, Isacoff EY, Callaway EM, Siegel RM (2010) Two-photon imaging of calcium in virally transfected striate cortical neurons of behaving monkey. PLoS One 5:e13829
66. Mayrhofer JM, Skreb V, von der Behrens W, Musall S, Weber B, Haiss F (2013) Novel two-alternative forced choice paradigm for bilateral vibrotactile whisker frequency discrimination in head-fixed mice and rats. J Neurophysiol 109:273–284
67. Dombeck DA, Graziano MS, Tank DW (2009) Functional clustering of neurons in motor cortex determined by cellular resolution imaging in awake behaving mice. J Neurosci 29: 13751–13760
68. Keller GB, Bonhoeffer T, Hübener M (2012) Sensorimotor mismatch signals in primary visual cortex of the behaving mouse. Neuron 74:809–815
69. Niell CM, Stryker MP (2010) Modulation of visual responses by behavioral state in mouse visual cortex. Neuron 65:472–479
70. Royer S, Zemelman BV, Losonczy A, Kim J, Chance F, Magee JC, Buzsaki G (2012) Control of timing, rate and bursts of hippocampal place cells by dendritic and somatic inhibition. Nat Neurosci 15:769–775
71. Nimmerjahn A, Mukamel EA, Schnitzer MJ (2009) Motor behavior activates Bergmann glial networks. Neuron 62:400–412
72. Harvey CD, Collman F, Dombeck DA, Tank DW (2009) Intracellular dynamics of hippocampal place cells during virtual navigation. Nature 461:941–946
73. Haiss F, Schwarz C (2005) Spatial segregation of different modes of movement control in the whisker representation of rat primary motor cortex. J Neurosci 25:1579–1587
74. Hentschke H, Haiss F, Schwarz C (2006) Central signals rapidly switch tactile processing in rat barrel cortex during whisker movements. Cereb Cortex 16:1142–1156
75. Gentet LJ, Avermann M, Matyas F, Staiger JF, Petersen CC (2010) Membrane potential dynamics of GABAergic neurons in the barrel cortex of behaving mice. Neuron 65:422–435
76. Greenberg DS, Kerr JN (2009) Automated correction of fast motion artifacts for two-photon imaging of awake animals. J Neurosci Methods 176:1–15
77. Grewe BF, Langer D, Kasper H, Kampa BM, Helmchen F (2010) High-speed in vivo calcium imaging reveals neuronal network activity with near-millisecond precision. Nat Methods 7:399–405
78. O'Connor DH, Huber D, Svoboda K (2009) Reverse engineering the mouse brain. Nature 461:923–929
79. Lütcke H, Margolis DJ, Helmchen F (2013) Steady or changing? Long-term monitoring of neuronal population activity. Trends Neurosci 36(7):375–384
80. Petreanu L, Mao T, Sternson SM, Svoboda K (2009) The subcellular organization of neocortical excitatory connections. Nature 457:1142–1145
81. Hasan MT, Friedrich RW, Euler T, Larkum ME, Giese G, Both M, Duebel J, Waters J, Bujard H, Griesbeck O, Tsien RY, Nagai T, Miyawaki A, Denk W (2004) Functional fluorescent Ca^{2+} indicator proteins in transgenic mice under TET control. PLoS Biol 2:e163
82. Heim N, Garaschuk O, Friedrich MW, Mank M, Milos RI, Kovalchuk Y, Konnerth A, Griesbeck O (2007) Improved calcium imaging in transgenic mice expressing a troponin C-based biosensor. Nat Methods 4:127–129
83. Han X, Qian X, Bernstein JG, Zhou HH, Franzesi GT, Stern P, Bronson RT, Graybiel AM, Desimone R, Boyden ES (2009) Millisecond-timescale optical control of neural dynamics in the nonhuman primate brain. Neuron 62:191–198
84. Wilt BA, Burns LD, Wei Ho ET, Ghosh KK, Mukamel EA, Schnitzer MJ (2009) Advances in light microscopy for neuroscience. Annu Rev Neurosci 32:435–506
85. Scanziani M, Häusser M (2009) Electrophysiology in the age of light. Nature 461:930–939
86. Cheng A, Goncalves JT, Golshani P, Arisaka K, Portera-Cailliau C (2011) Simultaneous two-photon calcium imaging at different depths with spatiotemporal multiplexing. Nat Methods 8:139–142
87. Kawakami R, Sawada K, Sato A, Hibi T, Kozawa Y, Sato S, Yokoyama H, Nemoto T (2013) Visualizing hippocampal neurons with in vivo two-photon microscopy using a 1030 nm picosecond pulse laser. Sci Rep 3:1014
88. Mittmann W, Wallace DJ, Czubayko U, Herb JT, Schaefer AT, Looger LL, Denk W, Kerr JN (2011) Two-photon calcium imaging of evoked activity from L5 somatosensory neurons in vivo. Nat Neurosci 14:1089–1093
89. Horton NG, Wang K, Kobat D, Clark CG, Wise FW, Schaffer CB, Xu C (2013) In vivo three-photon microscopy of subcortical structures within an intact mouse brain. Nat Photonics 7:205–209

Chapter 11

Neocortical Circuit Interrogation with Optogenetics

Lief E. Fenno and Karl Deisseroth

Abstract

Optogenetic tools have gained popularity for enabling manipulation of specified populations of neurons on a precise temporal scale. These opsins, as the proteins are known, utilize retinal (Vitamin A) as a co-factor—in a form related to the same molecule used by the human retina as a light-sensing moiety; opsin function does not require (at least in vertebrates) exogenous supplementation of this molecule [1]. Taken together, these properties of optogenetic tools situate them ideally for the interrogation of neural circuits in the settings of acute slice preparations, *in vivo* single and multiunit recordings, and behaving animals. Robust and diverse tools now exist to allow for both gain and loss of function experiments in all of these settings; for example, recent engineering and genomic approaches have introduced diversity in the spectral properties of both excitatory and inhibitory opsins, opening the door to combinatorial optogenetic experiments and simultaneous manipulation of multiple populations of neurons in the same physical volume [2] (see Chapter 5 in this volume). Here we discuss circuit-targeting strategies with examples from the recent literature, focusing particularly on approaches and results in the study of mammalian neocortex.

Key words Optogenetics, Neural circuit, Channelrhodopsin, Gamma, Parvalbumin, Excitation/inhibition

1 Viral Targeting of Neuronal Populations

Using optogenetics to understand the contribution of defined neuronal populations to circuit function requires the specific expression of a given optogenetic tool exclusively within those neurons (Fig. 1). As with any genetic system, there are a number of modalities available to introduce optogenetic tools into neuronal populations; virally mediated transduction is perhaps the most widespread, cost-effective, and efficient. Current mainstays of viral delivery include lentivirus (LV) and adeno-associated virus (AAV), which can be produced quickly as well as in titers that allow for a high infectivity and copy number, and thus potent efficacy *in vivo* [3]. An additional advantage of LV is that virus may be produced using standard tissue culture techniques and an ultracentrifuge; a single batch of LV may be produced within 48 h [3].

Bruno Weber and Fritjof Helmchen (eds.), *Optical Imaging of Neocortical Dynamics*, Neuromethods, vol. 85,
DOI 10.1007/978-1-62703-785-3_11,

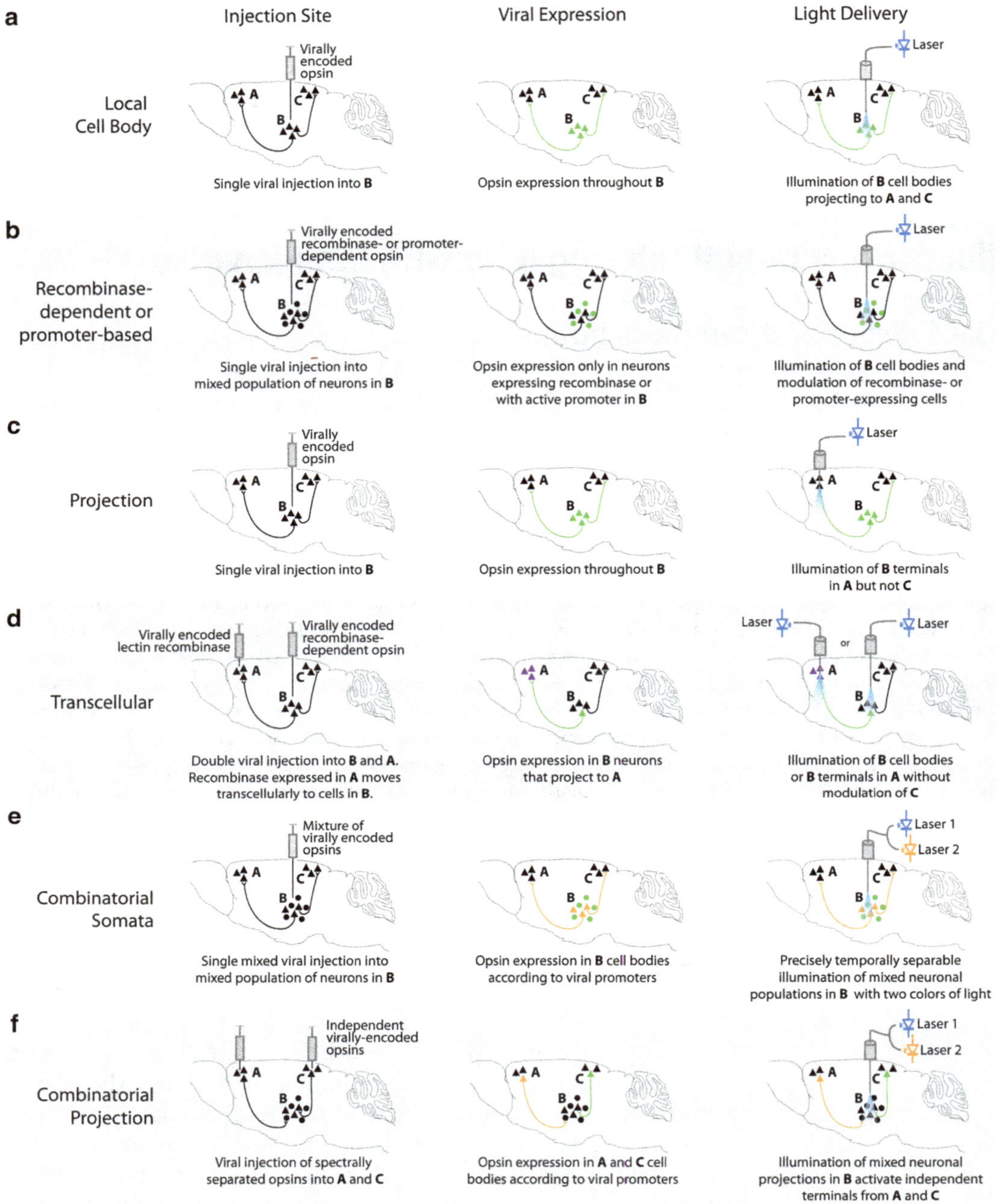

Fig. 1 *In vivo* optogenetic targeting strategies. (**a**) Viral targeting of a neuron population based on promoter specificity followed by stimulation of the same area as the injection. (**b**) Viral targeting of a neuron population based on expression of Cre recombinase or the restricted activity of a specific promoter followed by stimulation of the same area as the viral injection. (**c**) Viral targeting of a neuron population followed by stimulation of a downstream target region; noted that this will likely be a subset of the total population of neurons expressing opsin. (**d**) Dual viral targeting of a downstream population with lectin-recombinase (e.g., WGA-Cre) and upstream injection of cre-dependent DIO virus followed by stimulation of either the somata or axons of opsin-expressing population. (**e**) Injection of multiple opsins expressed in separate populations within the same space based on promoter or recombinase-mediated specificity followed by multi-wavelength stimulation at the somata. (**f**) Injection of multiple opsins in two separate locations followed by central multi-wavelength stimulation location to differentially modulate somata of neurons based on their projections (Adapted from [4]).

AAV may be made either in the laboratory or obtained from viral core facilities (for instance, the University of North Carolina viral core). Both LV and AAV have been used with success in multiple model organisms, including mouse, rat, and primate; additionally, high levels of expression are sustained over long periods with no reported toxicity [3, 5]. While neither AAV nor LV is expressed well in Zebrafish, both Sindbis and rabies viruses are reported to have efficacy in this setting [6].

Gene expression from the virus may be enabled in specific neuronal subtypes using promoters unique to the targeted population in the virus-injected region. Such promoters must be specific as well as sufficiently strong to drive high expression and small enough to fit within the viral backbone while still allowing efficient capsid packaging [4]. The promoter strategy has been successful in hypothalamus [7], subthalamic nucleus [8], brainstem [9, 10], hippocampus [11, 12], amygdala [13, 14], and neocortex [2, 15–23], among other targets. In neocortex, pure promoter-based strategies have most notably enabled the optogenetic targeting of local CaMKIIα (typically excitatory) neurons with the CaMKIIα promoter fragment [2, 16, 18–22].

Expression may be further refined by spatial restriction of the virus to a specific region of the brain by control of the injection volume and location, or by using a virus with functional expression conditional upon the presence of another molecule (e.g., in conjunction with a recombinase driver line; discussed below) [2, 3, 23] (Fig. 1). We recommend use of a sham virus encoding only the fluorophore (e.g., parallel cohorts of ChR2-YFP viral injections and YFP viral injections). This process controls for nonspecific effects of surgery, viral injection, viral expression, and the introduction of the activation light. It may also be important to conduct opsin-expressed but no-light controls to address any nonspecific effects of opsin expression itself.

2 Transgenic Animals

Although the virus/promoter combination strategy is an economical and rapid method of delivery for optogenetic constructs, this strategy suffers from the fact that there exists a relatively small subset of neurons with characterized strong and specific minimal promoters that fit the constraints of viral packaging. Fortunately, identifying such a promoter is no longer a prerequisite to targeting specific neuronal subtypes. To take advantage of the transgenic rodent recombinase-driver lines, viral constructs bearing optogenetic tools have been developed that employ the safe and potent elongation factor 1-α (EF1α) promoter fragment but are also Cre recombinase-dependent for expression [2, 20, 24].

This doublefloxed inverted open-reading-frame (DIO) expression system (Fig. 2) is initially inert with the opsin gene situated in

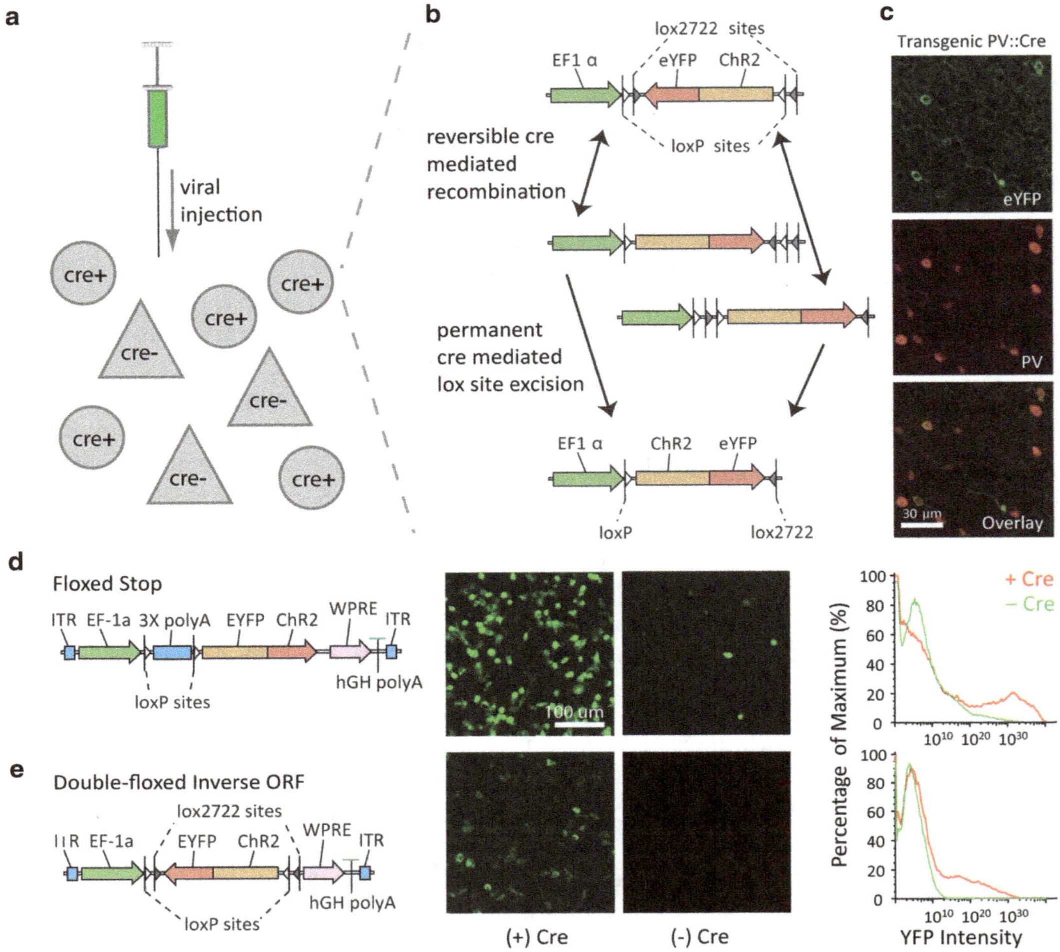

Fig. 2 Double inverted open-reading-frame strategy. (**a**) DIO-containing virus is injected into a transgenic mouse expressing Cre within a subset of neurons or in combination with exogenous Cre delivery (e.g., in concert with WGA-Cre injection). (**b**) Cre mediates the inversion of the reading frame and a subset of the lox recombinase recognition sites. Further Cre activity either returns the cassette to its original inverted position (reversible) or excises a pair of lox sites (permanent), thereby locking the reading frame in the sense position. (**c**) A sample from a parvalbumin PV::Cre transgenic mouse expressing DIO-ChR2-eYFP (*green*) counterstained against PV (*red*). (**d, e**) Alternative "Floxed Stop" (**d**) strategy compared to DIO (**e**) strategy. Floxed Stop gene expression is not as tightly regulated as DIO both qualitatively (*middle* fluorescence image) and quantitatively (note *green* trace intensities) (Adapted from [25]).

the antisense direction. Virus is injected into an animal expressing Cre recombinase in a subset of neurons (Fig. 2a). The opsin reading frame is flanked by an alternating inverted pair of incompatible Cre recombinase recognition sequences (e.g., loxP and lox2722 sites; Fig. 2b), which allows the gene direction to be reversed in the presence of Cre; thus functional expression is only able to occur in cells expressing the recombinase (Fig. 2c). Lox sites may be situated in either the reverse complement orientation relative to each other, which leads to the intervening sequence to be excised and then ligated in the antisense direction (as in the DIO [20]) or in the same orientation, which leads Cre recombinase to excise the intervening sequence and one of the recognition sites. Another common iteration of the Cre/lox system positions a number of STOP codons between two lox sites and upstream of a gene; these STOP codons are then removed in cells expressing Cre and expression commences. Comparing this system (Fig. 2d) and the DIO system (Fig. 2e) indicates that the DIO system has fewer cells expressing viable opsin in the absence of Cre (Fig. 2d, e middle, right panels).

Unlike the use of the alternative lox-STOP-lox strategy (where the STOP motif decreases expression of a downstream gene), this system excludes expression of the gene by its associated promoter, as the sense sequence will not be available as a transcription template in the absence of Cre activity. As functional expression in a neuron transduced by a DIO virus then becomes determined by the presence of Cre and not by a specific promoter, strong (but not too strong) promoters such as EF1α can be used. This versatile strategy allows the immediate and widespread use of current and new optogenetic tools [2] in the growing number of mouse (and rat) lines expressing Cre recombinase [26–30]. Of special note for neocortex in this regard are PV::Cre drivers [20, 26], a set of recent lines published by the laboratory of Josh Huang at Cold Spring Harbor Laboratory potentially useful for investigating specific subsets of cortical inhibitory neurons [26], and sets of neocortical layer-specific Cre driver mice [30–33]. Projections to neocortex of course can also be targeted with the Cre driver approach, including through use of TH::Cre and ChAT::Cre mice [24, 34, 35] and rats [36].

An alternative to the combined use of virus and transgenic recombinase lines is the direct production of transgenic or knock-in animals harboring specific optogenetic tools. The first opsin-expressing transgenic mouse lines were produced using the Thy1 promoter; line 18 was noted to have strong and widespread expression through layer 5 of neocortex and some expression in subcortical structures [37, 38], and has been used in diverse experimental paradigms including to examine the mechanism of action of deep brain stimulation in Parkinson's disease [8] and information processing in cortex [20]. Advantages of transgenic systems expressing a specific opsin include the ability to utilize these animals at an earlier age than those requiring injection (our lab typically performs viral injection at 6 weeks of age or older and waits an average of 3 weeks before performing experiments, depending on the axonal trafficking distance required). Transgenic animals also obviate the necessity to produce, handle, or inject virus, and allow for expression patterns that may not be readily reproduced using viral injection.

The main drawbacks to using transgenic systems center on loss of spatial specificity (with associated confounds linked to long-range projections) and the lack of flexibility: a separate line must be produced for each tool, whereas different viruses may be used with the same Cre transgenic line. Even so, a number of pioneering and highly useful transgenic lines have been produced expressing different opsins [38–43].

3 Layer-Specific Targeting

The ability to interrogate a circuit by perturbing a genetically defined element may not be sufficient to describe information processing in instances where the same genetically defined class of neuronal unit is used in multiple distinct computational steps within the local microcircuit. In the context of neocortical processing, this is indeed the case (for instance, pyramidal neurons of different layers play different roles in the circuit; [44]). To address this caveat, multiple studies [44, 45] have described the introduction of optogenetic tools using *in utero* electroporation (IUE). IUE takes advantage of distinct temporal windows during which the pyramidal neurons of different cortical layers are born. With knowledge of when the neurons of a layer are produced, one may introduce an expression construct with a given optogenetic tool to target only that layer. This has been used with success in rodents [32, 44–48] to analyze the contribution of layer-specific pyramidal populations to neocortical information processing. Similar to transgenic mouse lines, constructs electroporated in utero are expressed at birth, allowing for acute slice preparation of young animals. One drawback of this strategy is lower expression level of the opsin, possibly due to reduced expression construct copy number.

A recent use of IUE-introduced ChR2 by Adesnik and Scanziani [44] was integral to a description of the role of excitatory/inhibitory (E/I) interactions in layer-specific communication between and across cortical columns. Their approach to examining cortical circuitry involved introduction of ChR2 into layer 2/3 as well as use of a stimulating light ramp that increased in intensity over time. This group initially found that driving excitation in layer 2/3 pyramidal neurons in this way with ChR2 was sufficient to induce gamma-band oscillations in non-expressing cells from both inhibitory and excitatory classes, assessed by patch clamp. Using simultaneous four-cell patch clamp, this group found that transmission of ChR2-driven excitation and inhibition across layers (vertically) in cortex was clearest from layer 2/3 to 5, and less so to layers 4 or 6.

Additionally, checks and balances involving inhibition and excitation were found to be operative when spatially distinct subsets of cells expressing ChR2 were stimulated. Strikingly, when single cells in layer 2/3 were stimulated with current injection, photostimulation of nearby layer 2/3 cortical columns decreased induced spiking activity of the patched cell. In contrast, cells in layer 5 induced to spike with current injection saw a significant increase in firing rate during layer 2/3 photostimulation, which was reduced by lateral 2/3 excitation. In combination, it was found that layer 2/3 cells compete with layer 2/3 neighbors to drive cortical output via layer 5. These experiments utilized optogenetics to ask questions that could not be answered with other available neuroscience tools lacking this spatial and temporal specificity.

A distinct recent direct examination of the E/I ratio in cortex developed and employed novel ChR variants, including a "stabilized step function opsin" (SSFO) modified to remain in the activated state for >30 min after the introduction of blue light (but still inactivateable using yellow light) [2]. Like other step function opsins (SFOs), the photocurrent generated by SSFO is subthreshold and thus does not directly generate action potentials by driving a neuron above threshold. This paper [2] combined the novel opsin variants and viral delivery to manipulate E/I cortical balance through expression of SSFO in either pyramidal neurons (via the specific promoter CamKIIα to increase excitation E) or in parvalbumin expressing inhibitory interneurons (using a DIO-SSFO virus in combination with PV::Cre transgenic mouse to increase inhibition I). It was found that baseline gamma-band activity was increased in response to E/I elevation, even in freely moving mice (Fig. 3); moreover, both social behaviors and episodic learning (but not anxiety, locomotion, or novel object exploration) were nearly abolished in response to E/I elevation but not reduction; all of these features match signs and symptoms seen in autism and schizophrenia [49–52].

Further investigation in slice to examine potential mechanisms underlying this deficit revealed that information processing by pyramidal neurons became saturated at a much lower input frequency during SSFO activation of pyramidal neurons, whereas activation of PV neurons decreased the gain of the transfer curve, but without changing the slope; indeed, E/I balance elevation but not reduction was found to actually reduce information throughput in neocortical pyramidal neurons, assessed in bits per second. Lastly, combinatorial optogenetics (discussed below) was employed to examine the effect of simultaneously increasing excitation and inhibition (with a resultant E/I more similar to baseline than during manipulation of either E or I in isolation); partial restoration of social function was observed.

These two studies utilized optogenetic tools to investigate basic questions in neocortical microcircuitry. The combination of an expanding optogenetic toolbox with decades of rigorous experience in developing experimental paradigms and framing basic questions in neuroscience (across many laboratories) is just beginning to pay dividends in unlocking mysteries that have long been out of reach.

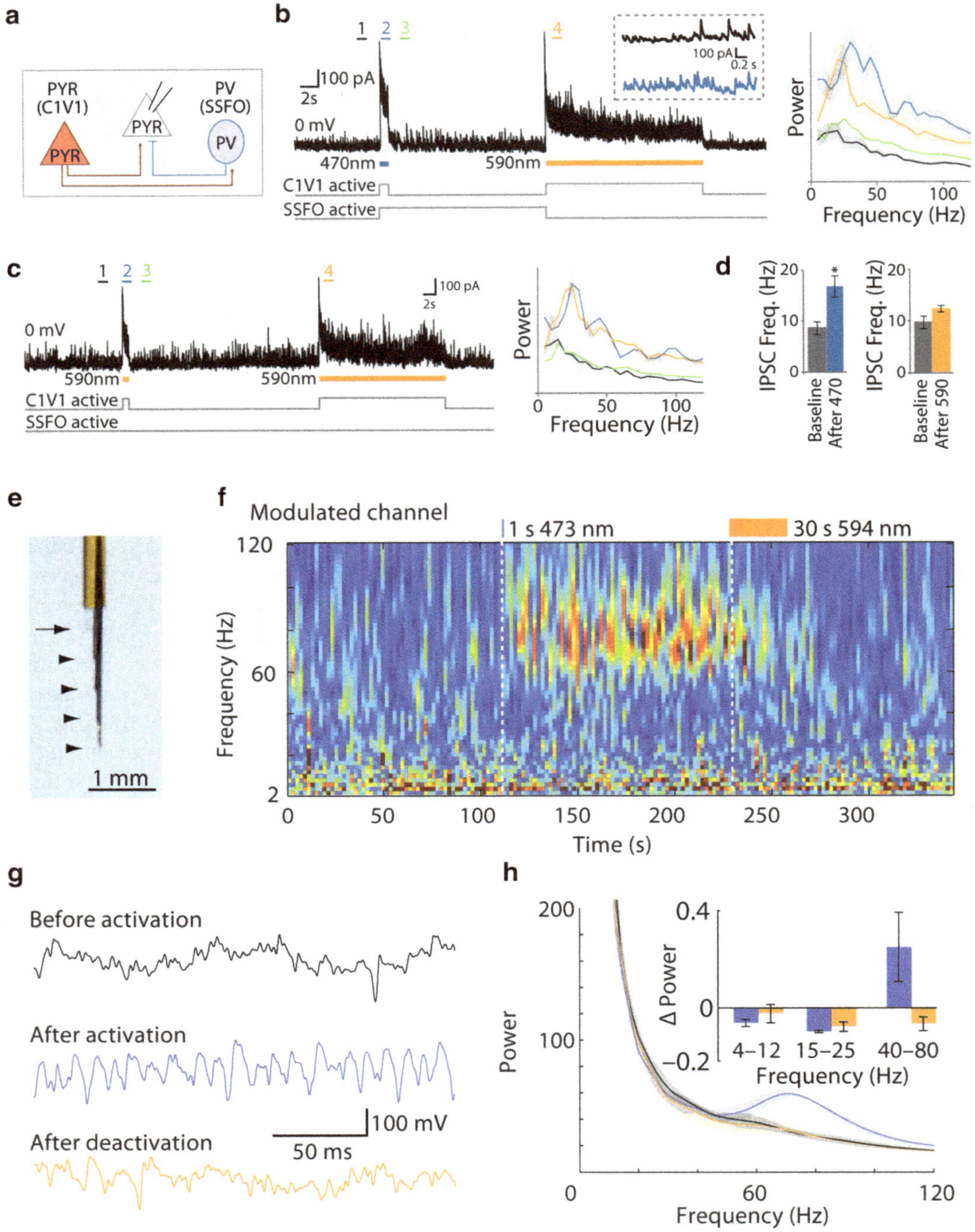

Fig. 3 Modulation of gamma-frequency activity using stabilized step function opsin (SSFO) in slice and freely behaving mice. (**a**) Experimental setup for combinatorial optogenetic slice recordings. *Blue* light activates both SSFO and C1V1 while *yellow* light activates only C1V1. (**b**) IPSC power within the gamma range is more highly increased during costimulation of PV and pyramidal neurons (2) than during stimulation of pyramidal neurons alone (4). Note SSFO remains active after *blue* light pulse cessation and inactivates during *yellow* light pulse (see *bottom* traces for predicted activity of both opsins). (**c**) No prior history dependence is noted in the shift

4 Topological Targeting

In addition to genetic identity and layer location, circuit elements may be defined based on synaptic partners. For instance, an intermingled population of neocortical pyramidal neurons may project to different downstream partners. Strategies to manipulate a single projection by restricting the intersection of opsin expression and activation light are described in the accompanying chapter (Chapter 5 in this volume). Now, additional strategies for targeting neurons based on specific connectivity have been achieved using variants of the Cre/lox-targeting system. These strategies rely upon the infection of downstream structures with one of a few viruses that work differently to achieve a similar result.

The first strategy involves the infection of downstream neurons with a virus encoding Cre recombinase fused to a trans-cellular tracer peptide (such as the wheat germ agglutinin (WGA) sequence [5]). The fusion WGA-Cre is trafficked trans-cellularly and moves retrogradely up the axons of impinging afferents. A second (Cre-dependent/DIO) virus encoding the opsin of interest is injected upstream at the origin of the putative afferents to be controlled. Either or both viruses can be designed in conjunction with a specific promoter to increase the intersectional specificity of the experimental paradigm. Although all neurons of a given genetic identity will express the DIO virus, only those making synaptic connections with the downstream WGA-Cre expressing neurons will express functional opsin; this strategy has been shown to be effective in cortico-cortical connections between primary sensory and motor cortex [5].

A variation on this theme replaces virus that encodes WGA-Cre with a virus known to transduce axon terminals, including herpes simplex virus (HSV) family members, pseudo-typed lentiviruses, or specific subsets of the AAV or CAV family. In this case, the virus encodes Cre in the absence of any trafficking peptide, but otherwise the coincidence detection mechanism is the same: a Cre-dependent DIO virus is injected in the upstream location and the intersection of Cre and the DIO determine the subpopulation that expresses a functional optogenetic tool.

Fig. 3 (continued) of gamma peak frequency during PV/pyramidal neuron costimulation seen in (**b**) (compare 2 and 4). (**d**) IPSC frequency in (**b**) is significantly increased over baseline after 470 nm light pulse has ended, indicating sustained activity of SSFO. (**e**) Image of chronically implanted quad electrode (*arrowheads*) array coupled to optical fiber (*arrow*). (**f**) Recording during free behavior of mouse injected with CK-SSFO in medial prefrontal cortex. Optical stimulation parameters are noted on *top* of power-time heat map. Note sustained gamma activity during activation of SSFO. Gamma is decreased to background levels after inactivation of SSFO. (**g**, **h**) Example traces from baseline, activation, and deactivation (**g**) periods are shown with summary data (**h**) showing increase in gamma frequency activity during SSFO activation in pyramidal neurons (Adapted from [2]).

These approaches are sufficient for experiments performed *in vivo* or in slice. Two other approaches ideally suited for acute neocortical slice preparations are also able to delineate the synaptic partners of a neuron using creative strategies. A promising technique [53, 54] utilizes single cell electroporation to transduce a single neocortical neuron with two plasmids that, together, encode a complete rabies virus. The viral package, however, is deficient in the glycoprotein necessary for infection. The consequence of this deficiency is that competent virus is only produced in the single electroporated neuron. As rabies virus is able to traverse the synapse, the partners of this single neuron will be infected, but the virus will not be able to travel further. Encoding a separate fluorophore on each of the electroporated constructs allows for the identification of the transduced neuron and its synaptic partners [54].

Finally, the DIO strategy has been combined [55] with a transgenic reporter line to identify neocortical–striatal projections to dopamine receptor-2 expressing cells by crossing a D2-GFP reporter line with a mouse expressing Emx1-cre and injecting DIO-ChR2-mCherry into neocortex. In this case, optogenetics was combined with two-photon glutamate uncaging and pharmacology to precisely describe the role of D2 receptors in glutamatergic synaptic transmission in the particular circuit [55].

5 Combinatorial Optogenetics

As the number of optogenetic tools has grown, delineations and distinctions have become more precisely defined [56]. Integral to operation in the context of neuroscience are the identity of the ions that are transmitted and the wavelength of light necessary for activity. For instance, depolarizing channelrhodopsins have been described that are activated by blue (ChR2) and red (C1V1) light. In addition, the hyperpolarizing halorhodopsin chloride pump is activated by orange light (NpHR) while the proton efflux pump Arch is activated by slightly more blueshifted light. It was noted in the original description of the neuroscience application of NpHR [57] that it may be used in combination with ChR2 by exploiting the spectral separation of each to achieve bimodal control of a single neuron utilizing separate light inputs. These two opsins were subsequently combined into single constructs using a variety of strategies [5, 58].

Although a red-shifted opsin (VChR1) was described soon after the birth of optogenetics [59], it was not until recently that combinatorial optogenetic excitation of multiple neuronal populations *in vivo* became possible in mammals. The chimeric opsin C1V1, which, like VChR1 is activated by red light, has the added property of large photocurrents and a shorter inactivation time constant, allowing for higher frequency stimulation [2]. The larger photocurrent of C1V1 allows stimulation of neurons using even further red-shifted non-peak frequencies of light, which, in the context of combinatorial control decreases crosstalk with ChR2. In the course of describing C1V1 and examining the excitatory/inhibitory balance of prefrontal cortical neurons, Yizhar et al. [2] used C1V1 in combination with both ChR2 and the SSFO in anesthetized and awake, behaving animals to simultaneously modulate excitatory pyramidal neurons and PV-expressing inhibitory interneurons as described above. In this way combinatorial control over somata in neocortex *in vivo* was achieved [2].

Moreover, combinatorial control over projections (including neocortical projections) also could be achieved [2]. In the same paper [2], the integration of corticothalamic and thalamocortical inputs to reticular neurons of the thalamus was examined by the separate activation of these afferents impinging upon a single RT neuron by separately expressing ChR2 and C1V1 in these two populations. Due to potential cross activation of C1V1 at the blue end of the spectra (using 405 nm light), these experiments were controlled to eliminate potentially confounding results, and cell bodies were removed from the slice to ensure that combinatorial control over axons was the mechanism of action [2]. Although a nascent technique, combinatorial optogenetics promises to be one of the most powerful experimental paradigms available to investigate the interaction of circuit elements to produce complex behavioral results.

6 Conclusion

As a genetically encoded tool functioning on a precise temporal scale, optogenetics enables interrogation of the causal contribution of individual components of neural circuitry, including systems as complex as the mammalian neocortex. By combining multiple viruses, transgenic animal lines, and Cre-dependent expression constructs, circuit elements may be defined based on genetic as well as topological and synaptic properties. The recent advent of the potent, red-shifted channelrhodopsin C1V1 has also allowed for the manipulation of multiple populations of neurons and axons within the same physical volume. Taken together, the diversity of the diversity available optogenetic tools and their delivery mechanisms combine to form a potent toolbox for neural circuit interrogation.

References

1. Zhang F, Wang LP, Boyden ES, Deisseroth K (2006) Channelrhodopsin-2 and optical control of excitable cells. Nat Methods 3:785–792
2. Yizhar O, Fenno LE, Prigge M, Schneider F, Davidson TJ, O'Shea DJ, Sohal VS, Goshen I, Finkelstein J, Paz JT, Stehfest K, Fudim R, Ramakrishnan C, Huguenard JR, Hegemann P, Deisseroth K (2011) Neocortical excitation/inhibition balance in information processing and social dysfunction. Nature 477:171–178
3. Zhang F, Gradinaru V, Adamantidis AR, Durand R, Airan RD, de Lecea L, Deisseroth K (2010) Optogenetic interrogation of neural circuits: technology for probing mammalian brain structures. Nat Protoc 5:439–456

4. Yizhar O, Fenno LE, Davidson TJ, Mogri M, Deisseroth K (2011) Optogenetics in neural systems. Neuron 71:9–34
5. Gradinaru V, Zhang F, Ramakrishnan C, Mattis J, Prakash R, Diester I, Goshen I, Thompson KR, Deisseroth K (2010) Molecular and cellular approaches for diversifying and extending optogenetics. Cell 141:154–165
6. Zhu P, Narita Y, Bundschuh ST, Fajardo O, Scharer YP, Chattopadhyaya B, Bouldoires EA, Stepien AE, Deisseroth K, Arber S, Sprengel R, Rijli FM, Friedrich RW (2009) Optogenetic dissection of neuronal circuits in Zebrafish using viral gene transfer and the Tet system. Front Neural Circuits 3:21
7. Adamantidis AR, Zhang F, Aravanis AM, Deisseroth K, de Lecea L (2007) Neural substrates of awakening probed with optogenetic control of hypocretin neurons. Nature 450:420–424
8. Gradinaru V, Mogri M, Thompson KR, Henderson JM, Deisseroth K (2009) Optical deconstruction of parkinsonian neural circuitry. Science 324:354–359
9. Tan W, Janczewski WA, Yang P, Shao XM, Callaway EM, Feldman JL (2008) Silencing preBotzinger complex somatostatin-expressing neurons induces persistent apnea in awake rat. Nat Neurosci 11:538–540
10. Abbott SB, Stornetta RL, Fortuna MG, Depuy SD, West GH, Harris TE, Guyenet PG (2009) Photostimulation of retrotrapezoid nucleus phox2b-expressing neurons in vivo produces long-lasting activation of breathing in rats. J Neurosci 29:5806–5819
11. Goshen I, Brodsky M, Prakash R, Wallace J, Gradinaru V, Ramakrishnan C, Deisseroth K (2011) Dynamics of retrieval strategies for remote memories. Cell 147:678–689
12. Dittgen T, Nimmerjahn A, Komai S, Licznerski P, Waters J, Margrie TW, Helmchen F, Denk W, Brecht M, Osten P (2004) Lentivirus-based genetic manipulations of cortical neurons and their optical and electrophysiological monitoring in vivo. Proc Natl Acad Sci U S A 101:18206–18211
13. Tye K, Prakash R, Kim S-Y, Fenno LE, Grosenick L, Zarabi H, Thompson KR, Gradinaru V, Ramakrishnan C, Deisseroth K (2011) Amygdala circuitry mediating reversible and bidirectional control of anxiety. Nature 471:358–362
14. Stuber GD, Sparta DR, Stamatakis AM, van Leeuwen WA, Hardjoprajitno JE, Cho S, Tye KM, Kempadoo KA, Zhang F, Deisseroth K, Bonci A (2011) Excitatory transmission from the amygdala to nucleus accumbens facilitates reward seeking. Nature 475:377–380
15. Covington HE III, Lobo MK, Maze I, Vialou V, Hyman JM, Zaman S, Laplant Q, Mouzon E, Ghose S, Tamminga CA, Neve RL, Deisseroth K, Nestler EJ (2010) Antidepressant effect of optogenetic stimulation of the medial prefrontal cortex. J Neurosci 30: 16082–16090
16. Paz JT, Bryant AS, Peng K, Fenno L, Yizhar O, Frankel WN, Deisseroth K, Huguenard JR (2011) A new mode of corticothalamic transmission revealed in the Gria4(−/−) model of absence epilepsy. Nat Neurosci 14:1167–1173
17. Cruikshank SJ, Urabe H, Nurmikko AV, Connors BW (2010) Pathway-specific feedforward circuits between thalamus and neocortex revealed by selective optical stimulation of axons. Neuron 65:230–245
18. Lee JH, Durand R, Gradinaru V, Zhang F, Goshen I, Kim DS, Fenno LE, Ramakrishnan C, Deisseroth K (2010) Global and local fMRI signals driven by neurons defined optogenetically by type and wiring. Nature 465:788–792
19. Aravanis AM, Wang LP, Zhang F, Meltzer LA, Mogri MZ, Schneider MB, Deisseroth K (2007) An optical neural interface: in vivo control of rodent motor cortex with integrated fiberoptic and optogenetic technology. J Neural Eng 4:S143–S156
20. Sohal VS, Zhang F, Yizhar O, Deisseroth K (2009) Parvalbumin neurons and gamma rhythms enhance cortical circuit performance. Nature 459:698–702
21. Han X, Qian X, Bernstein JG, Zhou HH, Franzesi GT, Stern P, Bronson RT, Graybiel AM, Desimone R, Boyden ES (2009) Millisecond-timescale optical control of neural dynamics in the nonhuman primate brain. Neuron 62:191–198
22. Tonnesen J, Sorensen AT, Deisseroth K, Lundberg C, Kokaia M (2009) Optogenetic control of epileptiform activity. Proc Natl Acad Sci U S A 106:12162–12167
23. Diester I, Kaufman MT, Mogri M, Pashaie R, Goo W, Yizhar O, Ramakrishnan C, Deisseroth K, Shenoy KV (2011) An optogenetic toolbox designed for primates. Nat Neurosci 14:387–397
24. Tsai HC, Zhang F, Adamantidis A, Stuber GD, Bonci A, de Lecea L, Deisseroth K (2009) Phasic firing in dopaminergic neurons is sufficient for behavioral conditioning. Science 324:1080–1084
25. Fenno L, Yizhar O, Deisseroth K (2011) The development and application of optogenetics. Annu Rev Neurosci 34:389–412
26. Taniguchi H, He M, Wu P, Kim S, Paik R, Sugino K, Kvitsani D, Fu Y, Lu J, Lin Y, Miyoshi G, Shima Y, Fishell G, Nelson SB,

Huang ZJ (2011) A resource of Cre driver lines for genetic targeting of GABAergic neurons in cerebral cortex. Neuron 71:995–1013
27. Geschwind D (2004) GENSAT: a genomic resource for neuroscience research. Lancet Neurol 3:82
28. Heintz N (2004) Gene expression nervous system atlas (GENSAT). Nat Neurosci 7:483
29. Gong S, Zheng C, Doughty ML, Losos K, Didkovsky N, Schambra UB, Nowak NJ, Joyner A, Leblanc G, Hatten ME, Heintz N (2003) A gene expression atlas of the central nervous system based on bacterial artificial chromosomes. Nature 425:917–925
30. Gong S, Doughty M, Harbaugh CR, Cummins A, Hatten ME, Heintz N, Gerfen CR (2007) Targeting Cre recombinase to specific neuron populations with bacterial artificial chromosome constructs. J Neurosci 27:9817–9823
31. Madisen L, Zwingman TA, Sunkin SM, Oh SW, Zariwala HA, Gu H, Ng LL, Palmiter RD, Hawrylycz MJ, Jones AR, Lein ES, Zeng H (2010) A robust and high-throughput Cre reporting and characterization system for the whole mouse brain. Nat Neurosci 13:133–140
32. Petreanu L, Mao T, Sternson SM, Svoboda K (2009) The subcellular organization of neocortical excitatory connections. Nature 457:1142–1145
33. O'Connor DH, Huber D, Svoboda K (2009) Reverse engineering the mouse brain. Nature 461:923–929
34. Witten IB, Lin SC, Brodsky M, Prakash R, Diester I, Anikeeva P, Gradinaru V, Ramakrishnan C, Deisseroth K (2010) Cholinergic interneurons control local circuit activity and cocaine conditioning. Science 330:1677–1681
35. Lindeberg J, Usoskin D, Bengtsson H, Gustafsson A, Kylberg A, Soderstrom S, Ebendal T (2004) Transgenic expression of Cre recombinase from the tyrosine hydroxylase locus. Genesis 40:67–73
36. Witten IB, Steinberg EE, Lee SY, Davidson TJ, Zalocusky KA, Brodsky M, Yizhar O, Cho SL, Gong S, Ramakrishnan C, Stuber GD, Tye KM, Janak PH, Deisseroth K (2011) Recombinase-driver rat lines for optogenetics. Neuron 72:721–733
37. Arenkiel BR, Peca J, Davison IG, Feliciano C, Deisseroth K, Augustine GJ, Ehlers MD, Feng G (2007) In vivo light-induced activation of neural circuitry in transgenic mice expressing channelrhodopsin-2. Neuron 54:205–218
38. Zhao S, Cunha C, Zhang F, Liu Q, Gloss B, Deisseroth K, Augustine GJ, Feng G (2008) Improved expression of halorhodopsin for light-induced silencing of neuronal activity. Brain Cell Biol 36:141–154
39. Hagglund M, Borgius L, Dougherty KJ, Kiehn O (2010) Activation of groups of excitatory neurons in the mammalian spinal cord or hindbrain evokes locomotion. Nat Neurosci 13:246–252
40. Katzel D, Zemelman BV, Buetfering C, Wolfel M, Miesenbock G (2010) The columnar and laminar organization of inhibitory connections to neocortical excitatory cells. Nat Neurosci 14:100–107
41. Thyagarajan S, van Wyk M, Lehmann K, Lowel S, Feng G, Wassle H (2010) Visual function in mice with photoreceptor degeneration and transgenic expression of channelrhodopsin 2 in ganglion cells. J Neurosci 30:8745–8758
42. Wang H, Peca J, Matsuzaki M, Matsuzaki K, Noguchi J, Qiu L, Wang D, Zhang F, Boyden E, Deisseroth K, Kasai H, Hall WC, Feng G, Augustine GJ (2007) High-speed mapping of synaptic connectivity using photostimulation in Channelrhodopsin-2 transgenic mice. Proc Natl Acad Sci U S A 104:8143–8148
43. Zhao S, Ting JT, Atallah HE, Qiu L, Tan J, Gloss B, Augustine GJ, Deisseroth K, Luo M, Graybiel AM, Feng G (2011) Cell type-specific channelrhodopsin-2 transgenic mice for optogenetic dissection of neural circuitry function. Nat Methods 8:745–752
44. Adesnik H, Scanziani M (2010) Lateral competition for cortical space by layer-specific horizontal circuits. Nature 464:1155–1160
45. Huber D, Petreanu L, Ghitani N, Ranade S, Hromadka T, Mainen Z, Svoboda K (2008) Sparse optical microstimulation in barrel cortex drives learned behaviour in freely moving mice. Nature 451:61–64
46. Lewis TL Jr, Mao T, Svoboda K, Arnold DB (2009) Myosin-dependent targeting of transmembrane proteins to neuronal dendrites. Nat Neurosci 12:568–576
47. Petreanu L, Huber D, Sobczyk A, Svoboda K (2007) Channelrhodopsin-2-assisted circuit mapping of long-range callosal projections. Nat Neurosci 10:663–668
48. Gradinaru V, Thompson KR, Zhang F, Mogri M, Kay K, Schneider MB, Deisseroth K (2007) Targeting and readout strategies for fast optical neural control in vitro and in vivo. J Neurosci 27:14231–14238
49. Hashimoto T, Volk DW, Eggan SM, Mirnics K, Pierri JN, Sun Z, Sampson AR, Lewis DA (2003) Gene expression deficits in a subclass of GABA neurons in the prefrontal cortex of subjects with schizophrenia. J Neurosci 23:6315–6326

50. Kehrer C, Maziashvili N, Dugladze T, Gloveli T (2008) Altered excitatory-inhibitory balance in the NMDA-hypofunction model of schizophrenia. Front Mol Neurosci 1:6
51. Orekhova EV, Stroganova TA, Nygren G, Tsetlin MM, Posikera IN, Gillberg C, Elam M (2007) Excess of high frequency electroencephalogram oscillations in boys with autism. Biol Psychiatry 62:1022–1029
52. Rubenstein JL, Merzenich MM (2003) Model of autism: increased ratio of excitation/inhibition in key neural systems. Genes Brain Behav 2:255–267
53. Wickersham IR, Finke S, Conzelmann KK, Callaway EM (2007) Retrograde neuronal tracing with a deletion-mutant rabies virus. Nat Methods 4:47–49
54. Wickersham IR, Lyon DC, Barnard RJ, Mori T, Finke S, Conzelmann KK, Young JA, Callaway EM (2007) Monosynaptic restriction of transsynaptic tracing from single, genetically targeted neurons. Neuron 53:639–647
55. Higley MJ, Sabatini BL (2010) Competitive regulation of synaptic Ca^{2+} influx by D2 dopamine and A2A adenosine receptors. Nat Neurosci 13:958–966
56. Mattis J, Tye KM, Ferenczi EA, Ramakrishnan C, O'Shea DJ, Prakash R, Gunaydin LA, Hyun M, Fenno LE, Gradinaru V, Yizhar O, Deisseroth K (2011) Principles for applying optogenetic tools derived from direct comparative analysis of microbial opsins. Nat Methods 9:159–172
57. Zhang F, Wang LP, Brauner M, Liewald JF, Kay K, Watzke N, Wood PG, Bamberg E, Nagel G, Gottschalk A, Deisseroth K (2007) Multimodal fast optical interrogation of neural circuitry. Nature 446:633–639
58. Kleinlogel S, Terpitz U, Legrum B, Gokbuget D, Boyden ES, Bamann C, Wood PG, Bamberg E (2011) A gene-fusion strategy for stoichiometric and co-localized expression of light-gated membrane proteins. Nat Methods 8(12):1083–1088
59. Zhang F, Prigge M, Beyriere F, Tsunoda SP, Mattis J, Yizhar O, Hegemann P, Deisseroth K (2008) Red-shifted optogenetic excitation: a tool for fast neural control derived from Volvox carteri. Nat Neurosci 11:631–633

Part III

Elucidating Metabolic States of Cortical Circuits

Chapter 12

Imaging of Astrocytic Activity in Living Rodents

Norio Takata, Yoshiaki Shinohara, Masamichi Ohkura, Tsuneko Mishima, Junichi Nakai, and Hajime Hirase

Abstract

Mounting evidence from in vitro experiments supports bidirectional interactions between astrocytes and neurons, suggesting glial involvement in neuronal information processing in the brain. Elevation of the cytosolic calcium ion (Ca^{2+}) concentration has been suggested to be important for gliotransmission; however, the study of Ca^{2+} dynamics in cerebral cortical astrocytes in vivo became possible only recently. Here, we describe a set of procedures to monitor Ca^{2+} concentration fluctuations in a population of astrocytes or in astrocytic processes using in vivo two-photon microscopy. Simultaneous recording of neuronal activity by electrodes and astrocytic activity by imaging is a promising way to reveal the nature of neuron–glia interactions in the brain of living rodents.

Key words Bolus loading, Adenovirus, G-CaMP, Astrocytes, Astrocytic processes, Cerebral cortex

1 Introduction

Since Virchow's seminal description of *Nervenkitt* (nerve-glue: neuroglia) in the mid-nineteenth century, glia gained citizenship in the central nervous system as major cellular components. In fact, glial cells constitute more than 50 % of the cell population in the human cerebral cortical gray matter (as reviewed in [1]) and in basal ganglia [2]. In rats, the neuron:glia ratio in the cerebral cortical gray matter is estimated to be 2:1, indicating the neuronal dominance in cell number [3]. Despite the variability of neuron:glia ratio in different species, it appears that the overall volume fraction of the glial cells remains more or less constant across species, as glia occupy approximately 50 % of the nervous system in volume [4]. Glia are subdivided into four cell types in the central nervous system: astrocytes, oligodendrocytes, NG2-positive glia, and microglia. In the peripheral nervous system, Schwann cells serve glial functions. In the rat cortical gray matter, astrocytes, oligodendrocytes, NG2-positive glia, and microglia represent approximately 21, 4, 2, and 9 %, leaving astrocytes to be the most numerous glial cell type [3].

Bruno Weber and Fritjof Helmchen (eds.), *Optical Imaging of Neocortical Dynamics*, Neuromethods, vol. 85,
DOI 10.1007/978-1-62703-785-3_12,

Although the physical presence of glial cells in large numbers has long been noted, they have classically been recognized as support cells in brain operation. For instance, oligodendrocytes provide myelination of axons thereby enabling saltatory conduction of action potentials [5, 6], while NG2 cells are oligodendrocyte precursor cells [7]. Microglia serve as immune cells and become phagocytic upon inflammation [8, 9]. Astrocytes are multifarious in their function. Astrocytes maintain extracellular milieu by buffering excess potassium and taking up glutamate upon neuronal activity such as action potential generation and synaptic transmission. This homeostatic function is critical in continued neuronal circuit operation, as accumulation of extracellular potassium or glutamate leads to undesirable membrane potential depolarization, which will eventually lead to cell death. Another important aspect of astrocytic function is metabolic support for neurons. Astrocytes receive glucose from the blood circulation, which in turn is metabolized to lactate. Lactate is then shuttled to neighboring neurons by monocarboxylate transporters to provide the source of energy for the neurons thereby establishing the "lactate shuttle" [10], a concept that has mainly been based on in vitro research with in vivo evidence emerging only recently (e.g., [11]).

Gray matter "protoplasmic" astrocytes possess perhaps the most complex cell morphology of all brain cells. The cell body of a protoplasmic astrocyte is usually small (~7 μm in diameter) in rodents and can be as large as 10–20 μm in humans. A few to several main processes branch out from the soma and these branches ramify into multitudinous astrocytic processes. Overall, the morphological appearance of protoplasmic astrocytes is somewhat similar to broccoli, although astrocytic main processes branch out radially. Electron microscopic investigations revealed that astrocytic processes enwrap excitatory synapses such as dendritic spines. Notably, in the cerebellar cortex, Purkinje cell synapses formed by the parallel fibers are completely covered by the Bergmann glia [12], which are specialized astrocytes whose cell bodies are located in the Purkinje cell layer. In the cerebral cortex and hippocampus, about a half of the synaptic perimeters is covered by astrocytic processes [13, 14]. Each protoplasmic astrocyte has its own territory of synapse coverage ("domain") [15]. A typical rodent astrocyte has a domain of 50 μm in diameter. It is estimated that a rodent astrocyte covers 20,000–120,000 synapses and a human astrocyte may cover up to two million synapses [16]. Such morphology of astrocytes hints to close relationship of astrocyte function and neuronal network dynamics, in particular, synaptic functions.

About a decade and a half ago, this intimate relationship of synapses and astrocytic processes was conceptually advanced by the discovery of gliotransmission—astrocytes were found to secrete chemical substances such as glutamate [17] and adenosine triphosphate (ATP) [18] that can directly influence synaptic transmission or signal reception. Hence the concept of the "tripartite synapse" was proposed [19], a view that a functional synaptic complex is composed of pre- and postsynaptic neuronal processes together with the surrounding astrocytic processes as a third partner. Remarkably, many of the gliotransmission phenomena observed with in vitro culture assays have been reported to be dependent on the astrocyte's intracellular Ca^{2+} concentration (reviewed in [20]). Interestingly, astrocytes express G protein-coupled neurotransmitter receptors such as the metabotropic glutamate receptors (mGluRs) or the P2Y purinergic receptors. When these receptors are activated, the astrocytic Ca^{2+} concentration is elevated via a release of Ca^{2+} from internal stores (e.g., endoplasmic reticulum). Therefore, astrocytic Ca^{2+} concentration reflects both the input and output aspects of astrocytic activity.

As many of the initial discoveries on astrocytic Ca^{2+} dynamics and substance release come from in vitro experiments, the extent to which such neuron–astrocyte interactions apply in the intact brain remains to be evaluated. Only in the recent decade, has it become feasible to monitor Ca^{2+} dynamics of cerebral cortical astrocytes in the living rodent brain by use of either organic or fluorescent protein-based Ca^{2+} indicators with the help of two-photon laser scanning microscopy. This chapter covers recent progress in the study of neuron–astrocyte interactions as assessed by optical imaging methods.

2 Methods for Acute Two-Photon Imaging

2.1 Surgical Preparations

Either mice or rats can be used for in vivo investigation of cerebral cortical astrocyte imaging. In general, more transgenic or knockout strains are available in mice, whereas cellular physiology is somewhat easier to perform in rats presumably due to larger cell size. We focus on the procedures for mature rats in this section with occasional precautions for mice experiments where appropriate. All experimental procedures described below were by the RIKEN Institutional Animal Care and Use Committee.

Adult male Sprague–Dawley rats (4–6 weeks old) are deeply anesthetized with 1.7 g/kg urethane (1.40–1.65 g/kg for mice). It is recommended that the dose is given in two or more intraperitoneal administrations at least 10 min apart. We use 30 % urethane solution in physiological saline (0.9 % NaCl) for rats. For mice, urethane solution should be further diluted (e.g., 7.5–15 %) to achieve more accurate injection volume measurements. Liquid paraffin is applied onto the animal's eyes to prevent from drying. Tracheotomy is performed for mice to avoid choking as urethane facilitates saliva secretion [21]. The body temperature of the rat is maintained at 37.5 °C with a regulated heating pad (TR-200, FST, CA, USA) throughout the surgery and following imaging experiment.

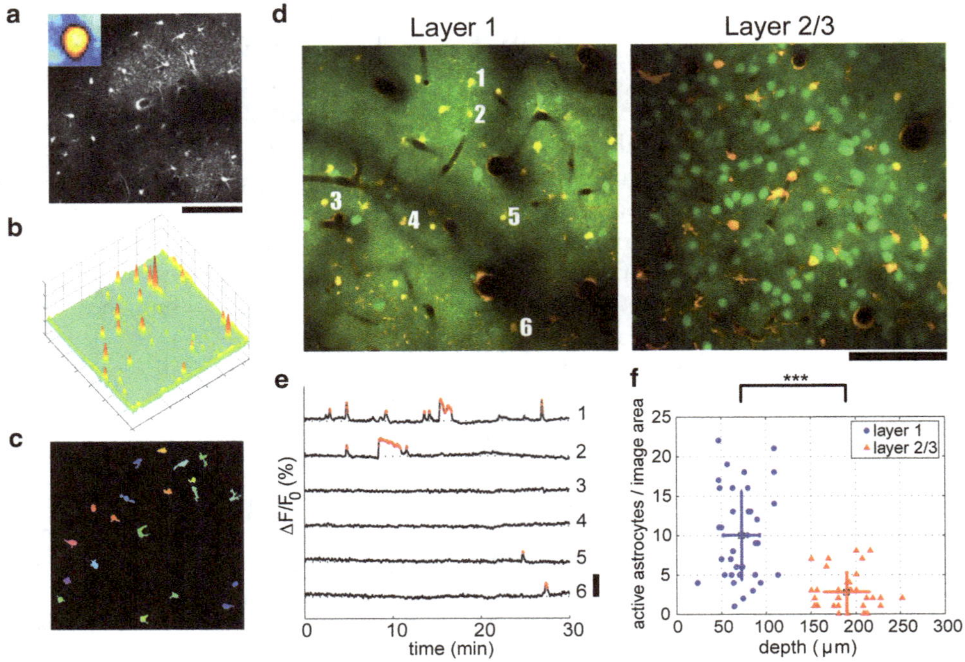

Fig. 1 Spontaneous astrocytic calcium signals in superficial layers of neocortex following bolus loading of an organic Ca^{2+} indicator. (**a**) An example two-photon image of OGB-1-labeled astrocytes for semiautomatic ROI selection. *Inset*: a feature vector expanded by four times to improve its visibility. The feature vector was obtained using library images for cells and non-cells. Scale bar 100 μm. (**b**) Convolved image of the fluorescence image and feature vector in (**a**). Each peak in the image is a candidate for soma location. (**c**) ROIs for somata were automatically selected based on the presumed somatic location and criteria for fluorescence intensity and soma size. (**d**) Representative in vivo two-photon images of L1 and L2/3. The Ca^{2+} indicator OGB-1 (*green*) labeled both neurons and astrocytes while SR101 (*red*) labeled astrocytes. Most of the labeled cells were astrocytes in L1, whereas neurons (i.e., SR101-negative) outnumber astrocytes (SR101-positive) in L2/3. Scale bar 100 μm. (**e**) Representative examples of spontaneous Ca^{2+} surges of astrocytes in L1. Numbers (*1–6*) correspond to the astrocytes marked in (**d**). Scale bar 50 %. *Red dots* represent Ca^{2+} events detected by an automatic computer program. (**f**) Number of astrocytes with at least one spontaneous Ca^{2+} surge ("active astrocyte") per imaged area is plotted against the depth from the cortical pia. L1 had significantly more active astrocytes than L2/3 (Modified from [22]).

This is particularly important as deeply anesthetized animals have attenuated autonomic nervous functions and the body temperature drops to the level of hypothermia if no external supply of heat is provided. Physiological saline containing dextrose (5 % w/v) is injected subcutaneously as necessary (up to 10 mL/kg/h) to maintain fluid balance.

2.2 Cranial Window Preparation and Bolus Loading

After the skull is exposed, a metal frame is attached to the skull using a dental acrylic (Fuji LUTE BC, GC, Tokyo, Japan). For bolus loading of organic Ca^{2+} indicator or other fluorescent indicators (see details below; Fig. 1), a craniotomy and/or a durotomy has to be performed. Transgenic animals with genetically encoded sensor protein can be imaged through a thinned-skull preparation (e.g., [23]). Thinned-skull preparations have the advantage of undamaged nervous tissue, which is ideal for glial imaging as glial cells are sensitive to tissue damage. Indeed, blood–brain barrier dysfunction has been reported following craniotomy [24]. On the other hand, thinned-skull preparations often have compromised depth penetration and/or spatial resolution due to inferior quality of optical access through the skull.

For imaging in the somatosensory cortex, a craniotomy (diameter 1.0–1.5 mm), centered at the stereotaxic coordinate of anterior–posterior approximately −2.5 mm and medial–lateral ~3.0 mm from the bregma, is prepared using a dental drill. We recommend to use a carbon-hardened metal drill-bit of diameter of 0.4 mm (No. 23013, Nakanishi, Japan). It is absolutely essential to perform the craniotomy and the following durotomy with greatest care as any damage to the nervous tissue is critical for the outcome of bolus loading or surface application of membrane-permeable dyes. It is also important to keep the dura and pia from drying, lest they should adhere. Use of a surgical microscope or stereo microscope is a must. It is important to keep the craniotomy diameter as small as possible, because a larger craniotomy will result in edema in the course of 30 min and longer. Bone chips can be removed by fine forceps (be careful!) or by mild suction using an aspirator while applying physiological saline onto the craniotomy.

Durotomy dramatically increases the visibility of the cranial window especially for rats and higher animals as the dura mater is translucent and highly scattering. The dura mater can be surgically removed. We use a specialized commercial tool for dura removal (10066–15, FST, CA, USA), but a tungsten sharp electrode for unit recording or a 26-gauge needle with the tips bent is fine, too. Every care should be taken to avoid any damage to the pial vessel and cortical surface. Indeed, if damage (even a small one) is visible, one should consider starting with a new animal. While such sensitivity to the quality of durotomy (or the whole surgical preparation for that matter) might not need to be imposed on neuronal imaging, astrocytes and other glial cell types are known to change their morphology and physiology in insulted nervous tissue ("reactive glia").

For multi-cell bolus loading (MCBL) a dye mixture of the Ca^{2+}-sensitive fluorescence indicator Oregon Green 488 BAPTA-1 (OGB-1) AM ester and astrocyte-specific fluorescence marker Sulforhodamine 101 (SR101) is typically used in our experiments [25, 26]. The dye mixture is composed of the following reagents: OGB-1 AM (50 μg, O-6807, Invitrogen), Pluronic F-127 (5 μL, P3000MP, Invitrogen), phosphate-buffered saline (42 μL, pH 7.4, PBS), and SR101 (3 μL of 1 mM solution, S-359, Molecular Probes). The dye mixture may be sonicated (Misonix XL-2000, Qsonica) briefly, and should be syringe-filtered with a 0.22 μm pore size Millex-GV (Millipore). We use a quartz glass micropipette (QF100-50-10, Sutter Instruments) for dye-loading. As quartz glass has a higher rigidity than borosilicate glass, the tip bows less in the brain tissue, allowing more accurate positioning of the pipette tip. As quartz glass has a higher melting temperature than borosilicate glass, a laser puller (e.g., P-2000, Sutter Instruments) is usually needed to pull a quartz glass pipette. However, the use of quartz glass is not an absolute requirement. The dye-loading pipette has a tip diameter of approximately 1–2 μm which corresponds to a typical whole-cell patch electrode and contains typically 3–5 μL of the dye mixture. The pipette is held in a patch-electrode holder to allow exertion of positive air pressure and is slowly inserted into the cortex (200 μm below the pia targeting cortical L2/3 for our experiments) using an electronic fine manipulator (EMM-3, Narishige, Tokyo, Japan). The dye mixture is introduced to the tissue by air pressure application to the pipette (70 kPa, 1 min) under two-photon observation to confirm dye diffusion among neuropil avoiding somata. A successful injection of the fluorescent dye to the brain tissue can be confirmed by two-photon imaging of the site with a 40× water immersion objective 10–15 min after the injection. The pipette is then slowly retracted and the craniotomy is covered with agarose (1.5 % w/v in PBS) and sealed by placing a thin glass coverslip on top (3 × 3 mm, thickness <0.12 mm, Matsunami Glass Ind., Osaka, Japan). The cranial window is secured by dental cement. As stated explicitly in [27], one of the key factors for successful loading is freshness of the reagents, especially the Pluronic. When AM-ester fluorescent dyes do not get loaded into cells, it is advisable to purchase a fresh Pluronic. The craniotomy is continuously superfused with warmed HEPES-Ringer solution (~35 °C) to maintain the surface temperature of the brain during experiments. This seems to be important as neural responses and spontaneous local field potential (LFP) patterns change with brain temperature.

Another small craniotomy (diameter 0.5 mm) is made approximately 2 mm posterior to the imaging window for LFP recording so that the electrode will not make a physical intrusion to the imaged area, while measuring the population activity in the vicinity of the imaged area. We advise to position the electrode to the location of interest well before the imaging experiment (~2 h), as it takes a considerable amount of time before LFP responses recover (Box 1).

2.3 Adenovirus Preparation and Inoculation

As an alternative to organic fluorescent Ca^{2+} indicators, genetically expressed fluorescent Ca^{2+}-sensitive proteins can be used. Since the invention of the Cameleon Ca^{2+} sensor protein [28], many Ca^{2+} sensor proteins of various affinities and excitation/emission wavelengths are available to date. Recent updates of Ca^{2+} sensors are covered elsewhere in this volume (see Chapter 1.4 in this volume). As far as astrocytes are concerned, the resting intracellular Ca^{2+} concentration $[Ca^{2+}]_i$ is around 100 nM and upon Ca^{2+} surges, $[Ca^{2+}]_i$ can be as high as several μM. Therefore, a Ca^{2+} sensor that shows large changes of its fluorescence within such Ca^{2+} concentration

Box 1 Recovery of Field Potential Response After Electrode Insertion

Quite often, imaging is performed with electrophysiological recording from the brain. However, the fact that electrode insertion results in a transient attenuation of neuronal response is often overlooked. Here, we demonstrate that even gentle insertion of a silicon probe microelectrode requires at least 1 h of field response recovery period. The LFP from the mouse barrel cortex layer 4 was continuously recorded from just after the electrode insertion. The peak amplitude shows a time-dependent increase in the first 60 min, and stabilizes thereafter. This increase of amplitude is likely to be due to a recovery of brain tissue damage from surgery and electrode insertion. Thus, the recovery period of at least 1 h is strongly recommended. The variability of evoked responses becomes larger after 60 min as indicated by the size of the error bar. This variability reflects the amplitude of slow oscillations that also recovers in parallel (Fig. B1).

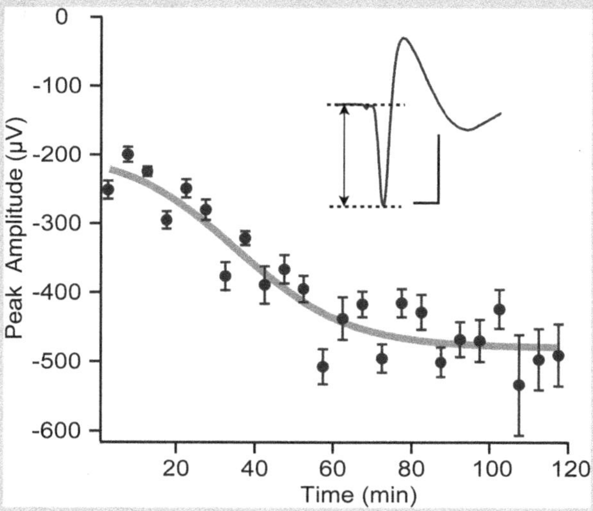

Fig. B1 LFP response amplitude to single air puff whisker stimulation (0.05 Hz) plotted vs. time after electrode insertion. The peak amplitude is defined as the negative peak amplitude of the evoked response (*inset*), and averaged in 5 min bins. Scale bars: 200 μV, 200 ms.

range is desirable for astrocytic imaging. We synthesized replication-deficient recombinant adenovirus to express Ca^{2+} sensor G-CaMP-HS [29] under the control of the CMV promoter using a commercial kit (AdEasy XL, Stratagene), as the adenovirus has a high tropism for astrocytes in the postnatal rodent brain. We find that purification of adenovirus by ultracentrifuge using CsCl discontinuous gradient yields in general higher titer virus (10^{10-11} pfu), compared with commercial affinity column-based purification. Experimental animals are anesthetized with ketamine–xylazine cocktail (9 % ketamine and 1 % xylazine in physiological saline; dosage 1.0 mg/kg for rats). Similar to the bolus loading procedure described above, up to 300 nL of virus-containing solution (10^{9-10} pfu) is pressure-injected into the area of interest with a Nanoject II (Drummond Scientific). The animals are recovered for 1–2 weeks before imaging experiments. The expression of G-CaMP-HS could be observed within 1 week after infection and lasted for at least 3 weeks. For optimal expression, we recommend injecting virus 2 weeks prior to imaging.

2.4 Two-Photon Imaging and LFP Recording

Imaging is performed with a custom-modified Olympus FV-1000-based two-photon microscope. A Chameleon XR laser (Coherent) is used as the excitation light source. The group velocity dispersion is compensated by a prism-coupled pre-chirper and the beam diameter is adjusted by a Kepler telescope. The laser output power is adjusted by a motorized graded neutral density filter or a polarizing prism (e.g., Glan-Thompson prism). The laser power measured after the objective lens (LUMPLANFL40XW, N.A. 0.8, Olympus, Tokyo, Japan) is <10 mW for both L1 (up to 100 μm from the pia) and L2/3 (~150–200 μm) with a pixel dwell time of 2 μs/pixel. Typically, an area of 320 × 320 μm is imaged for 30 min at 0.5 Hz with 1 μm^2 pixel resolution. For LFP monitoring, a glass pipette (resistance 2–3 MΩ) containing physiological saline is inserted 300–400 μm below the pial surface at an insertion angle of 30°. LFP is measured by taking a differential signal between the microelectrode and a reference signal at the neck of the animal or a screw attached to the interparietal bone. The LFP signal is amplified by a Multiclamp 700B amplifier (Molecular Devices, CA, USA) and digitized with a PCI-6259 data acquisition board (National Instruments) using custom software written in LabView (National Instruments).

2.5 Post Hoc Immunohistochemistry

After imaging experiments, the rats are transcardially perfused with physiological saline, followed by 100 mL of 0.1 M phosphate buffer containing 4 % paraformaldehyde. Coronal sections corresponding to the imaged area are cut at 60 μm thickness in 0.1 M phosphate buffer (pH 7.4) using a microslicer (Pro-7 Linear Slicer, Dosaka EM, Kyoto, Japan). For the detection of cytoarchitectural difference of cortical layers, slices are immunostained using primary antibodies of anti-S100β rabbit polyclonal antibody (1:1,000 dilution, Product ID 37, Swant) and anti-NeuN mouse monoclonal antibody (1:1,000 dilution, Product ID MAB377, Chemicon). For the assessment of craniotomy condition, slices are immunostained using anti-glial fibrillary acidic protein (GFAP) mouse monoclonal antibody (1:1,000 dilution, Product ID G3893, Sigma-Aldrich) and anti-ionized Ca^{2+}-binding adaptor molecule 1 (Iba-1) rabbit polyclonal antibody (1:1,000 dilution, Product ID 019–19741, Wako Pure Chemicals). Subsequently, slices are incubated with fluorescent secondary antibodies (Alexa Fluor 488-conjugated anti-mouse antibody, 1:2,000 dilution, Molecular Probes-Invitrogen; Cy3-conjugated anti-rabbit antibody, 1:2,000 dilution, Jackson Immunoresearch Laboratories).

2.6 Data Analysis

Fluorescence signal is quantified by measuring the mean pixel value of a semiautomatically selected somatic area for each frame of the image stack using MATLAB (Mathworks, MA). The semiautomatic selection of regions of interests ("ROIs") is based on Fisher's linear discriminant method [30]. First, soma locations are obtained from local maxima in the image obtained by convolving a fluorescence image with a feature vector f (Fig. 1a, b). For the reference fluorescence image, the averaged image of the time series worked well with OGB-1 experiments while the maximum intensity projection image works well with Fluo-4 experiments. The feature vector f satisfies the equation: $1/2 * (\Sigma_C + \Sigma_{NC}) \cdot f = m_C - m_{NC}$, where m_C and m_{NC} are the mean of manually collected library images of cells and non-cells, respectively, and Σ_C and Σ_{NC} are the corresponding covariances. Because the matrix $\Sigma_C + \Sigma_{NC}$ was not always invertible, the matrix is diagonalized numerically and the first largest eigenvalues accounting for 90–95 % of the variance are retained to obtain the feature vector f [30]. After soma positions are detected, ROIs are selected based on the fluorescence intensity and size (Fig. 1c). Ca^{2+} surges of astrocyte are detected when the fluorescence intensity of OGB-1 reached values higher than ~120 % of the baseline fluorescence intensity (Note that threshold for Ca^{2+} surge should be adjusted according to your measurements. In our experiments, we've used ~120 % for OGB-1 and 50 % for Fluo-4, which correspond to ~3× SD [22, 31]). The baseline fluorescence intensity is defined as the mean of the lower two-third of the fluorescence intensity histogram for each astrocyte during the 30 min imaging. The duration of each Ca^{2+} surge is defined as duration of fluorescence intensity above 50 % of its peak intensity during a Ca^{2+} surge (also called "full width at half maximum"). To analyze the synchrony of Ca^{2+} dynamics among the processes of an astrocyte, the Pearson correlation coefficients are computed for each pair of the selected main branches. All data are expressed as mean ± standard deviation (SD).

3 Imaging of Astrocytic Activity

3.1 Spontaneous Glial Activities Observed with Organic Ca^{2+} Indicator

The presence of astrocytes dominates over neurons in layer 1 (L1) of the cerebral cortex, whereas the neuronal population is predominant in layer 2/3. We asked whether the spontaneous astrocytic Ca^{2+} activity reflects the cytoarchitectural difference. MCBL of OGB-1 and SR101 typically labeled cells in an area within 350 μm from the tip of the injection electrode. As expected, the composition of labeled cells was substantially different between L1 and L2/3. In L1, most of the OGB-1-loaded cells were also labeled by SR101, indicating that they were astrocytes (Fig. 1d, left). In contrast, a mixture of astrocytes and OGB-1-loaded neurons without SR101 labeling was observed in L2/3 (Fig. 1d, right).

L1 and L2/3 were monitored with a field of view of 320 × 320 μm at 1 μm² pixel resolution microscope for 30 min. Figure 1e shows representative traces of somatic Ca^{2+} level fluctuations for the L1 astrocytes marked in Fig. 1d. We termed astrocytes that gave rise to at least one spontaneous Ca^{2+} surge as "active astrocytes," and astrocytes without a Ca^{2+} surge during the 30 min imaging session as "inactive astrocytes." We found that the average number of active astrocytes in L1 (10.0 ± 5.6 per imaging field) was significantly higher than that in L2/3 (2.8 ± 2.4, $p < 0.001$, *t*-test) (Fig. 1f), which may reflect cytoarchitectural differences between L1 and L2/3. The mean number of spontaneous Ca^{2+} events per astrocytes (active and inactive) in 30 min was calculated to be 1.5 ± 2.3 for L1 and 0.4 ± 1.1 for L2/3 and the difference was significant ($p < 0.001$, *t*-test). Active astrocytes in L1 had 3.0 ± 2.5 Ca^{2+} surges per astrocyte in 30 min and were significantly more active than those in L2/3 which had 2.0 ± 1.7 Ca^{2+} surges ($p < 0.001$, *t*-test). These astrocytic spontaneous Ca^{2+} events had an average duration (full width at half maximum) of 34.8 ± 35.8 s in L1 and 32.5 ± 34.6 s in L2/3. We showed that these spontaneous Ca^{2+} surges were independent from the surrounding neuronal activities [22]. These results suggest that astrocytes in L1 and L2/3 have distinct physiological properties in regard to spontaneous Ca^{2+} activities in the somata.

Because astrocytic processes surround and sometimes contact synapses, we next investigated the Ca^{2+} activity of astrocytic processes. With the pixel resolution of ~1 μm², only primary processes of astrocytes could reliably be detected in SR101-labeled images. We chose three clearly distinguishable primary branches of each astrocyte. For each chosen astrocytic process, two ROIs were placed—one being more proximal (p) and the other more distal (d) to the soma (Fig. 2a). At astrocytic processes, fluorescence signals were generally smaller and the noise levels of $\Delta F/F_0$ were relatively large compared to somatic signals (Fig. 2b). Therefore, we assessed the Ca^{2+} dynamics at astrocytic processes by taking the pair-wise Pearson product–moment correlation coefficient (*r*-value) as a measure of synchronization.

The correlation coefficient was calculated for pairs of ROIs from 30 processes of 10 astrocytes in 3 rats (Fig. 2c). Both in L1 and L2/3, ROIs within the same astrocyte process had a weak but positive association (e.g., proc1p vs. proc1d in L1 had $r \sim 0.34$, proc2p vs. proc2d in L2/3 had $r \sim 0.43$), indicating that the Ca^{2+} activity of within-processes is correlated. On the other hand, while ROIs placed at different processes had little association in L1 (e.g., proc1p vs. proc2p had $r \sim 0.19$), in the case of L2/3, Ca^{2+} traces of ROIs placed at different astrocyte processes had a positive association (e.g., proc1p vs. proc2p had $r \sim 0.39$). These data suggest that processes of L1 astrocytes operated independently, while L2/3 astrocytic processes tend to have synchronous Ca^{2+} activity.

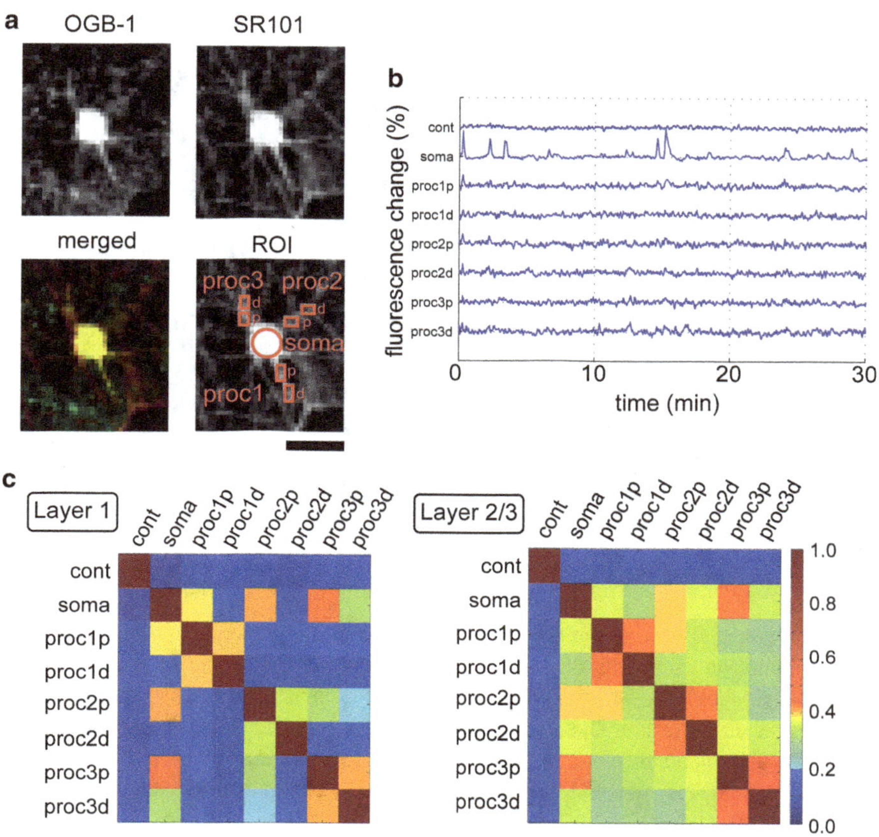

Fig. 2 Distinct modes of Ca^{2+} activity in astrocytic processes. (**a**) Astrocyte's soma and processes were identified based on the SR101 image. Three primary processes (proc) of a representative astrocyte in L1 were arbitrarily selected and two rectangular ROIs were placed at each primary process. One ROI was placed at proximal (p) part of the process and the other was at distal (d) part. As a control, one ROI was placed in the neuropil at least 100 μm away from the astrocyte. This control ROI had the same size as the ROI of astrocytic soma. Scale bar 20 μm. (**b**) Ca^{2+} activity was monitored throughout the imaged session for each of the ROIs selected in (**a**). Scale bar 100 %. (**c**) Pearson's linear correlations (*r*-values) between Ca^{2+} traces of ROIs were calculated for L1 and L2/3. The correlation coefficient of "within-process" (e.g., proc1p vs. proc1d) ROIs was high in both L1 and L2/3. Note that the between-processes correlation was high only in L2/3 (e.g., proc1p vs. proc2p in L2/3) (Modified from [22]).

3.2 Sensory-Driven Glial Responses Measured with a Ca^{2+}-Sensitive Fluorescent Protein

Analysis of Ca^{2+} activity in astrocytic processes with MCBL using OGB-1 is limited to the primary branches of astrocytes due to the low S/N ratio. To achieve Ca^{2+} surge detection in astrocytic processes, we expressed the Ca^{2+}-sensitive protein G-CaMP-HS using a recombinant adenovirus. Injection of 100 nL virus-containing solution resulted in selected G-CaMP-HS expression in astrocytes. The inoculated area did not show any noticeable inflammatory reactions, which was confirmed by post hoc immunohistochemistry for a microglia marker Iba-1 (not shown). G-CaMP-HS

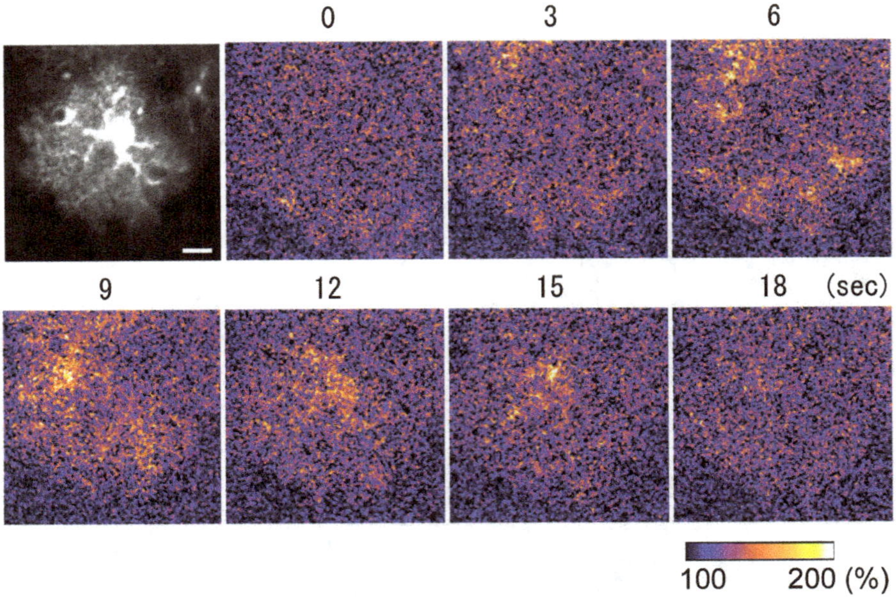

Fig. 3 Use of a Ca^{2+}-sensitive fluorescent protein revealed Ca^{2+} surge propagation from the processes to the soma of a single astrocyte upon whisker stimulation. A representative image of an astrocyte expressing G-CaMP-HS is shown in the *upper left* corner (scale bar 10 μm). The astrocyte was located 105 μm below the pial surface. Color images show time-lapse astrocytic Ca^{2+} responses upon whisker stimulation (5 Hz, 1 min). The numeral above each image indicates the elapsed time in second from the onset of whisker stimulation. Ca^{2+} activity is expressed by F/F_0 of G-CaMP-HS. Note the sensory stimulation-induced astrocytic Ca^{2+} propagation from the processes converging to the soma.

fluorescence of a single astrocyte visualized the whole morphology including the fine processes in an area distant from the virus injection where labeling was sparse (Fig. 3, upper left). Whisker stimulation (5 Hz, 1 min) triggered Ca^{2+} surges initially in astrocytic processes within a few seconds. These Ca^{2+} surges at processes propagated into the soma at a speed of 3–4 μm/s. The precedence of Ca^{2+} surges at processes upon neuronal activity is in accordance with the idea that astrocytes detect synaptic activities with their fine processes, presumably sensing spilled-over glutamate from the synaptic clefts. We saw Ca^{2+} surges that are limited to within astrocytic processes more frequently than Ca^{2+} surges that involve the soma.

4 Discussion and Outlook

4.1 Spatiotemporal Aspects of Astrocytic Ca^{2+} Dynamics

We find that somatic elevation of astrocytic $[Ca^{2+}]_i$ is large in magnitude and long-lasting (mode was 10 ~ 14 s) in the cerebral cortex in vivo. The dynamics of these events is similar to those reported in in vitro experiments [32, 33]. Under urethane anesthesia, such astrocytic $[Ca^{2+}]_i$ elevations occur spontaneously, albeit at low frequency (1–3 times in 10 min). Most of these large somatic Ca^{2+} events are likely to be mediated by activation of IP_3 receptors, as such large Ca^{2+} events were not observed in mice deficient of IP_3R2 in hippocampal slice [34] or in the cortex in vivo [31]. These astrocytic $[Ca^{2+}]_i$ activities are spatially organized in two ways. First, layer 1 astrocytes have more frequent spontaneous $[Ca^{2+}]_i$ events than layer 2/3 astrocytes. As pharmacological blockades of various receptors and channels did not have much effect on the spontaneous astrocyte $[Ca^{2+}]_i$ elevation, it is likely that these events occur independent of the surrounding neuronal activity [22]. It is noted, however, that GABA-B receptor blockade has not been performed to date and it remains possible that spontaneous astrocyte Ca^{2+} events are driven by some GABAergic activity. Another source of drive to the spontaneous activity could be enzymatic activity in the extracellular space. Recently, it has been reported that astrocytes express the protease-activated receptors type 1 (PAR-1), activation of which triggers large cytosolic Ca^{2+} elevations [35].

Second, we find that coactive astrocytes are located close to each other [22, 36]. Similar findings have been reported for the hippocampal astrocytes in vivo [37], suggesting that spatially and temporally clustered Ca^{2+} elevation of astrocytes is a general phenomenon in cortical structure. Interestingly, in vitro studies using acute hippocampal slices show that glutamate is secreted from astrocytes as $[Ca^{2+}]_i$ elevates [37, 38], suggesting that astrocytic Ca^{2+} events depolarize neighboring dendrites and hence enhance synchronization of local neuronal activity. In our studies and others, the astrocytic Ca^{2+} events are discrete and intercellular traveling wave propagation was hardly observed. However, in the cerebellar cortex, radially propagating Ca^{2+} waves are observed in Bergman glia networks [39]. Interestingly, a recent study reports intercellular astrocytic Ca^{2+} waves in the hippocampus in vivo under a somewhat low body temperature condition [40].

As spontaneous astrocytic Ca^{2+} events become more frequent with decreased levels of anesthesia [41], an immediate next step forward is to monitor and characterize astrocytic Ca^{2+} activities in unanesthetized conditions. Such imaging procedures are now actively employed in neuronal imaging (e.g., Chapter 2.5 in this volume), and some studies report astrocytic activity in unanesthetized condition as well. For example, in the cerebellar cortex, mass-synchronized Ca^{2+} occurs among Bergmann glia as the mouse initiates motor movement [42]. Similarly, mass-synchronized Ca^{2+} elevations occur in the somatosensory cortex during the period of active motor movements [43]. It has been reported that sensory activation and subcortical neurotransmitters can give rise to astrocytic $[Ca^{2+}]_i$ elevations [41, 44–46]. In waking states, subcortical neurotransmitter systems, such as acetylcholine and noradrenalin, are tonically active. As G protein-coupled receptors for subcortical neurotransmitters are abundantly expressed by astrocytes, astrocytic $[Ca^{2+}]_i$ elevations in waking states may play more active roles.

4.2 Subcellular Ca^{2+} Events

Aside from control of local blood flow (see Chapters 14, 15 and 16 in this volume), the exact function of astrocytic Ca^{2+} elevation is still being explored [47]. One intriguing proposed role of astrocytes in neuronal information processing is their involvement in synaptic plasticity, although this is highly controversial [31, 48–51]. If astrocytes are capable of controlling plasticity of individual synapses via a $[Ca^{2+}]_i$ signaling, imaging subcellular compartment of astrocytes, such as processes, should yield some insights into synapse–astrocyte interactions. Interestingly, in vitro experiments show that astrocytic Ca^{2+} is critically important for neurite growth [52] and intimate physical interactions between astrocytic processes and neuronal spines take place for spine maturation[53, 54]. It remains an open question whether such neuron–glia interactions occur in the intact brain during brain development and synaptic plasticity. It should be noted that astrocytic processes are prone to photo-damage by excitation light for multiphoton microscopy. Spurious astrocytic process $[Ca^{2+}]_i$ elevations are produced [44] and intercellular Ca^{2+} waves cease in high laser [41] power conditions (Box 2). We showed that $[Ca^{2+}]_i$ elevations in astrocytic processes in the somatosensory cortex are driven by sensory stimulation, suggesting that these events are not merely induced by photo-damage. Recently reported membrane-bound Ca^{2+} indicators may further reveal the Ca^{2+} dynamics of fine structures [55]. Future studies should address the detailed mechanisms for astrocytic process $[Ca^{2+}]_i$ elevations and their implications to the neuronal circuit dynamics in vivo.

Box 2 Potential Laser-Induced Astrocyte Ca^{2+} Signals

It has become increasingly evident that overexposure to light induces adverse effects on living cells. For instance, the laser beam for two-photon imaging can trigger action potentials in neurons, when applied with a high intensity and focused on the membrane [56]. It turns out that astrocytes are more sensitive to laser light excitation than neurons, causing artificially induced Ca^{2+} elevations in astrocytic processes above an intensity threshold (Fig. B2). According to Kuga et al. [40], intercellular astrocytic Ca^{2+} waves are observable in the hippocampus in vivo only when the excitation laser power is below 10 mW. The exact condition beyond which astrocytes give rise to photo-induced Ca^{2+} elevations depends on many parameters such as the laser pulse width, repetition rate, objective lens properties, and imaging depth. Determination of such condition and the use of minimal intensity excitation beam are recommended for astrocyte Ca^{2+} imaging.

(continued)

Box 2 (continued)

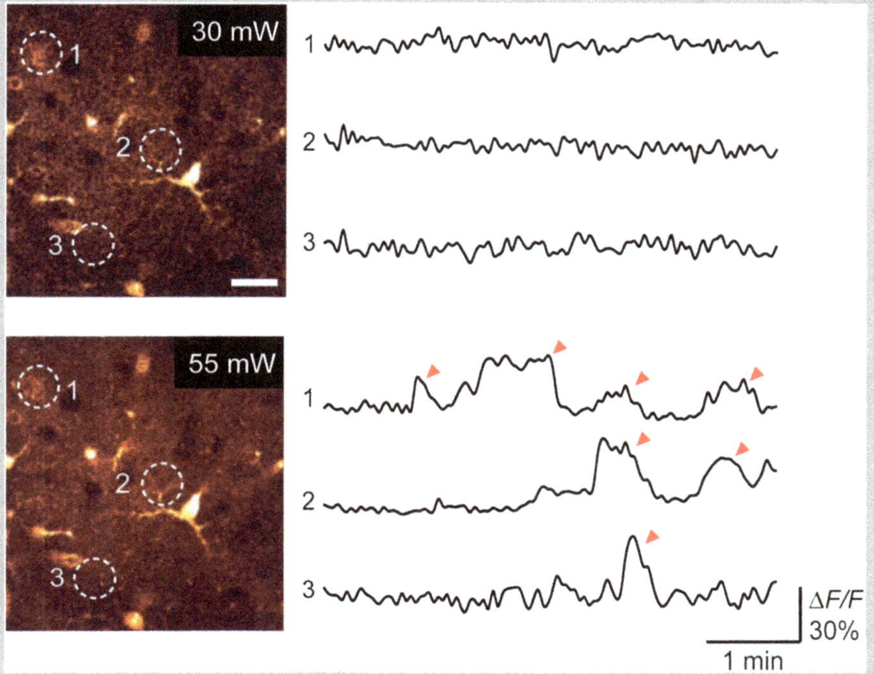

Fig. B2 *Top row*: Barrel cortex astrocytes loaded with Fluo-4 AM located 120 μm below the pial surface. One frame was imaged every 3 s for 5 min using 30 mW laser power (at the exit of the 20× objective, 0.9 NA, Olympus). Traces depict the spontaneous astrocytic Ca^{2+} activity in areas indicated by *white dashed circles*. *Bottom row*: Same field imaged with 55 mV laser power (1 frame every 3 s). *Red arrowheads* indicate spontaneous Ca^{2+} increases. Note that no spontaneous Ca^{2+} increases are evident during the first minute of recording (Modified from [44]).

References

1. Nedergaard M, Ransom B, Goldman SA (2003) New roles for astrocytes: redefining the functional architecture of the brain. Trends Neurosci 26:523–530
2. Karlsen AS, Pakkenberg B (2011) Total numbers of neurons and glial cells in cortex and basal ganglia of aged brains with Down syndrome—a stereological study. Cereb Cortex 21:2519–2524
3. Mishima T, Sakatani S, Hirase H (2005) In vivo intracellular recording and subsequent morphological visualization of rat neocortical astrocytes. Soc Neurosci. Abstract, 385.4
4. Verkharatsky A, Butt A (2007) Glial neurobiology. Wiley, Chichester, West Sussex
5. Tasaki I (1939) The electro-saltatory transmission of the nerve impulse and the effect of narcosis upon the nerve fiber. Am J Physiol 127: 211–227
6. Huxley AF, Stampfli R (1949) Evidence for saltatory conduction in peripheral myelinated nerve fibres. J Physiol 108:315–339
7. Nishiyama A, Chang A, Trapp BD (1999) NG2+ glial cells: a novel glial cell population in the adult brain. J Neuropathol Exp Neurol 58:1113–1124
8. Aloisi F (2001) Immune function of microglia. Glia 36:165–179
9. Hanisch UK, Kettenmann H (2007) Microglia: active sensor and versatile effector cells in the normal and pathologic brain. Nat Neurosci 10: 1387–1394
10. Magistretti PJ, Pellerin L, Rothman DL, Shulman RG (1999) Energy on demand. Science 283:496–497
11. Wyss MT, Jolivet R, Buck A, Magistretti PJ, Weber B (2011) In vivo evidence for lactate as

a neuronal energy source. J Neurosci 31: 7477–7485

12. Grosche J, Matyash V, Moller T, Verkhratsky A, Reichenbach A, Kettenmann H (1999) Microdomains for neuron-glia interaction: parallel fiber signaling to Bergmann glial cells. Nat Neurosci 2:139–143
13. Chao TI, Rickmann M, Wolff JR (2002) The synapse-astrocyte boundary: anatomical basis for an integrative role of glia in synaptic transmission. In: Volterra A, Magistretti P, Haydon PG (eds) Tripartite synapses: synaptic transmission with glia. Oxford University Press, New York, pp 3–23
14. Ventura R, Harris KM (1999) Three-dimensional relationships between hippocampal synapses and astrocytes. J Neurosci 19:6897–6906
15. Bushong EA, Martone ME, Jones YZ, Ellisman MH (2002) Protoplasmic astrocytes in CA1 stratum radiatum occupy separate anatomical domains. J Neurosci 22:183–192
16. Oberheim NA, Takano T, Han X, He W, Lin JH, Wang F, Xu Q, Wyatt JD, Pilcher W, Ojemann JG, Ransom BR, Goldman SA, Nedergaard M (2009) Uniquely hominid features of adult human astrocytes. J Neurosci 29:3276–3287
17. Parpura V, Basarsky TA, Liu F, Jeftinija K, Jeftinija S, Haydon PG (1994) Glutamate-mediated astrocyte-neuron signalling. Nature 369:744–747
18. Guthrie PB, Knappenberger J, Segal M, Bennett MV, Charles AC, Kater SB (1999) ATP released from astrocytes mediates glial calcium waves. J Neurosci 19:520–528
19. Araque A, Parpura V, Sanzgiri RP, Haydon PG (1999) Tripartite synapses: glia, the unacknowledged partner. Trends Neurosci 22:208–215
20. Volterra A, Meldolesi J (2005) Astrocytes, from brain glue to communication elements: the revolution continues. Nat Rev Neurosci 6:626–640
21. Moldestad O, Karlsen P, Molden S, Storm JF (2009) Tracheotomy improves experiment success rate in mice during urethane anesthesia and stereotaxic surgery. J Neurosci Methods 176:57–62
22. Takata N, Hirase H (2008) Cortical layer 1 and layer 2/3 astrocytes exhibit distinct calcium dynamics in vivo. PLoS One 3:e2525
23. Yang G, Pan F, Parkhurst CN, Grutzendler J, Gan WB (2010) Thinned-skull cranial window technique for long-term imaging of the cortex in live mice. Nat Protoc 5:201–208
24. Olesen SP (1987) Leakiness of rat brain microvessels to fluorescent probes following craniotomy. Acta Physiol Scand 130:63–68
25. Stosiek C, Garaschuk O, Holthoff K, Konnerth A (2003) In vivo two-photon calcium imaging of neuronal networks. Proc Natl Acad Sci USA 100:7319–7324
26. Nimmerjahn A, Kirchhoff F, Kerr JN, Helmchen F (2004) Sulforhodamine 101 as a specific marker of astroglia in the neocortex in vivo. Nat Methods 1:31–37
27. Sullivan MR, Nimmerjahn A, Sarkisov DV, Helmchen F, Wang SS (2005) In vivo calcium imaging of circuit activity in cerebellar cortex. J Neurophysiol 94:1636–1644
28. Miyawaki A, Llopis J, Heim R, McCaffery JM, Adams JA, Ikura M, Tsien RY (1997) Fluorescent indicators for Ca^{2+} based on green fluorescent proteins and calmodulin. Nature 388:882–887
29. Muto A, Ohkura M, Kotani T, Higashijima S, Nakai J, Kawakami K (2011) Genetic visualization with an improved GCaMP calcium indicator reveals spatiotemporal activation of the spinal motor neurons in zebrafish. Proc Natl Acad Sci USA 108:5425–5430
30. Metzner W, Koch C, Wessel R, Gabbiani F (1998) Feature extraction by burst-like spike patterns in multiple sensory maps. J Neurosci 18:2283–2300
31. Takata N, Mishima T, Hisatsune C, Nagai T, Ebisui E, Mikoshiba K, Hirase H (2011) Astrocyte calcium signaling transforms cholinergic modulation to cortical plasticity in vivo. J Neurosci 31:18155–18165
32. Pasti L, Zonta M, Pozzan T, Vicini S, Carmignoto G (2001) Cytosolic calcium oscillations in astrocytes may regulate exocytotic release of glutamate. J Neurosci 21:477–484
33. Fiacco TA, McCarthy KD (2006) Astrocyte calcium elevations: properties, propagation, and effects on brain signaling. Glia 54:676–690
34. Petravicz J, Fiacco TA, McCarthy KD (2008) Loss of IP3 receptor-dependent Ca^{2+} increases in hippocampal astrocytes does not affect baseline CA1 pyramidal neuron synaptic activity. J Neurosci 28:4967–4973
35. Shigetomi E, Bowser DN, Sofroniew MV, Khakh BS (2008) Two forms of astrocyte calcium excitability have distinct effects on NMDA receptor-mediated slow inward currents in pyramidal neurons. J Neurosci 28:6659–6663
36. Hirase H, Qian L, Bartho P, Buzsaki G (2004) Calcium dynamics of cortical astrocytic networks in vivo. PLoS Biol 2:E96
37. Sasaki T, Kuga N, Namiki S, Matsuki N, Ikegaya Y (2011) Locally synchronized astrocytes. Cereb Cortex 21:1889–1900
38. Fellin T, Pascual O, Gobbo S, Pozzan T, Haydon PG, Carmignoto G (2004) Neuronal synchrony mediated by astrocytic glutamate

through activation of extrasynaptic NMDA receptors. Neuron 43:729–743

39. Hoogland TM, Kuhn B, Gobel W, Huang W, Nakai J, Helmchen F, Flint J, Wang SS (2009) Radially expanding transglial calcium waves in the intact cerebellum. Proc Natl Acad Sci USA 106:3496–3501
40. Kuga N, Sasaki T, Takahara Y, Matsuki N, Ikegaya Y (2011) Large-scale calcium waves traveling through astrocytic networks in vivo. J Neurosci 31:2607–2614
41. Schummers J, Yu H, Sur M (2008) Tuned responses of astrocytes and their influence on hemodynamic signals in the visual cortex. Science 320:1638–1643
42. Nimmerjahn A, Mukamel EA, Schnitzer MJ (2009) Motor behavior activates Bergmann glial networks. Neuron 62:400–412
43. Dombeck DA, Khabbaz AN, Collman F, Adelman TL, Tank DW (2007) Imaging large-scale neural activity with cellular resolution in awake, mobile mice. Neuron 56:43–57
44. Wang X, Lou N, Xu Q, Tian GF, Peng WG, Han X, Kang J, Takano T, Nedergaard M (2006) Astrocytic Ca2+ signaling evoked by sensory stimulation in vivo. Nat Neurosci 9:816–823
45. Bekar LK, He W, Nedergaard M (2008) Locus coeruleus alpha-adrenergic-mediated activation of cortical astrocytes in vivo. Cereb Cortex 18:2789–2795
46. Petzold GC, Albeanu DF, Sato TF, Murthy VN (2008) Coupling of neural activity to blood flow in olfactory glomeruli is mediated by astrocytic pathways. Neuron 58: 897–910
47. Takata N, Nagai T, Ozawa K, Oe Y, Mikoshiba K, Hirase H (2013) Cerebral blood flow modulation by Basal forebrain or whisker stimulation can occur independently of large cytosolic Ca^{2+} signaling in astrocytes. PLoS One 8:e66525
48. Yang Y, Ge W, Chen Y, Zhang Z, Shen W, Wu C, Poo M, Duan S (2003) Contribution of astrocytes to hippocampal long-term potentiation through release of D-serine. Proc Natl Acad Sci USA 100:15194–15199
49. Perea G, Araque A (2007) Astrocytes potentiate transmitter release at single hippocampal synapses. Science 317:1083–1086
50. Henneberger C, Papouin T, Oliet SH, Rusakov DA (2010) Long-term potentiation depends on release of D-serine from astrocytes. Nature 463:232–236
51. Agulhon C, Fiacco TA, McCarthy KD (2010) Hippocampal short- and long-term plasticity are not modulated by astrocyte Ca^{2+} signaling. Science 327:1250–1254
52. Kanemaru K, Okubo Y, Hirose K, Iino M (2007) Regulation of neurite growth by spontaneous Ca2+ oscillations in astrocytes. J Neurosci 27:8957–8966
53. Haber M, Zhou L, Murai KK (2006) Cooperative astrocyte and dendritic spine dynamics at hippocampal excitatory synapses. J Neurosci 26:8881–8891
54. Nishida H, Okabe S (2007) Direct astrocytic contacts regulate local maturation of dendritic spines. J Neurosci 27:331–340
55. Shigetomi E, Kracun S, Sofroniew MV, Khakh BS (2010) A genetically targeted optical sensor to monitor calcium signals in astrocyte processes. Nat Neurosci 13:759–766
56. Hirase H, Nikolenko V, Goldberg JH, Yuste R (2002) Multiphoton stimulation of neurons. J Neurobiol 51:237–247

Chapter 13

Imaging Morphology and Function of Cortical Microglia

Bianca Brawek and Olga Garaschuk

Abstract

Microglial cells are the innate immune cells of the central nervous system. In the healthy adult brain "resting" ramified microglia continuously palpate their environment to monitor the integrity of and to react to any disturbance of brain homeostasis. During injury, inflammation, and in the course of neurodegenerative diseases microglia become activated, proliferate, and release a plethora of cytokines as well as reactive oxygen species. In addition to their well known role in disease, it has become increasingly clear that "resting" microglia also contribute to normal brain physiology, both during postnatal development and in the mature adult brain.

Functional in vivo imaging of microglia first of all captures the morphological changes accompanying microglial transition between "resting" and activated states. In addition, intracellular Ca^{2+} homeostasis of microglia is believed to be altered between the two states [1–3]. So far, however, microglial Ca^{2+} signaling was predominantly studied in reduced preparations like brain slices or cell cultures, in which microglia are found in a rather activated state. In this chapter we describe a technique for studying microglial Ca^{2+} signaling in vivo. Furthermore, we discuss a new approach for visualization of morphological dynamics of microglial cells in vivo at high resolution. This approach utilizes a lectin-based staining technique and is applicable to any deliberate mouse strain at any developmental stage.

Key words Microglia, Plant lectins, In vivo imaging, Single-cell electroporation, Ca^{2+} signaling, Neurodegenerative diseases

1 Introduction

Microglia are the resident immune cells of the central nervous system (CNS). The majority of these cells are likely to originate from the yolk sac macrophages which engraft in the brain during early embryogenesis [4]. Microglia are abundantly present in all regions of the CNS. Their number in the human brain (100–200 billion cells, depending on the condition, e.g., healthy, infected, or diseased brain [5]) approximately equals to that of neurons (85–120 billion cells [5, 6]). Their density varies in different brain regions with the highest density found in the grey matter [7].

In the healthy adult brain microglia is found in a so-called resting state. "Resting" microglial cells have ramified morphology [8, 9].

Bruno Weber and Fritjof Helmchen (eds.), *Optical Imaging of Neocortical Dynamics*, Neuromethods, vol. 85, DOI 10.1007/978-1-62703-785-3_13, © Springer Science+Business Media New York 2014

They continuously palpate their vicinity with their fine processes [10, 11] and frequently contact synapses of neighboring neurons [12, 13]. Microglial surveillance seems to be necessary for normal synaptic maturation as these cells are actively involved in pruning of surplus synapses during postnatal development [14]. Furthermore, by removal of dysfunctional or apoptotic neurons as well as elimination of nonfunctional synaptic elements (for a review, see [15]), microglia contribute to adult circuit plasticity.

In response to certain cues like cytokines, chemokines, neurotransmitters, complement proteins, and many other factors microglia adopt an activated phenotype [9, 16], which is characterized by enlarged somata and retracted, thickened processes [7, 9]. Activation of microglia accompanies a large number of CNS insults like acute CNS injury, brain tumors, apoptosis, infection, ischemia, as well as neurodegenerative diseases like Alzheimer's disease, Parkinson's disease, and HIV dementia [7, 8, 17, 18]. Whether activated microglia are harmful or beneficial for the disease progression is still a matter of debate. On one hand, it has been suggested that fully activated microglia that resemble macrophages are neurotoxic as they release reactive oxygen species, nitric oxide, or pro-inflammatory cytokines [7–9, 18, 19]. On the other hand, microglia can be neuroprotective, for example, by degrading cell debris or misfolded proteins like amyloid β, which accumulates in the course of Alzheimer's disease [7, 9] or by releasing anti-inflammatory and neuroprotective factors [20].

There are several mouse lines in which microglia are genetically labeled with enhanced green fluorescent protein (GFP) [21–23]. So far two of those were used for in vivo studies of microglial function. In one mouse line the gene for the fractalkine receptor CX_3CR1 is replaced by a gene coding for GFP [21]. In the second line GFP is expressed under control of ionized Ca^{2+}-binding adaptor molecule 1 (Iba1) promoter [22]. In the CNS both of these proteins, CR_3CR1 as well as Iba1, are exclusively expressed in microglia thus allowing cell type-specific labeling of these cells. The aforementioned mouse strains were used for in vivo studies revealing the surveillance function of ramified microglia in the intact brain [10, 11], as well as their important role in remodeling synaptic circuits [12]. However, the use of these mouse lines has also disadvantages. CX_3CR1 is the only receptor for the chemokine fractalkine, which is known to mediate communication between neurons and microglia [24]. Although microglia in heterozygous $CX_3CR1^{GFP/+}$ mice seem to behave alike their wild-type counterparts [21, 25], a reduced level of the protein might impact microglial function in the aged and/or diseased brain [26, 27]. Iba1-GFP animals, in turn, suffer from lower expression levels of GFP compared to $CX_3CR1^{GFP/+}$ mice (B.B. and O.G., unpublished data). In addition, the GFP expression level in Iba1-GFP mice (e.g., brightness of labeled cells) seems to vary a lot across experimental animals. Furthermore, the use of genetically labeled mouse strains requires time-, mice-, and resources-consuming breeding of animals, especially when analyzing aged mice with additional pathological mutations.

As in many other non-excitable cells, changes in the intracellular Ca^{2+} concentration represent one of the major pathways for signal transduction in microglia. The intracellular Ca^{2+} concentration ($[Ca^{2+}]_i$) in microglia is tightly regulated by diverse mechanisms [8, 28]. For example, the cells express a multitude of different ionotropic and metabotropic receptors in the plasma membrane, which cause either an influx of Ca^{2+} through the plasma membrane or its release from the intracellular Ca^{2+} stores [1, 29]. Moreover, elevation of $[Ca^{2+}]_i$ has been shown to be involved in the activation process of microglia [2] and to be coupled to executive functions of microglial cells such as release of pro- and anti-inflammatory cytokines, NO, or trophic factors [1, 9, 16, 30]. So far, studies of microglial Ca^{2+} signaling were conducted in cultured cells or acute brain slices [1, 16, 28, 31], preparations in which those cells are present in a rather activated state due to isolation or slicing procedures. To get an insight into the mechanisms underlying microglial Ca^{2+} signaling under physiological conditions, these cells have to be studied in vivo. However, simple in vivo methods for labeling the cells with small molecule Ca^{2+} indicators as, for example, multicell bolus loading (MCBL, [32]) fail to label microglia [25, 33]. In principle, in vivo microglia can also be labeled with genetically encoded Ca^{2+} sensors (e.g., Yellow Cameleon 3.60, TN-XXL, GCaMP2, GCaMP3, Yellow Cameleons-Nano, and GECOs [34–39]). Interestingly, when expressing Yellow Cameleon 3.60 under glia-specific S100β promoter, Atkin et al. [40] also labeled Iba1-positive cells (microglia). In these mice, however, the percentage of labeled microglial cells was rather low (10–40 %, depending on the brain region and the mouse strain analyzed). In addition, the labeling was not cell type-specific because many other cells (e.g., astrocytes, NG2-cells, oligodendrocyte progenitor cells, motor neurons in the brain stem) were also labeled. Recently, another genetically encoded Ca^{2+} sensor (GCaMP2) was effectively expressed in microglia using a retroviral transduction system [33]. One drawback of this approach is that only proliferating microglial cells are labeled. To trigger proliferation of microglia a stab wound has to be induced before injection of a retrovirus. Although labeled microglia returned to a morphologically ramified state 42 days after the injury, it remains unclear if the labeled cells were altered or "primed" by the initial activation process.

Here we discuss the techniques recently developed in our laboratory for in vivo studies of "resting" and activated microglia. We describe how to visualize individual microglial cells in any mouse strain and at any developmental stage and how to study intracellular Ca^{2+} signaling in these cells.

2 In Vivo Labeling of Microglia

2.1 Visualization of Microglia by Means of Plant Lectins

Plant lectins are highly specific tools recognizing carbohydrates on cell surfaces. On the basis of their specificity, lectins can be classified into five groups according to the monosaccharide for which they have the highest affinity (e.g., mannose, galactose/*N*-acetylgalactosamine, *N*-acetylglucosamine, fucose, and *N*-acetylneuraminic acid). Of the various monosaccharides that can be found in nature, only those mentioned above are typical constituents of surfaces of eukaryotic cells [41].

Isolectin IB_4 is a 114 kDa glycoprotein isolated from the seeds of the tropical African legume *Griffonia simplicifolia* [42] which binds specifically to α-D-galactose [43], whereas the 100 kDa glycopolypeptide Tomato lectin is isolated from *Lycopersicon esculentum* and binds to poly-*N*-acetylglucosamine [44]. Both lectins are effective markers of blood vessels and microglial cells in the CNS and have extensively been used for in vitro studies. For instance, Isolectin IB_4 has been used as a microglial marker in rat culture [45, 46] or fixed brain slices [47, 48]. Tomato lectin has been used for identification of microglial cells, for example, prior to electrophysiological recordings, in acute human hippocampal slices [49], acute rat brain slices [45], acute mouse cortical slices [33], in fixed murine hippocampal slices [50], and in cell culture [46]. Furthermore, both lectins have been shown to label macrophages in the lung, [51] as well as in and around brain tumors in the CNS [52].

Commercially available lectins conjugated with fluorophores allow fast, one-step detection and visualization of microglia. Therefore, we established a protocol enabling the use of these fluorescent markers in vivo. The stock solution (500 μg/mL) of Isolectin IB_4 conjugated to Alexa 594 (from Invitrogen, Eugene, OR, USA) was prepared by dissolving the dye in the standard pipette solution of the following composition (in mM): 150 NaCl, 2.5 KCl, and 10 HEPES, pH 7.4, supplemented with 0.1 mM $CaCl_2$. Stock solutions of Tomato lectin conjugated to DyLight 594 (1 mg/mL) or fluorescein (2 mg/mL) were purchased directly from Vector Laboratories (Burlingame, CA, USA). On the day of experiment, lectins were further diluted in the standard pipette solution to obtain the desired concentration of 75 μg/mL for Isolectin IB_4 and 25 μg/mL for Tomato lectin, respectively. This solution was filtered using a Millipore Ultrafree®-MC Centrifugal Filter with a pore diameter of 0.45 μm (Millipore, Billerica, MA, USA) and kept in a dark place until use, to prevent photobleaching. For in vivo staining of microglia, a glass micropipette (tip diameter 1 μm) was filled with the lectin solution and inserted into the cortex via a craniotomy with a help of a micromanipulator (see [32, 53] for detailed description of mice surgery). In the case of Tomato lectin, the dye was pressure-injected (at 60 kPa) into the cortex for 30 s. In the case of Isolectin IB_4 a

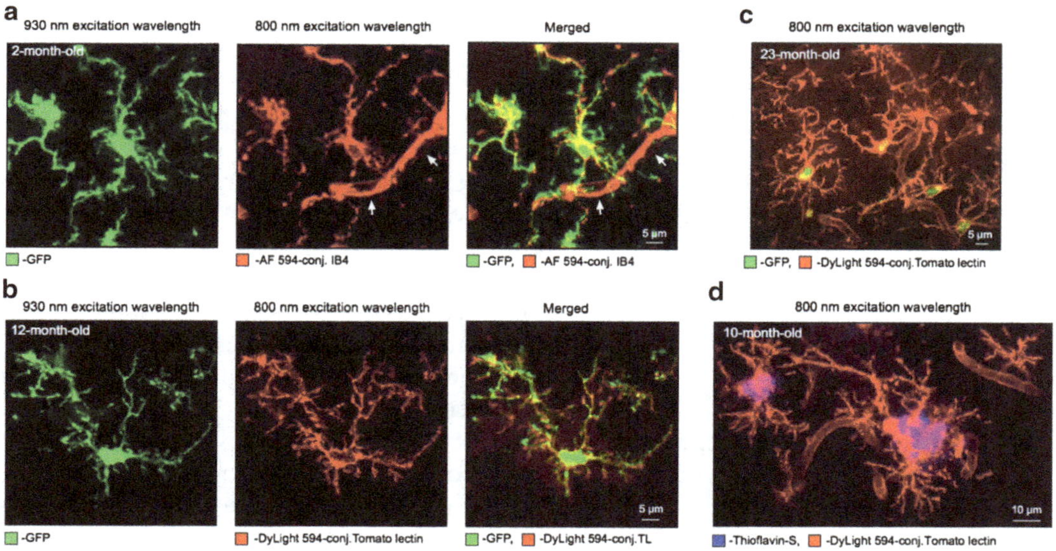

Fig. 1 Visualization of microglia by means of plant lectins. (**a**) Maximum intensity projection (MIP) two-photon excited fluorescence images (±10 μm, step 1 μm) of microglial cells in the cortex of a juvenile $CX_3CR1^{GFP/+}$ mouse labeled with Isolectin IB_4 conjugated to Alexa 594. *Left*: GFP fluorescence imaged at an excitation wavelength of 930 nm. *Middle*: fluorescence of Isolectin IB_4 conjugated to Alexa 594 imaged at an excitation wavelength of 800 nm. *Right*: An overlay of the two images. Besides microglia endothelial cells lining blood vessels were stained (*arrows*). (**b**) MIP images of a microglial cell in the cortex of an adult $CX_3CR1^{GFP/+}$ mouse labeled with Tomato lectin conjugated with DyLight 594. *Left*: GFP fluorescence; *middle*: fluorescence of Tomato lectin conjugated with DyLight 594; *right*: an overlay of the two images. (**c**) MIP image of microglial cells in the cortex of an aged $CX_3CR1^{GFP/+}$ mouse labeled with Tomato lectin conjugated with DyLight 594. Because GFP is poorly excited at 800 nm, only cell bodies are visible in the *green* channel. (**d**) MIP image of microglial cells surrounding amyloid plaques in a mouse model of Alzheimer's disease (APP23PS45$CX_3CR1^{GFP/+}$ mouse). Only *blue* and *red* channels are merged. Thioflavin-S is a *blue* dye which binds to amyloid β and thus specifically labels senile plaques. Panel (**a**) is reproduced from [25] with permission from Elsevier; panels (**b**)–(**d**) are reproduced from [54].

longer injection time (90 s) was necessary. Thereafter, the experiment was interrupted for 30–50 min to allow for equilibration of the dye and for the washout of surplus dye solution. Especially when labeling aged brain tissue (>15-month-old mice) longer washout periods were required to obtain the best possible image quality. So far we have tested the above protocol with DyLight 594- and fluorescein-conjugated Tomato lectin as well as Alexa 594-conjugated Isolectin IB_4. We assume, however, that similar protocols can be used for labeling microglial cells with lectins conjugated to other fluorophores (e.g., DyLight 594- or fluorescein-conjugated Isolectin IB_4 or Texas Red-conjugated Tomato lectin).

Individual lectin-labeled cells were visualized in vivo by means of two-photon microscopy. In addition to microglial cells also endothelial cells lining the interior surface of blood vessels were labeled by both lectins. These structures, however, were easily distinguished from microglia due to their long and tubular appearance (see arrows in Fig. 1a). Single injection labeled a spherical tissue volume with a diameter of about 100–150 μm. Several equidistant injections were used to increase the size of the labeled area.

2.1.1 Isolectin IB_4 Conjugated to Alexa 594

Figure 1a (middle panel) illustrates the image quality obtained when labeling microglia with Isolectin IB_4 (conjugated to Alexa 594). Isolectin IB_4 allows visualization of microglial somata in juvenile, adult, and aged wild-type (WT) mice. However, labeling of cell processes is discontinuous (compare GFP labeling and Isolectin IB_4 staining of the very same cells in Fig. 1a). This patchy, discontinuous labeling of cell processes precluded studies on their motility and dynamics. Labeling remained stable for at least 5 h (B.B. and O.G., unpublished data). Surprisingly, Isolectin IB_4 cannot be used for visualization of microglial cells in mouse models of Alzheimer's disease (Fig. S2 in [54]). In these mice injection of Isolectin IB_4 resulted in more or less homogeneous staining of the entire brain parenchyma, making identification of an individual microglial cell impossible. These data suggest that the surface expression of glycoconjugates bearing terminal α-D-galactose, to which Isolectin IB_4 binds, is upregulated in Alzheimer's disease in a cell type-nonspecific manner.

2.1.2 Tomato Lectin Conjugated to DyLight 594 or Fluorescein

Injection of Tomato lectin conjugated to DyLight 594 resulted in a high-quality labeling of microglial cells (Fig. 1b). Of note, it also reliably labeled microglial processes (Fig. 1b and Fig. 1 in [54]). Similar image quality was obtained when using Tomato lectin conjugated to fluorescein (see Fig. 1 in [54]). To test the reliability of Tomato lectin staining we counted all labeled microglia in GFP-expressing $CX_3CR1^{GFP/+}$ mice. Within stained cortical areas, 100 % of GFP-positive cells were also positive for Tomato lectin [54].

Similar to Isolectin IB_4, Tomato lectin also allowed labeling of microglial cells in animals of any experimental age, from juvenile to adult and aged (Fig. 1b, c). In contrast to Isolectin IB_4, however, Tomato lectin enabled labeling of microglial cells in mouse models of Alzheimer's disease. In these mice, Tomato lectin labeled microglial cells surrounding senile plaques (one of the major histological markers of the disease) as well as microglia in plaque-free areas (Fig. 1d). Once labeled with Tomato lectin, microglial cells remain visible for the whole experimental day (~6 h) thus enabling in vivo investigation of microglial behavior during prolonged imaging sessions. Of note, Tomato lectin labeling is preserved after fixation of the tissue with formaldehyde [54].

The use of Tomato lectin can easily be combined with the use of other cell type-specific markers, like sulforhodamine 101, or ion-sensitive dyes (e.g., Oregon Green BAPTA-1 (OGB-1) or Fluo-4) enabling simultaneous identification and functional analyses of astrocytes, neurons, and microglia in one preparation. Conveniently, green- (conjugated to fluorescein) or red fluorescent (conjugated to DyLight 594 or Texas Red) Tomato lectin is

Table 1
Labeling approaches for in vivo imaging of microglia.

In vivo labeling technique	Labeling of cell body	Labeling of microglial processes	Overall image quality	Chronic imaging	Applicable to WT mice	Applicable to AD mice	Versatility of the approach
Chronic expression of GFP	Yes	Yes	Excellent	Feasible	Yes	Yes	Requires breeding of experimental animals
Acute application of Tomato lectin	Yes	Yes	Very good	Requires re-labeling	Yes	Yes	Applicable to any mouse strain at any experimental age
Acute application of Isolectin IB_4	Yes	Poor	Medium	Requires re-labeling	Yes	No	Works in juvenile, adult, and aged WT but not AD mice

available, thus enabling multicolor imaging in combination with different fluorophores.

However, there is accumulating evidence that fluorescent markers introduced into the living tissue to label a specific cell type may significantly modify the network under study. The astrocytic marker sulforhodamine 101 [55], for example, can induce long-lasting changes in synaptic transmission and neuronal network activity [56–58]. Therefore, we tested whether the use of Tomato lectin according to our protocol influences network activity of surrounding neurons. To do so we monitored frequencies and amplitudes of spontaneous Ca^{2+} transients before and 20 min after staining of microglia with Tomato lectin. We did not observe any Tomato lectin-induced changes in these parameters (Fig. S5 in [54]).

In conclusion, both Isolectin IB_4 and Tomato lectin are suitable for visualization of microglial cell bodies in juvenile, adult, and aged WT mice. The processes of these cells, however, are better visualized using Tomato lectin because Isolectin IB_4 results in a patchy, discontinuous process staining (see Table 1). Furthermore, as Isolectin IB_4 strongly and unspecifically labels the brain parenchyma in mouse models of Alzheimer's disease, Tomato lectin should be used to stain microglial cells in these mice.

2.2 In Vivo Labeling of Microglia with Ca^{2+} Indicator Dyes

Ca^{2+} signaling of neurons and astrocytes can easily be studied in vivo using simple loading techniques like MCBL, which provides the basis for monitoring activity of tens to hundreds of cells at the same time [32]. Unfortunately, microglia do not take up small molecule Ca^{2+} indicators like OGB-1 or Fluo-4 when these molecules are delivered only for a short time (seconds-to-minutes). This characteristic has so far precluded investigations of microglial Ca^{2+} signaling in vivo. Another possibility for loading cells with Ca^{2+} indicators is the single-cell electroporation technique. The application of an electric field results in the transient formation of small pores in the cell membrane. Simultaneously, charged molecules are moved within the field and thus are loaded into the cells. In vivo this approach was first used for loading neurons with Ca^{2+} indicators like OGB-1 [59]. We modified this protocol for labeling of microglia [25].

For loading, GFP- or lectin-labeled cells are approached by a glass micropipette containing 10 mM OGB-1 (Molecular Probes, Eugene, OR, USA) using a micromanipulator. OGB-1 is dissolved in the intracellular solution containing (in mM): 140 K-gluconate, 14 KCl, 4 NaCl, and 10 HEPES, pH 7.3. To prevent damage of the cell membrane, size of the pipette tip should not exceed 1 μm (pipette resistance: 5–7 MΩ, when filled with the pipette solution mentioned above). As soon as the pipette touches the cell membrane, a negative current (−600 nA) is applied once for 10 ms to move negatively charged Ca^{2+} indicator molecules into a cell. After staining, the pipette is immediately withdrawn. The electroporation procedure should be followed by a 10–15 min recovery period to allow equilibration of the dye and filling of microglial processes (Fig. 2a).

Several lines of evidence suggest that the majority of microglial cells are not damaged during the electroporation procedure. First, electroporated cells do not change their morphology for at least 2 h after electroporation, as it would be expected if the cells were activated or damaged. Second, labeled microglial cells are able to move their processes towards the tip of an ATP-containing pipette [25]. Third, in contrast to on-purpose-damaged cells [25] and tissues [11], the electroporated cells do not attract their intact neighbors. If occasionally damaged, microglial cell somata turn extremely bright and processes become fragmented. Such cells can be clearly distinguished from non-damaged counterparts to be excluded from further studies. In our hands, only a small fraction (~10 %) of cells is damaged in the course of pipette positioning/electroporation procedure [25].

3 Application of the Method to the Mammalian Cortex

3.1 Monitoring Microglial Function in Lectin-Labeled Cells

Microglial cells respond to changes in their microenvironment by (1) undergoing morphological changes and (2) increasing the intracellular concentration of second messengers (for example, Ca^{2+}). Monitoring of morphological changes is possible in cells labeled

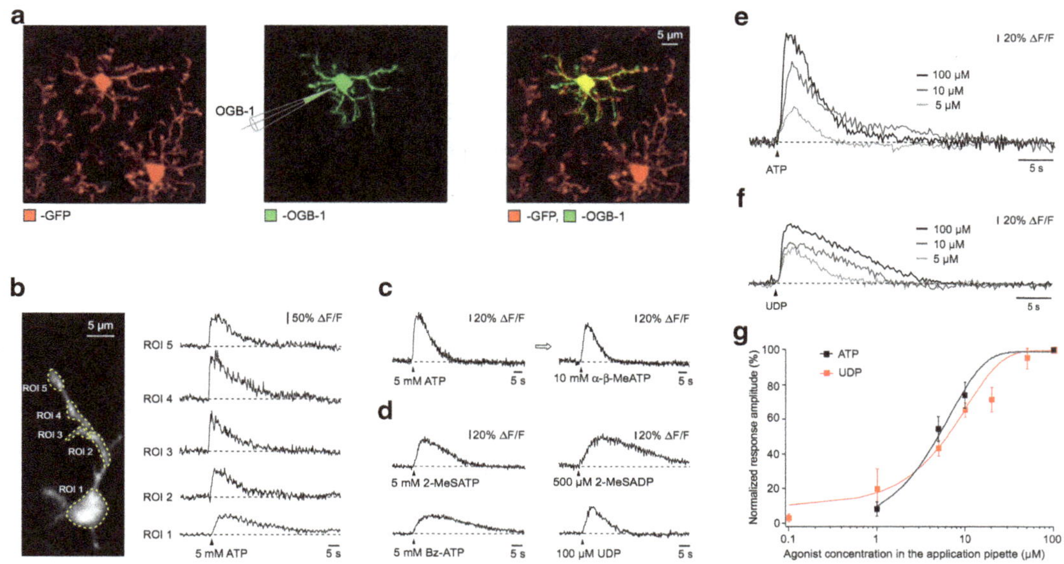

Fig. 2 In vivo labeling of microglia with calcium indicator dyes. (**a**) Single-cell electroporation of microglia. *Left*: MIP two-photon excited fluorescence image of GFP-positive cells in the cortex of a $CX_3CR1^{GFP/+}$ mouse. *Middle*: the upper microglial cell was stained with OGB-1 by means of single-cell electroporation (electroporation pulse: 10 ms, −600 nA; 10 mM OGB-1 in the micropipette). Schematic drawing indicates the position of the micropipette used for the loading procedure. *Right*: an overlay of the two images. (**b**) *Right*: Ca^{2+} transients evoked by a pressure application of 5 mM ATP (20 kPa, 50 ms) recorded from the soma and processes of a microglial cell. The corresponding regions of interest (ROIs) are marked with respective numbers in the *left* panel. (**c**, **d**) Ca^{2+} signals evoked by an application of the respective purinergic agonist. (**c**) Here and below each cell was first challenged with a brief application of ATP (*left*; 5 mM, 50 ms, 20 kPa) followed by the drug tested (*right*; α-β-MeATP, 10 mM). (**d**) Representative traces of Ca^{2+} signals evoked by 2-MeSATP (*top left*; 5 mM), 2-MeSADP (*top right*; 500 μM), Bz-ATP (*bottom left*; 5 mM), and UDP (*bottom right*; 100 μM). (**e**, **f**) Representative examples of Ca^{2+} transients evoked by applications of different concentrations of ATP and UDP, respectively. (**g**) Dose–response curves for ATP (*black*, $n = 11$ cells) and UDP (*red*, $n = 11$ cells). The response amplitude is normalized to the maximum change in fluorescence caused by an application of 100 μM ATP or UDP, respectively. *Solid lines* represent fits to respective sets of data using the Hill equation. Reproduced from [25] with permission from Elsevier.

with lectins alone. For the reasons explained above (see 2.1.2) we would suggest to use Tomato lectin for such studies. For combined monitoring of morphology and Ca^{2+} signals electroporation with a Ca^{2+} indicator dye has to be performed in addition (see 2.2).

In our lab we used Tomato lectin to study directed motility of microglial processes towards the pipette filled with 5 mM ATP (Fig. 2d in [54]). As was shown by Davalos et al., such experimental settings faithfully mimic local damage of brain tissue [11]. Because the entire microglial cells were labeled, it was possible to visualize an initial thickening of the tips of the processes in response to the pipette insertion, formation of a spherical containment, and continuous decrease of its diameter due to the movement of the contributing microglial processes. So far in vivo such image quality could only be achieved when using GFP-expressing microglial cells (see Supplementary Video 5 in [11]). Neither the velocity of process movement nor the final diameter of the spherical containment was altered in Tomato lectin-labeled cells compared to their GFP-labeled counterparts (Fig. 2e–g in [54]).

To test whether microglial Ca^{2+} signaling is altered due to binding of Tomato lectin to microglial membranes, cells were loaded with OGB-1 by means of single-cell electroporation (as described in 2.2) and Ca^{2+} transients evoked by purinergic agonists ATP, UDP, and Bz-ATP (all obtained from Sigma-Aldrich, St. Louis, MO, USA) were compared to those evoked in GFP-expressing cells. To estimate the amount of agonist pressure-applied to a given cell, we co-injected a cell-impermeant red fluorescent dye Alexa 594 and measured the area under the transient in the Alexa 594-channel. This value was then used to normalize agonist-evoked Ca^{2+} signals (area under the respective agonist-evoked Ca^{2+} transient). There was no significant difference between the normalized areas under the Ca^{2+} transients recorded from GFP-expressing and Tomato lectin-labeled microglial cells (Fig. 2j in [54]). We concluded, therefore, that functional properties of microglial cells (e.g., process motility and Ca^{2+} signaling) are not altered when labeling these cells with Tomato lectin.

3.2 In Vivo Ca^{2+} Imaging of Microglial Cells

Using single-cell electroporation procedure for loading microglial cells with the Ca^{2+} indicator OGB-1 enabled for the first time analyses of spontaneous as well as agonist-evoked microglial Ca^{2+} signals in vivo.

Cells in the mouse cortex were identified based on the expression of GFP (in $CX_3CR1^{GFP/+}$ mice) or lectin binding (WT mice, Fig. 2a and [25]). To check for cell viability after the electroporation procedure, cells were challenged with 5 mM ATP, pressure-applied from a glass micropipette into the extracellular space 30–40 μm away from the electroporated cell. Application of ATP evoked large Ca^{2+} transients invading the soma as well as microglial processes (Fig. 2b).

Studies using cell cultures and tissue slices have shown that in vitro microglia express a wide range of ionotropic and metabotropic receptors for ATP, glutamate, acetylcholine, etc. as well as voltage-gated Ca^{2+} channels [29, 60–63]. In vivo, however, the pharmacological profile of microglial cells was much narrower (see Fig. 4c in [25]). The cells readily responded with Ca^{2+} transients to ATP as well as to other agonists of ionotropic (P2X) and metabotropic (P2Y) ATP receptors (Fig. 2c, d), but not to other agonists reported to cause microglial Ca^{2+} transients in vitro. In vivo we have tested the following ATP agonists: α-β-MeATP for $P2X_{1,3,5,6}$, 2-MeSATP for $P2X_{1,2,4,5,7}$ and $P2Y_1$, 2-MeSADP for $P2Y_{1,12,13}$, Bz-ATP for $P2X_7$, and UDP for $P2Y_6$. All of them readily evoked Ca^{2+} transients in microglia (Fig. 2c, d; Fig. 4c in [25]). It is worth to note that in vivo microglial cells were able to respond to very low concentrations of UDP and ATP (1–5 μM in the application pipette; Fig. 2e–g). Taking into account the dilution of substance that occurs during its diffusion in the extracellular space, the effective concentration reaching the microglial cell is estimated to be in the nanomolar range. Such high sensitivity is unique to intact cells in vivo, because it has never been observed in cultured microglial cells. In contrast, cultured cells do not respond to exogenous ATP applied at a concentration lower than 10 μM [64].

In addition to evoked Ca^{2+} signals, we also studied spontaneously occurring microglial Ca^{2+} transients. Interestingly, when compared to other cell types like neurons or astrocytes, ramified microglia in the juvenile mouse cortex were rather silent under resting conditions. Only 20 % of microglial cells (6/27) showed a spontaneous Ca^{2+} transient during 15-min-long recording period. Even when neuronal activity was increased in the presence of bicuculline, the spontaneous activity of microglia remained low [25].

Further, we tested whether microglia are able to sense and to react to neuronal damage. To this end, a patch-like pipette was inserted into the cortex and located on a surface of a single neuron. By moving the pipette 10 μm forward in one step, neuronal membrane was ruptured but the pipette remained within the damaged cell (see Fig. 3 in [25]). Remarkably, this procedure reliably evoked large Ca^{2+} transients in microglia located in the immediate vicinity of the damaged neuron. We termed these Ca^{2+} signals damage-induced Ca^{2+} transients (DICTs; [25]). Subsequent pharmacological experiments revealed an involvement of Ca^{2+} release from the intracellular Ca^{2+} stores in the generation of DICTs. Taken together, these data suggest that in vivo Ca^{2+} transients in microglial cells do not occur spontaneously. Rather, they reliably signal a minor damage in the microglial microenvironment.

4 Discussion and Outlook

The generation of transgenic mice harboring genetically labeled microglia facilitated numerous morphological studies characterizing microglial behavior in the healthy and diseased brain [10–12, 65, 66]. Our laboratory has contributed a complementary lectin-based technique for in vivo labeling of microglia. The latter is applicable for fast in vivo labeling of microglial cells in any mouse strain and at any developmental stage from juvenile to aged. In addition, it is also applicable for labeling microglial cells in mouse models of neurodegenerative diseases. Microglia become chronically activated during aging and neurodegeneration, being potentially toxic to their microenvironment [9, 17–19, 67, 68]. Therefore, visualization of these cells in the aging and diseased brain and understanding of their in vivo function represent an important direction of the future basic and clinical research.

The use of plant lectins (especially Tomato lectin) coupled to bright fluorophores allows high-quality labeling of microglial cells, comparable to that in transgenic GFP-expressing mice. Importantly, the established labeling protocol changes neither network activity of surrounding neurons nor the functional characteristics of labeled microglial cells. Therefore, we consider this technique to be a valuable tool for in vivo studies of microglial function.

By now, two alternative techniques are available for in vivo loading of microglial cells with Ca^{2+} indicators. Either the cells can be loaded by single-cell electroporation [25] or a genetically encoded Ca^{2+} sensor can be expressed in microglial cells constitutively [40] or with the help of a viral vector [33]. So far, however, only retroviral vectors have been used [33]. Because retroviral vectors only infect dividing cells, cell proliferation was required for sensor expression. Because only activated microglial cells proliferate (this process is called reactive microgliosis [69]), a stab wound was required to trigger proliferation. To label microglial cells, the virus was then injected 2 days post-injury. Under these conditions one also labels other macrophages attracted to the site of injury in addition to microglia [33]. Of note, no cell labeling was observed when injecting the virus at the day of injury. Thus, the retroviral vector-based technique in principle allows the expression of genetically encoded Ca^{2+} sensors in microglia, but has serious drawbacks: (1) "resting" microglia cannot be labeled, (2) even at the site of injury this protocol only labels proliferating cells, (3) even when imaged long after the injury the properties of sensor-expressing cells may still be modified [9, 33], and (4) this protocol does not provide the desired cell type specificity because cells other than microglia are also labeled [33]. Although it is foreseeable that other viral vectors may be better suited for expressing genetically encoded Ca^{2+} sensors in microglia, the demonstration of their effectiveness still belongs to the future. The same applies to genetically encoded Ca^{2+} sensors expressed constitutively under a microglia-specific promoter.

At present we are using single-cell electroporation technique for in vivo Ca^{2+} imaging of microglia. Although electroporation procedure per se also might cause cell damage, using the mild protocol described above this does not seem to be the main issue [25]. In addition, damaged cells can be clearly distinguished from their intact counterparts based on their morphology and diminished responsiveness to low concentrations of ATP or UDP. Furthermore, damaged cells quickly attract the processes of their intact neighbors. Therefore, the main drawback of single-cell electroporation procedure is that it is labor-intensive. Fortunately once labeled, the cells can be studied for several hours.

In conclusion, the emerging toolkit enables for the first time high resolution in vivo imaging of microglial structure and function in a deliberate mouse strain and at any developmental stage. Similar techniques may also be applicable to other experimental animals. This will accelerate analyses of the role of "resting" microglia for brain's development, plasticity, and other vital functions. Moreover, it will enable analyses of its function during aging and neurodegenerative diseases as well as during acute brain injury.

References

1. Färber K, Kettenmann H (2006) Functional role of calcium signals for microglial function. Glia 54:656–665
2. Hoffmann A, Kann O, Ohlemeyer C, Hanisch UK, Kettenmann H (2003) Elevation of basal intracellular calcium as a central element in the activation of brain macrophages (microglia): suppression of receptor-evoked calcium signaling and control of release function. J Neurosci 23:4410–4419
3. Light AR, Wu Y, Hughen RW, Guthrie PB (2006) Purinergic receptors activating rapid intracellular Ca increases in microglia. Neuron Glia Biol 2:125–138
4. Prinz M, Priller J, Sisodia SS, Ransohoff RM (2011) Heterogeneity of CNS myeloid cells and their roles in neurodegeneration. Nat Neurosci 14:1227–1235
5. Soulet D, Rivest S (2008) Microglia. Curr Biol 18:R506–R508
6. Herculano-Houzel S (2009) The human brain in numbers: a linearly scaled-up primate brain. Front Hum Neurosci 3:31
7. Block ML, Zecca L, Hong JS (2007) Microglia-mediated neurotoxicity: uncovering the molecular mechanisms. Nat Rev Neurosci 8:57–69
8. Kettenmann H, Hanisch UK, Noda M, Verkhratsky A (2011) Physiology of microglia. Physiol Rev 91:461–553
9. Hanisch UK, Kettenmann H (2007) Microglia: active sensor and versatile effector cells in the normal and pathologic brain. Nat Neurosci 10:1387–1394
10. Nimmerjahn A, Kirchhoff F, Helmchen F (2005) Resting microglial cells are highly dynamic surveillants of brain parenchyma in vivo. Science 308:1314–1318
11. Davalos D, Grutzendler J, Yang G, Kim JV, Zuo Y, Jung S, Littman DR, Dustin ML, Gan WB (2005) ATP mediates rapid microglial response to local brain injury in vivo. Nat Neurosci 8:752–758
12. Wake H, Moorhouse AJ, Jinno S, Kohsaka S, Nabekura J (2009) Resting microglia directly monitor the functional state of synapses in vivo and determine the fate of ischemic terminals. J Neurosci 29:3974–3980
13. Tremblay ME, Lowery RL, Majewska AK (2010) Microglial interactions with synapses are modulated by visual experience. PLoS Biol 8:e1000527
14. Paolicelli RC, Bolasco G, Pagani F, Maggi L, Scianni M, Panzanelli P, Giustetto M, Ferreira TA, Guiducci E, Dumas L, Ragozzino D, Gross CT (2011) Synaptic pruning by microglia is necessary for normal brain development. Science 333:1456–1458
15. Tremblay ME, Stevens B, Sierra A, Wake H, Bessis A, Nimmerjahn A (2011) The role of microglia in the healthy brain. J Neurosci 31:16064–16069
16. McLarnon JG (2005) Purinergic mediated changes in Ca^{2+} mobilization and functional responses in microglia: effects of low levels of ATP. J Neurosci Res 81:349–356
17. Ransohoff RM, Perry VH (2009) Microglial physiology: unique stimuli, specialized responses. Annu Rev Immunol 27:119–145
18. Glass CK, Saijo K, Winner B, Marchetto MC, Gage FH (2010) Mechanisms underlying inflammation in neurodegeneration. Cell 140:918–934
19. Kreutzberg GW (1996) Microglia: a sensor for pathological events in the CNS. Trends Neurosci 19:312–318
20. Neumann H, Kotter MR, Franklin RJ (2009) Debris clearance by microglia: an essential link between degeneration and regeneration. Brain 132:288–295
21. Jung S, Aliberti J, Graemmel P, Sunshine MJ, Kreutzberg GW, Sher A, Littman DR (2000) Analysis of fractalkine receptor CX_3CR1 function by targeted deletion and green fluorescent protein reporter gene insertion. Mol Cell Biol 20:4106–4114
22. Hirasawa T, Ohsawa K, Imai Y, Ondo Y, Akazawa C, Uchino S, Kohsaka S (2005) Visualization of microglia in living tissues using

Iba1-EGFP transgenic mice. J Neurosci Res 81:357–362

23. Sasmono RT, Oceandy D, Pollard JW, Tong W, Pavli P, Wainwright BJ, Ostrowski MC, Himes SR, Hume DA (2003) A macrophage colony-stimulating factor receptor-green fluorescent protein transgene is expressed throughout the mononuclear phagocyte system of the mouse. Blood 101:1155–1163
24. Harrison JK, Jiang Y, Chen S, Xia Y, Maciejewski D, McNamara RK, Streit WJ, Salafranca MN, Adhikari S, Thompson DA, Botti P, Bacon KB, Feng L (1998) Role for neuronally derived fractalkine in mediating interactions between neurons and CX3CR1-expressing microglia. Proc Natl Acad Sci USA 95:10896–10901
25. Eichhoff G, Brawek B, Garaschuk O (2011) Microglial calcium signal acts as a rapid sensor of single neuron damage in vivo. Biochim Biophys Acta 1813:1014–1024
26. Rogers JT, Morganti JM, Bachstetter AD, Hudson CE, Peters MM, Grimmig BA, Weeber EJ, Bickford PC, Gemma C (2011) CX3CR1 deficiency leads to impairment of hippocampal cognitive function and synaptic plasticity. J Neurosci 31:16241–16250
27. Lee S, Varvel NH, Konerth ME, Xu G, Cardona AE, Ransohoff RM, Lamb BT (2010) CX3CR1 deficiency alters microglial activation and reduces beta-amyloid deposition in two Alzheimer's disease mouse models. Am J Pathol 177:2549–2562
28. Möller T (2002) Calcium signaling in microglial cells. Glia 40:184–194
29. Pocock JM, Kettenmann H (2007) Neurotransmitter receptors on microglia. Trends Neurosci 30:527–535
30. Inoue K (2002) Microglial activation by purines and pyrimidines. Glia 40:156–163
31. Re DB, Przedborski S (2006) Fractalkine: moving from chemotaxis to neuroprotection. Nat Neurosci 9:859–861
32. Stosiek C, Garaschuk O, Holthoff K, Konnerth A (2003) In vivo two-photon calcium imaging of neuronal networks. Proc Natl Acad Sci USA 100:7319–7324
33. Seifert S, Pannell M, Uckert W, Farber K, Kettenmann H (2011) Transmitter- and hormone-activated Ca^{2+} responses in adult microglia/brain macrophages in situ recorded after viral transduction of a recombinant Ca^{2+} sensor. Cell Calcium 49:365–375
34. Horikawa K, Yamada Y, Matsuda T, Kobayashi K, Hashimoto M, Matsu-ura T, Miyawaki A, Michikawa T, Mikoshiba K, Nagai T (2010) Spontaneous network activity visualized by ultrasensitive Ca^{2+} indicators, yellow Cameleon-Nano. Nat Methods 7:729–732
35. Mank M, Santos AF, Direnberger S, Mrsic-Flogel TD, Hofer SB, Stein V, Hendel T, Reiff DF, Levelt C, Borst A, Bonhoeffer T, Hubener M, Griesbeck O (2008) A genetically encoded calcium indicator for chronic in vivo two-photon imaging. Nat Methods 5:805–811
36. Tallini YN, Ohkura M, Choi BR, Ji G, Imoto K, Doran R, Lee J, Plan P, Wilson J, Xin HB, Sanbe A, Gulick J, Mathai J, Robbins J, Salama G, Nakai J, Kotlikoff MI (2006) Imaging cellular signals in the heart in vivo: cardiac expression of the high-signal Ca^{2+} indicator GCaMP2. Proc Natl Acad Sci USA 103:4753–4758
37. Nagai T, Yamada S, Tominaga T, Ichikawa M, Miyawaki A (2004) Expanded dynamic range of fluorescent indicators for Ca^{2+} by circularly permuted yellow fluorescent proteins. Proc Natl Acad Sci USA 101:10554–10559
38. Tian L, Hires SA, Mao T, Huber D, Chiappe ME, Chalasani SH, Petreanu L, Akerboom J, McKinney SA, Schreiter ER, Bargmann CI, Jayaraman V, Svoboda K, Looger LL (2009) Imaging neural activity in worms, flies and mice with improved GCaMP calcium indicators. Nat Methods 6:875–881
39. Zhao Y, Araki S, Wu J, Teramoto T, Chang YF, Nakano M, Abdelfattah AS, Fujiwara M, Ishihara T, Nagai T, Campbell RE (2011) An expanded palette of genetically encoded Ca^{2+} indicators. Science 333:1888–1891
40. Atkin SD, Patel S, Kocharyan A, Holtzclaw LA, Weerth SH, Schram V, Pickel J, Russell JT (2009) Transgenic mice expressing a cameleon fluorescent Ca^{2+} indicator in astrocytes and Schwann cells allow study of glial cell Ca^{2+} signals in situ and in vivo. J Neurosci Methods 181:212–226
41. Lis H, Sharon N (1998) Lectins: carbohydrate-specific proteins that mediate cellular recognition. Chem Rev 98:637–674
42. Murphy LA, Goldstein IJ (1977) Five alpha-D-galactopyranosyl-binding isolectins from Bandeiraea simplicifolia seeds. J Biol Chem 252:4739–4742
43. Streit WJ, Schulte BA, Balentine DJ, Spicer SS (1985) Histochemical localization of galactose-containing glycoconjugates in sensory neurons and their processes in the central and peripheral nervous system of the rat. J Histochem Cytochem 33:1042–1052
44. Slifkin M, Doyle RJ (1990) Lectins and their application to clinical microbiology. Clin Microbiol Rev 3:197–218
45. Boucsein C, Kettenmann H, Nolte C (2000) Electrophysiological properties of microglial

cells in normal and pathologic rat brain slices. Eur J Neurosci 12:2049–2058
46. Newell EW, Stanley EF, Schlichter LC (2007) Reversed Na^+/Ca^{2+} exchange contributes to Ca^{2+} influx and respiratory burst in microglia. Channels (Austin) 1:366–376
47. Streit WJ, Kreutzberg GW (1987) Lectin binding by resting and reactive microglia. J Neurocytol 16:249–260
48. Streit WJ (1990) An improved staining method for rat microglial cells using the lectin from Griffonia simplicifolia (GSA I-B4). J Histochem Cytochem 38:1683–1686
49. Bordey A, Spencer DD (2003) Chemokine modulation of high-conductance Ca^{2+}-sensitive K^+ currents in microglia from human hippocampi. Eur J Neurosci 18:2893–2898
50. Rappert A, Bechmann I, Pivneva T, Mahlo J, Biber K, Nolte C, Kovac AD, Gerard C, Boddeke HW, Nitsch R, Kettenmann H (2004) CXCR3-dependent microglial recruitment is essential for dendrite loss after brain lesion. J Neurosci 24:8500–8509
51. Bankston PW, Porter GA, Milici AJ, Palade GE (1991) Differential and specific labeling of epithelial and vascular endothelial cells of the rat lung by Lycopersicon esculentum and Griffonia simplicifolia I lectins. Eur J Cell Biol 54:187–195
52. Moffett JR, Els T, Espey MG, Walter SA, Streit WJ, Namboodiri MA (1997) Quinolinate immunoreactivity in experimental rat brain tumors is present in macrophages but not in astrocytes. Exp Neurol 144:287–301
53. Garaschuk O, Milos RI, Grienberger C, Marandi N, Adelsberger H, Konnerth A (2006) Optical monitoring of brain function in vivo: from neurons to networks. Pflugers Arch 453:385–396
54. Schwendele B, Brawek B, Hermes M, Garaschuk O (2012) High resolution in vivo imaging of microglia using a versatile non genetically-encoded marker. Eur J Immunol 42:2193–2196
55. Nimmerjahn A, Kirchhoff F, Kerr JN, Helmchen F (2004) Sulforhodamine 101 as a specific marker of astroglia in the neocortex in vivo. Nat Methods 1:31–37
56. Kang J, Kang N, Yu Y, Zhang J, Petersen N, Tian GF, Nedergaard M (2010) Sulforhodamine 101 induces long-term potentiation of intrinsic excitability and synaptic efficacy in hippocampal CA1 pyramidal neurons. Neuroscience 169:1601–1609
57. Garaschuk O (2013) Imaging microcircuit function in healthy and diseased brain. Exp Neurol 242:41–49
58. Fink S, Kovalchuk Y, Homma R, Schwendele B, Direnberger S, Cohen LB, Griesbeck O, Garaschuk O (2012) In vivo functional imaging of the olfactory bulb at single cell resolution. In: Fellin T, Halassa M (eds) Neuronal network analysis. Humana Press, New York, pp 21–43
59. Nevian T, Helmchen F (2007) Calcium indicator loading of neurons using single-cell electroporation. Pflugers Arch 454:675–688
60. Whittemore ER, Korotzer AR, Etebari A, Cotman CW (1993) Carbachol increases intracellular free calcium in cultured rat microglia. Brain Res 621:59–64
61. Zhang L, McLarnon JG, Goghari V, Lee YB, Kim SU, Krieger C (1998) Cholinergic agonists increase intracellular Ca^{2+} in cultured human microglia. Neurosci Lett 255:33–36
62. Biber K, Laurie DJ, Berthele A, Sommer B, Tolle TR, Gebicke-Harter PJ, van Calker D, Boddeke HW (1999) Expression and signaling of group I metabotropic glutamate receptors in astrocytes and microglia. J Neurochem 72:1671–1680
63. Colton CA, Jia M, Li MX, Gilbert DL (1994) K^+ modulation of microglial superoxide production: involvement of voltage-gated Ca^{2+} channels. Am J Physiol 266:C1650–C1655
64. Verderio C, Matteoli M (2001) ATP mediates calcium signaling between astrocytes and microglial cells: modulation by IFN-gamma. J Immunol 166:6383–6391
65. Meyer-Luehmann M, Spires-Jones TL, Prada C, Garcia-Alloza M, de Calignon A, Rozkalne A, Koenigsknecht-Talboo J, Holtzman DM, Bacskai BJ, Hyman BT (2008) Rapid appearance and local toxicity of amyloid-beta plaques in a mouse model of Alzheimer's disease. Nature 451:720–724
66. Hefendehl JK, Wegenast-Braun BM, Liebig C, Eicke D, Milford D, Calhoun ME, Kohsaka S, Eichner M, Jucker M (2011) Long-term in vivo imaging of beta-amyloid plaque appearance and growth in a mouse model of cerebral beta-amyloidosis. J Neurosci 31:624–629
67. Tambuyzer BR, Ponsaerts P, Nouwen EJ (2009) Microglia: gatekeepers of central nervous system immunology. J Leukoc Biol 85:352–370
68. Yong VW, Rivest S (2009) Taking advantage of the systemic immune system to cure brain diseases. Neuron 64:55–60
69. Streit WJ, Walter SA, Pennell NA (1999) Reactive microgliosis. Prog Neurobiol 57:563–581

Chapter 14

Functional Imaging of Cerebral Oxygenation with Intrinsic Optical Contrast and Phosphorescent Probes

Anna Devor, Sava Sakadžić, Mohammad A. Yaseen, Emmanuel Roussakis, Peifang Tian, Hamutal Slovin, Ivo Vanzetta, Ivan Teng, Payam A. Saisan, Louise E. Sinks, Anders M. Dale, Sergei A. Vinogradov, and David A. Boas

Abstract

Microscopic in vivo measurements of cerebral oxygenation are of key importance for understanding normal cerebral energy metabolism and its dysregulation in a wide range of clinical conditions. Relevant cerebral pathologies include compromised blood perfusion following stroke and a decrease in efficiency of single-cell respiratory processes that occurs in neurodegenerative diseases such as Alzheimer's and Parkinson's disease. In this chapter we review a number of quantitative optical approaches to measuring oxygenation of blood and cerebral tissue. These methods can be applied to map the hemodynamic response and study neurovascular and neurometabolic coupling, and can provide microscopic imaging of bio-markers in animal models of human disease, which would be useful for screening potential therapeutic approaches.

Key words O_2 sensing, Phosphorescence quenching, Intrinsic optical signals, Energy metabolism, In vivo imaging, Hemoglobin, Two-photon microscopy, CCD

1 Introduction

Cerebral oxygenation and O_2 metabolism are key parameters defining healthy brain function. Until recently, though, our knowledge of the microscopic dynamics of cerebral oxygenation has been limited by a lack of methods with the spatial and temporal resolution needed to follow oxygenation changes induced by varying neuronal activity. Most of what we know about the behavior of cerebral oxygenation during the hemodynamic response comes from measurements of light absorption by hemoglobin in the blood. This method, commonly referred to as "optical imaging of intrinsic signals (OIS)", is based on differential absorption spectra of oxy- and deoxy-hemoglobin (HbO and Hb) in the visible range and does not require

Bruno Weber and Fritjof Helmchen (eds.), *Optical Imaging of Neocortical Dynamics*, Neuromethods, vol. 85, DOI 10.1007/978-1-62703-785-3_14, © Springer Science+Business Media New York 2014

loading of extrinsic O_2-sensitive probes. Following the original demonstration [1], the intrinsic imaging method was widely used for investigation of neuro-hemodynamic coupling [2–11] and mapping of cortical neuronal responses [12–15] (Box 1).

The hemoglobin molecule is composed of four monomers, each containing an O_2-binding heme group. Hemoglobin tetramer

Box 1 Optical Imaging of Intrinsic Signals—A Historical Perspective

One hundred and twenty years ago, Roy and Sherrington [16] argued that neuronal activation causes the local vasculature to respond. While unequivocal confirmation of this claim had to wait nearly a century, until radioactive methods became available [17–20], large reflectance changes of brain tissue during localized seizure activity could be visualized already in the late 1930s [21]. A few decades later advances made it possible to detect and analyze the much smaller optical signals during activity of the normal cortex. These were accounted for by activity-associated changes in cerebral blood flow (CBF) and volume (CBV) [18, 22]; in addition, Chance [23] and Jöbsis [22] observed that neuronal activity is often accompanied by oximetric signals that can be detected optically by monitoring the absorption (and/or fluorescence) of hemoglobin and other intrinsic chromophores.

In the late 1980s, Grinvald et al. [1] showed that the small light absorption changes induced by these activity-evoked hemodynamic responses can be used to explore cortical functional architecture in vivo, by using a CCD camera to image the cortex upon illumination at specific wavelengths during the presentation of sensory stimuli. The resulting cortical images can then be used to produce functional maps at the spatial resolution of a few tens of microns, more than enough to image the columnar structure of the mammalian neocortex.

Since then, the interpretation of intrinsic signals in terms of neuronal activity—and thus their utility for functional brain mapping—has been tightly linked to our understanding of the mechanisms underlying neurovascular coupling. In fact, such knowledge is necessary to distinguish between optical signals resulting from changes in the activity of local neuronal populations and those resulting from the vascular effects of remote neuronal events (e.g., venous drain). To address this issue, Frostig et al. [4] imaged the sensory-evoked optical responses in the cortex at several wavelengths and tried to decompose them spectroscopically into CBV and O_2 saturation. A few years later, Malonek and Grinvald [6] used a continuous wavelength spectroscopic approach, optical imaging spectroscopy, to investigate the hemodynamic response in further detail. Both studies concluded that some hemodynamic events colocalize better with neuronal activity than others. In particular, an early increase in deoxy-hemoglobin concentration (Hb)—the so-called initial dip—was detected, and interpreted as resulting from local O_2 consumption induced by neuronal activity while the vasculature is still at rest. Thus, the initial dip is expected to colocalize more accurately with neuronal activity than the subsequent hemodynamic events, i.e., increased CBV and CBF, which are mediated by the complex spatiotemporal transfer function of the active vascular response [11, 24]. Which hemodynamic response component is best for functional mapping and how to optimally choose the imaging parameters (wavelength, timing of data acquisition, etc.) are therefore still the object of some debate [8, 25–31].

Whatever its conclusion, de facto, imaging the activity of visual stimuli in early visual cortex using oximetric wavelengths (600–630 nm, emphasizing changes in [Hb] over those in [HbO]) has allowed investigators to obtain high-quality single-condition functional maps, i.e., maps obtained by comparing stimulated conditions to rest. This has not been possible using isosbestic wavelengths, where Hb and HbO absorb with equal strength and which thus emphasize changes in CBV. At those wavelengths, differential approaches (the comparison of several stimulated conditions one to another) are needed to visualize functional architecture at the columnar level [4, 24], at least upon stimulation with large visual stimuli. Interestingly, in other cortices such as auditory [32–34] and rat somatosensory cortex [26], CBV-based optical signals have turned out to warrant at least as good mapping as oximetric ones. Moreover, in primate visual cortex [30] it has been shown that, upon stimulation with small

(continued)

Box 1 (continued)

retinotopic stimuli, the resolution of CBV-dominated signals is as precise as that obtained using oximetric ones.

Taken together, these accumulating results indicate that the optimal choice of the signal to be used for functional mapping might depend on the relation between the topology of the local vascular network and that of the functional architecture to be mapped, that is, on the specific cortical area. However, kind and duration of the stimulus, as well as the signals' etiology, must also be taken into account (e.g., the CBV response appears to have an early capillary component and a delayed macrovascular one), in rodents [27, 35, 36] and primates [30]. Finally, it has been shown that, under some circumstances, imaging at long wavelengths (>700 nm) is advantageous in spite of the small signal sizes, because it reduces the contribution of surface vasculature [37]. Yet, the precise composition of the optical signal at those wavelengths remains to be determined [38].

Irrespective of the above issues, detailed maps of the cortical functional architecture have been obtained in a large variety of preparations—from rodent through cat and ferret up to the primate—in several cortical areas, including both sensory (visual, auditory, somatosensory) and motor. In early visual cortex, intrinsic imaging of functional domains was first reported by Grinvald et al. [1] and later on by many others [39–42]. These studies have laid the groundwork for optical imaging of higher visual areas such as MT [43], V4 [44], and inferotemporal cortex [45]. Fine functional organization of auditory cortex revealed by Fourier optical imaging was reported by Kalatsky et al. [46], after earlier work by Frostig and co-workers [47] and other groups. In the somatosensory cortex, the visualization of whisker barrels in the rat has been a standard protocol for nearly 20 years now [48]. Moreover, the representation of functional domains for digits [49–51] and recently the representation of mechanical nociceptive stimuli within SI have been reported [52].

The elimination of anesthesia has been another critical step. High-resolution optical imaging in alert monkeys was first reported by Grinvald et al. [53], and later in trained monkeys by Roe's group [54]. However, the challenge of long-term imaging in the awake behaving primate could be met only with the development of an artificial, transparent dura substitute [14, 55, 56]. This has allowed investigation not only of the functional architecture but also of its stability over time [14], as well as its behavioral correlation with neurophysiology [15]. Optical imaging has also enabled long-term developmental studies such as imaging the development of orientation preference maps [57, 58] and the layout of functional domains in strabismic animals [59, 60].

Due to its high resolution and non-invasiveness, optical imaging of intrinsic signals has also been used to explore pathological situations, such as the exploration of epilepsy, in animal models as well as in humans [61, 62] and in intraoperative imaging in humans [63, 64], with the goal of finely localizing the site of specific cortical functions.

exists in equilibrium between two states (T and R), characterized by different affinity to O_2. The O_2 binding to hemoglobin is allosterically regulated in the sense that the binding itself shifts the equilibrium towards the higher affinity R state (for review see [65]). Conformationally the T and R states are different, but their optical absorption spectra are very similar. In contrast, binding of O_2 to a heme group induces significant spectroscopic changes. Therefore, each monomer in the hemoglobin molecule, whether it exists in the T or in the R state, can be considered an independent chromophoric unit.

During an increase in neuronal activity, the associated increase in flow and volume of blood results in a decrease of O_2 extraction fraction (from blood to tissue) producing a decrease in deoxyhemoglobin concentration ([Hb]) and an increase in oxyhemoglobin concentration ([HbO]). This effect can be detected, for example, by a decrease in light absorption at 580 nm and the corresponding changes in the absorption-based image (for a recent example from rat somatosensory cortex see [66]). Thus, mapping of neuronal activity can be achieved using a simple epi-illumination/reflectance setup and a CCD camera for detection. Multi-wavelength illumination [6] can be used to extend this method to quantify HbO and Hb changes (as % change from the baseline) [38, 67, 68]. Such absorption-based measurements are linearly related to the fraction of O_2 bound to hemoglobin, while their relationship to the blood pO_2 (dissolved O_2) is defined by the non-linear hemoglobin O_2 binding curve.

An alternative approach to quantifying intravascular O_2 changes in the brain is based on introducing extrinsic phosphorescent probes into the blood stream [25, 69–71,] and measuring phosphorescence lifetime. The method—namely the use of phosphorescence quenching to quantify O_2 tension—was originally developed by Wilson and colleagues [72, 73] and relies on the fact that molecular oxygen (O_2)—a triplet molecule in the ground state—is able to quench emission from excited triplet states of other molecules, such as phosphorescent metalloporphyrin-based probes. Thus, while the intrinsic imaging method measures hemoglobin oxygenation, the phosphorescence quenching technique detects O_2 by measuring kinetics of quenching. Therefore, it can be applied for both intravascular and tissue measurements, given an adequate strategy for delivery of the probe. The phosphorescence lifetime of a probe depends on the partial pressure of O_2 (pO_2) in the immediate vicinity of the probe, providing a spatially localized measurement. O_2 is the only effective dynamic quencher of phosphorescence present in biological systems in sufficient concentration, so measurements of phosphorescence report O_2 levels with high specificity. Importantly, the decay lifetime (rather than the phosphorescence intensity) is measured, which makes the technique insensitive to local variations in the probe concentration and, very importantly, to changes in tissue optical properties. As such, lifetime measurements are insensitive to the absorption and scattering changes associated with the hemodynamic response in vivo, as long as a sufficient number of photons reach the detector for the accurate fitting of the phosphorescent decay. A brief discussion of phosphorescent probes can be found in Box 2 below as well as in recent papers (see, for example, [74, 75]).

Imaging with intrinsic hemoglobin-based contrast or with phosphorescence in wide-field mode using a CCD camera has no true depth resolution: the signal at every pixel represents a weighted sum of the response through the whole depth of light penetration with the highest sensitivity to the cortical surface. This limitation might be acceptable for certain types of intravascular studies, in particular, studies of the "columnar" organization of circulation in cerebral cortex: the vertical orientation of diving arterioles and, more importantly, surfacing venules, transporting

Box 2 Phosphorescent O_2 Sensors

The ideal phosphorescent O_2 probe for in vivo applications consists of a phosphorescent chromophore of the maximal attainable brightness, encapsulated inside a protective jacket, the role of which is to regulate O_2 diffusion to the chromophore (and thereby constant k_q, Eq. 2) in order to maintain high O_2 sensitivity throughout the physiological range. Unrestricted O_2 access (too high k_q) leads to complete quenching of phosphorescence already at low O_2 concentrations, whereas overly restricted access (low k_q) results in a narrow dynamic range, i.e., the difference between the phosphorescence of the quenched and unquenched probes is too small. Similarly, too short triplet lifetimes result in low sensitivity because of low probability of collisions with O_2.

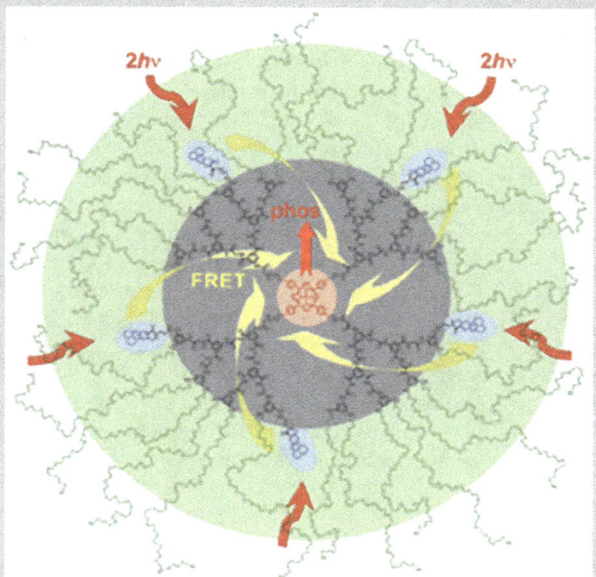

Fig. B1 Molecular structure of PtP-C343. PtP-C343 is a dendritic O_2 probe, in which phosphorescence of metalloporphyrins upon two-photon excitation is enhanced by intramolecular Förster resonance energy transfer (FRET) from two-photon coumarin-343 antennae chromophores (*blue*). The O_2 sensitivity of the probes is regulated by dendritic encapsulation of the core metalloporphyrin (*gray*), while peripheral polyethyleneglycol residue groups on the dendrimer isolate the probes from contact with biological macromolecules in the environment (*green*).

Only a few chromophores exhibit bright phosphorescence in solutions at ambient temperatures. Based on the optical spectra, emission quantum yields, and triplet lifetimes, porphyrins of Pt and Pd are by far the best suited for O_2 measurements in the physiological pO_2 range. The original probes were based on simple derivatives of Pd porphyrins [73, 76]. Such porphyrins are relatively hydrophobic and need to be prebound to serum albumin (usually bovine albumin) to improve their solubility and prevent aggregation in aqueous media. As a result, exogenous protein becomes a part of the injected material, which may lead to unwanted immunogenic reactions and possible toxicity.

Introduction of dendritic polyglutamic porphyrins [77, 78], known as Oxyphors R2 and G2, alleviated the necessity of pre-binding probes to albumin, thanks to polyglutamic branches which render the compounds highly water-soluble. Nevertheless, Oxyphors R2 and G2 still require binding to endogenous albumin, because only albumin complexes of these probes are suitable for measurements in the physiological O_2 range [79]. As a result, O_2 measurements with R2 and G2 can be performed only in albumin-rich environments (e.g., blood plasma). Even in the blood, though, incomplete binding, easily

(continued)

Box 2 (continued)

encountered at high probe concentrations (above ~10^{-5} M), can lead to heterogeneous decays and compromise measurement accuracy.

Keeping the calibration constants of probes unaltered in biological systems is a significant challenge, but also a stringent requirement for quantitative imaging. Indeed, it is not sufficient to have a luminescent probe whose signal simply changes in response to O_2. It is important that both k_q and τ_0 (Eq. 2) remain exactly the same as they were determined during the calibration. These probes' parameters are defined by the probes' immediate environment; to insure their stability the probe molecule must remain in aqueous phase, forming no complexes with bio-macromolecules, cellular membranes, or other biological objects. These requirements have been fulfilled in the so-called fully protected third-generation probes [74, 80]. Pt or Pd porphyrins in these probes are encapsulated inside hydrophobic polyarylglycine (AG) dendrimers, whose exteriors are modified with hydrophilic polyethyleneglycol (PEG) residues. PEGylated jackets regulate the sensitivity and dynamic range of the measurement and insure the probes' high aqueous solubility and inertness with respect to the biological system. This general design is inherent in both one- and two-photon (vide infra) O_2 probes.

To enable two-photon phosphorescence lifetime microscopy (2PLM), special phosphorescent probes were recently developed with enhanced two-photon absorption cross sections. One such probe is PtP-C343 [81], which has been successfully utilized in vivo in brain imaging studies [82–84]. The functional elements of PtP-C343 are Pt tetraarylporphyrin (PtP), which acts as a phosphorescent core, and several C343 units, which play the role of the two-photon antenna (Fig. B1). Near-infrared energy (e.g., at 840 nm) is absorbed by the C343-antenna and within fractions of a nanosecond channeled to PtP via intramolecular FRET, where it is utilized to populate the emissive triplet excited state. The redox potentials of PtP and C343 and the distances between them within the probe molecule are optimized to minimize the unwanted triplet quenching by electron transfer [80, 81]. The C343-antenna enhances phosphorescence in PtP-C343 approximately 25-fold compared to that or a regular one-photon probe. Because of the effect of the dendritic layer and peripheral PEG groups, PtP-C343 is pH-insensitive in the physiological range and its signal is independent of interactions with proteins in the blood.

the deoxygenated blood along the imaging axis. Sensing tissue oxygenation, however, has a stronger requirement for the resolution in depth. Indeed, metabolic differences between cortical layers have been demonstrated with other methods, e.g., cytochrome oxidase staining and incorporation of 2-deoxyglucose [85–87]. In recognition of the importance of depth-resolved tissue pO_2 measurements, numerous studies have employed point polarographic O_2 sensors, also called "O_2 electrodes" [88]. While measurements with O_2 electrodes have provided invaluable insight into tissue O_2 dynamics in cerebral [89, 90] and cerebellar cortex [91], the point nature of the measurement, blind positioning, and invasiveness of the method—the need for repetitive penetrations throughout the cortical tissue for mapping the signal at different locations—put significant constraints on throughput and data interpretations. Indeed, O_2 electrode measurements have demonstrated a significant degree of unaccounted for variability between nearby locations, which could result from differences in the electrode position relative to the blood vessels and/or tissue damage associated with electrode insertion [92, 93]. O_2 electrodes have been successfully used to map oxygenation on the cortical surface [94–96], circumventing the uncertainty of blind positioning. Other methods of measuring tissue O_2 include binding of nitroimidazole-based drugs [97] and electron paramagnetic resonance (EPR) oximetry [98, 99]. These methods differ in spatial and temporal resolution, but generally cannot be applied to real-time tissue O_2 microscopy.

Several new approaches to tissue oximetry have recently been disclosed. First, imaging modalities based on photo-acoustic effect (photo-acoustic tomography [PAT], photo-acoustic microscopy [PAM], optical resolution photo-acoustic microscopy [OR-PAM]) are capable of volumetric measurements of the Hb/HbO ratio [100] with an optical contrast and, except in the case of OR-PAM, ultrasound resolution. Due to the low scattering of sound waves, the imaging resolution of PAT is much higher than that of Diffuse Optical Tomography (DOT), e.g., sub-millimeter resolution at depths as high as 50 mm. However, direct comparisons of O_2 images obtained by the two methods have yet to be performed. Also, penetration depth of PAM is higher than that of current optical microscopy technologies (e.g., two-photon microscopy and optical coherence tomography). Second, a microscopic O_2 imaging method has been described by Warren and collaborators [101] who quantified the Hb/HbO ratio by measuring the excited state absorption of the two components using multiphoton absorption and excitation by shaped femtosecond infrared pulses. This technique is subject to the same depth limitations as all methods relying on ballistic photons and, like DOT, suffers from low sensitivity. A variant of luminescence quenching technique has recently been described by Ashkenazi et al. [102] who measured triplet kinetics using transient triplet–triplet absorption and acoustic detection. However, generation of adequate acoustic signals requires very high probe concentrations and/or near-saturation excitation regime in order to produce adequate amounts of the triplet state. Finally, a method is being developed, which is based on the dependence between the time-averaged fluorescence intensity of regular fluorescent dyes and population of dark triplet states of these dyes in O_2-dependent fashion [103, 104]. This approach also requires extremely high excitation intensities, unlikely to be sustainable by biological tissues.

Seeking to overcome the problem of spatial resolution in CCD-based wide-field imaging of pO_2 based on O_2-dependent phosphorescence quenching, some studies have attempted to combine the phosphorescent quenching method with two-photon excitation [71, 105]. Unfortunately, regular metalloporphyrin-based phosphorescent probes have very low two-photon absorption cross sections, and therefore require extremely high probe concentrations and high laser powers to observe adequate signals. Probes such as Oxyphor R2 [77, 79] and simple hydrophilic Pd porphyrins [73], which can operate only as complexes with albumin (vide infra), partially exist in unbound form at such high concentrations. As a result, the phosphorescence decays are characterized by complex lifetime distributions and cannot be interpreted quantitatively. To enable 2PLM of O_2, special two-photon-enhanced phosphorescent probes have recently been designed [80, 81, 106]. Here, two-photon-"antenna" chromophores are included in the probe molecules for the purpose of capturing excitation energy and transmitting the excitation to phosphorescent cores via intramolecular Förster resonance energy transfer (FRET).

The other major obstacle precluding widespread use of two-photon excitation of phosphorescent probes is a temporal mismatch between high repetition rates of femtosecond Ti:sapphire lasers, typically used for multiphoton excitation (~80 MHz), and long triplet lifetimes of porphyrin-based probes (tens to hundreds of microseconds). To circumvent this problem, the excitation must be gated to allow for acquisition of the phosphorescence decay. Commercial systems suitable for simultaneous measurement/imaging of phosphorescent lifetimes during the gate off-time and fluorescent lifetimes during the gate on-time are just starting to appear (Becker & Hickl data acquisition boards SPC-830 and SPC-150, http://www.becker-hickl.de). However, it is possible to modify or build a 2PLM system in a laboratory with optical expertise [71, 81–83]. 2PLM allows measurement of both intravascular and tissue pO_2 with unprecedented spatial resolution and is well suited to imaging of pO_2 changes during functional activation [84]. Furthermore, it has been recently demonstrated that 2PLM enables estimation of blood flow in individual capillaries simultaneously with intravascular pO_2 [83].

Despite its advantages in spatial resolution and applicability to measurements of tissue oxygenation, 2PLM does not replace measurements of intrinsic signals. Rather, it complements the suite of tools available for metabolic/hemodynamic measurements. Multiple measurement techniques need to be employed to obtain the full picture of the underlying physiological process and to cross-validate one another. For example, one can combine measurements of oxygenation with extrinsic measurements of blood flow to estimate the cerebral rate of O_2 metabolism: 2PLM with optical coherence tomography [107, 108] for microscopic resolution, or intrinsic imaging with either laser speckle [66] or wide-field red blood cell velocity measurements [109] for a larger "mesoscopic" view. Below we discuss the technical aspects of the multi-wavelength CCD-based method for imaging of intrinsic signals and the application of phosphorescence quenching for imaging of intravascular and tissue pO_2, and provide specific examples and references for further reading.

2 Optical Imaging of Hemoglobin Oxygenation

Researchers have long recognized intrinsic optical changes in fluorescence, absorption, and scattering associated with neuronal and metabolic activity. The sources of fluorescence changes are small, intrinsically fluorescent metabolites such as NADH and FAD, whose signals change during a metabolic response to increased neuronal activity or changes in O_2 availability (e.g., during hypoxia) [110]. These molecules have characteristic chromophore-specific excitation bands. The hemoglobin in blood absorbs visible light and is the main source of intrinsic contrast in brain tissue in the visible spectrum [4, 6, 25, 111–114]. At longer wavelengths, where Hb and HbO have negligible absorbance, light-scattering effects dominate. Near-infrared scattering changes, for example, underlie the contrast in optical coherence tomography [115, 116].

Since hemoglobin-based OIS was introduced to measure the functional architecture of the cortex in vivo [1], the method has been widely applied (Box 1). The popularity of OIS for in vivo studies is due largely to the robustness of the hemodynamic response, utilization of endogenous contrast agents, and moderate complexity of the optical instrumentation setup. In addition, intrinsic signals have a relatively large percentage change from the baseline ($\Delta I/I$) in response to neuronal activation, enabling data collection with minimally invasive procedures, such as imaging through the thinned skull.

Whereas OIS relies on functional hemoglobin changes to create contrast, increases in blood volume and the amount of hemoglobin during the hemodynamic response produce unwanted "artifacts" for fluorescence imaging techniques in vivo, such as calcium or NADH imaging. Absorption and scattering changes typically affect both the illuminating light on the way to the fluorophore and the emitted fluorescence. To correct the signal, one can employ algorithms based on the reflectance changes at an additional wavelength [117–119] or fluorescence signals from a physiologically inert fluorophore such as SR101 or GFP [120, 121].

OIS typically uses a CCD camera for signal detection and provides spatial resolution in the XY plane ranging from ~10 μm on the cortical surface to hundreds of micrometer in cortical layers 2–3, and no true depth resolution (see Chapter 2 in this volume). By measuring the absorption changes at multiple wavelengths of light, and given knowledge of the path length of light through the tissue, it is possible to quantify absolute hemoglobin concentration changes [38, 68]. Below we consider the quantitative estimation of Hb and HbO from multi-wavelength OIS data and the spatial resolution of common OIS instrumentation setups.

2.1 Multi-wavelength Imaging Method for Quantification of Oxy- and Deoxy-hemoglobin

Performing OIS with a number of different illumination wavelengths (referred to hereafter as "spectral" imaging) enables estimation of changes in oxy-hemoglobin (ΔHbO), deoxy-hemoglobin (ΔHb), and total hemoglobin (ΔHbT) [66–68, 122, 123]; for a recent review see [124].

For spectral imaging, one can employ a white illumination source filtered with a rotating filter wheel [9, 26, 66–68, 123, 125–127], a set of electronically controlled flashing LEDs optically chopped with appropriate filters [128], or a grating monochromator

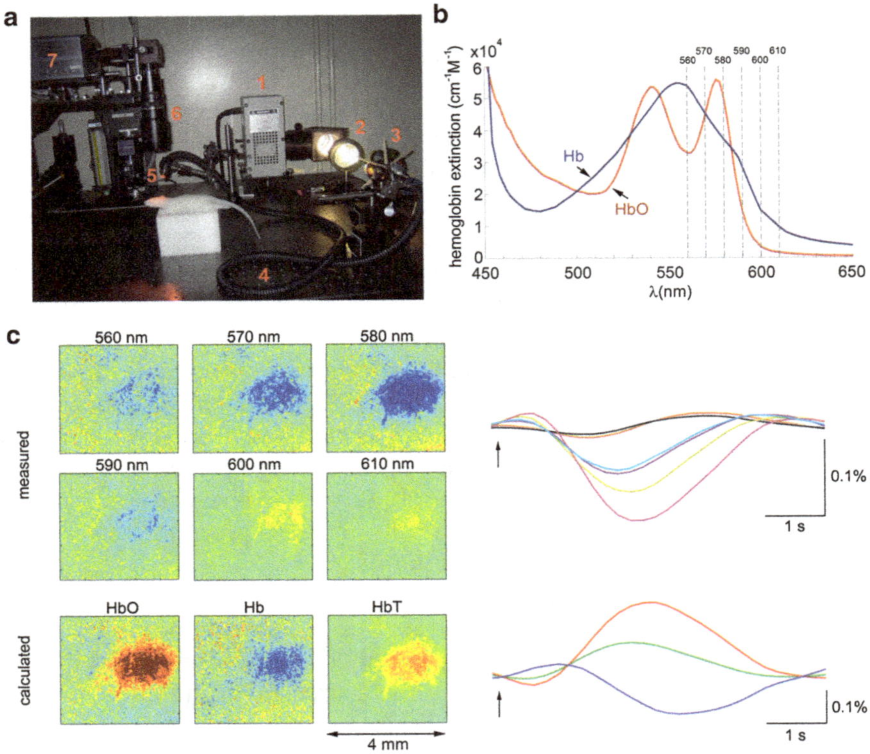

Fig. 1 Conversion of multi-wavelength signals to HbO, Hb, and HbT. (**a**) Optical instrumentation setup with a light source (*1*), focusing lens (*2*), filter wheel (*3*), fiber bundle (*4*) that brings illumination to the specimen (*5*), objective lens (*6*), and CCD camera (*7*). (**b**) Absorption spectra of hemoglobin in the visible range. (**c**) Single-filter maps of absorption changes (*left*) and the corresponding signal time-courses extracted from the active region (*right*). Signal time-courses for each of the six wavelengths are superimposed. (**d**) Estimated HbO, Hb, and HbT maps (*left*) and time-courses (*right*).

[129]. Figure 1a shows an example of a simple instrumentation setup used in our laboratory. Illuminating light (labeled "1" in the figure) from a tungsten–halogen light source (Oriel, Spectra-Physics) is directed by a lens ("2") through a 6-slot filter wheel ("3"), coupled to a 12-mm fiber bundle ("4") that brings the illuminating light to the specimen ("5"). In this filter wheel, the center wavelengths of the filters range between 560 and 610 nm at 10-nm intervals. The filter wheel is mounted on a DC motor and rotates at ~18 Hz, resulting in a ~3 Hz acquisition rate per wavelength. The light is collected by an objective lens ("6") and images are acquired by a cooled 16-bit CCD camera (Cascade 512B, Photometrics) ("7"). Image acquisition is triggered by individual filters in the filter wheel passing through an optic sensor. The image set at each wavelength is averaged across trials and the averaged data are converted to ΔHbO and ΔHb at each pixel using the modified Beer–Lambert relationship:

$$\Delta A(\lambda,t) = \left(\varepsilon_{HbO}(\lambda)\Delta C_{HbO}(t) + \varepsilon_{Hb}(\lambda)\Delta C_{Hb}(t)\right)D(\lambda) \quad (1)$$

Table 1
The pathlength correction factor and extinction coefficients for commonly used illumination wavelengths.

λ (nm)	560	570	580	590	600	610
$D(\lambda)$	0.0537	0.0500	0.0503	0.0909	0.1587	0.2044
$\varepsilon_{HbO}(\lambda)$, ×1.0e+05	0.7511	1.0247	1.1539	0.3316	0.0737	0.0346
$\varepsilon_{Hb}(\lambda)$, ×1.0e+05	1.2387	1.0380	0.8526	0.6523	0.3380	0.2174

The values for extinction coefficients are corrected for the concentration of hemoglobin in tissue.

where $\Delta A(\lambda,t) = \log(R_o/R(t))$ is the attenuation at each wavelength, R_o and $R(t)$ are the measured reflectance intensities at baseline and time t, respectively, ΔC_{HbO} and ΔC_{Hb} are the changes in concentrations of HbO and Hb, respectively, and ε_{HbO} and ε_{Hb} are the wavelength-dependent molar extinction coefficients (Fig. 1b). This equation is solved for ΔC_{HbO} and ΔC_{Hb} using a least-squares approach. The differential pathlength factor, $D(\lambda)$, accounts for the fact that each wavelength travels slightly different pathlengths through the tissue due to the wavelength dependence of scattering and absorption in the tissue. It is estimated using the approach described by Kohl et al. [38] through Monte Carlo simulations of light propagation in tissue (Table 1). Baseline concentrations of 60 and 40 μM are assumed for HbO and Hb, respectively [3, 8]. Results for relative hemoglobin changes during functional activation are only weakly sensitive to these assumed baseline values [66]. Figure 1c–d shows an example of ΔHbO/ΔHb estimation from the six-wavelength data in rat barrel cortex in response to a single deflection of one whisker. Larger hemodynamic changes induced by various pathological manipulations (e.g., middle cerebral artery occlusion, cortical spreading depression) require application of nonlinear fitting algorithms to estimate hemoglobin changes [130].

2.2 Optical Resolution and Depth Sensitivity Considerations

In two-dimensional (2D) camera-based OIS, a collimated beam in the visible spectrum incident onto the brain surface propagates inside the tissue. The reflected light from the brain is collected by the objective lens and imaged onto the CCD chip, forming a 2D image. Tissues at different depths of the cortex contribute to the image on the camera, such that the measured signal intensity of each pixel represents a weighted sum of the response through the whole depth of light penetration (with the highest sensitivity to the cortical surface) [131, 132]. Because of light scattering in the cortex, even a point source in the brain tissue will not result in an image corresponding to a diffraction-limited spot but to a much larger region. The size of this region is determined by the configuration of the optical imaging system (numerical aperture (NA) and the depth at which the image is in focus, referred to as "focal plane

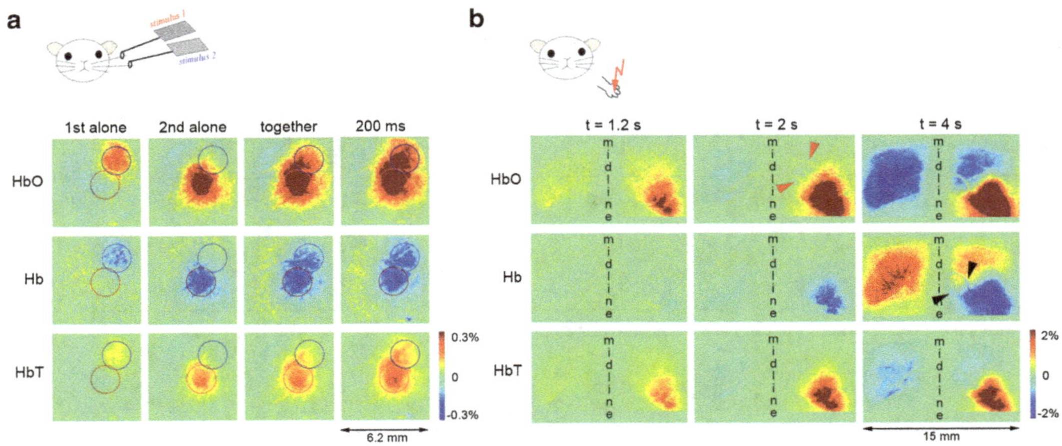

Fig. 2 Imaging of hemoglobin oxygenation in the rat primary somatosensory cortex. (**a**) Example HbO, Hb, and HbT maps in response to a tactile stimulus of single whiskers. The maps correspond to 2.5 s after the stimulus onset for four stimulus conditions: one deflection of a single whisker ("1st alone" and "2nd alone"), simultaneous deflection of both whiskers ("together"), and a sequential deflection with a 200-ms interval ("200 ms"). The signal for Hb and HbO is expressed in percent change relative to its own baseline concentration (40 and 60 μM, respectively). HbT was calculated as a sum of Hb and HbO. (**b**) Example HbO, Hb, and HbT maps in response to an electrical forepaw stimulus. *Red* and *black arrows* point to the surface arteries and veins, visible on HbO and Hb images, respectively.

depth") and the optical properties of the tissue (such as the scattering and absorption coefficients).

Propagation of light in a highly scattering medium such as cortical tissue is complicated and cannot be modeled analytically with adequate accuracy. Instead, numerical methods, such as Monte Carlo methods, are widely used to simulate light transport in tissues [133–135]. A systematic Monte Carlo study of spatial resolution and depth sensitivity for 2D optical imaging methods with configurations typically encountered in functional brain imaging demonstrated that more than 97 % of the signal comes from the top 500 μm of the tissue [134] (see Chapter 3 in this volume). Therefore, no significant contribution is expected to originate in cortical layers 4 or 5. The spatial resolution using the common NA and focal depth settings is less than or comparable to the size of cortical columns in rodent barrel cortex.

2.3 Example Application: Mapping of Rat Barrel Cortex Using Multi-wavelength Optical Imaging

OIS provides a sensitive tool for mapping the location of neuronal responses to sensory stimulation and for studying neurovascular coupling while varying the evoked hemodynamic response in the temporal and spatial domain [4, 11, 14, 48, 136]. Figure 2a shows an example from the rat barrel cortex. The skull over the barrel cortex contralateral to the stimulated whisker pad was thinned until transparent, and tactile stimuli were delivered to single whiskers at different time delays [137]. The hemodynamic response to tactile stimulation of individual whiskers is clearly localized (red and blue circles in Fig. 2a). Moreover, one can study the additive properties of the response by stimulating multiple whiskers, either simultaneously or sequentially.

Figure 2b shows a different example of ΔHb/ΔHbO/ΔHbT maps in response to a much stronger stimulus: an electrical stimulus delivered to a forepaw in the rat. In this case, the skull was thinned bilaterally over the forepaw region of the primary somatosensory cortex. Note that the amplitude of the signal change is almost an order of magnitude larger than in the previous case. The maps in both cases have a spatial structure, which can be exploited to study the transformation of neuronal circuit activity to hemodynamic signal changes [126, 127].

3 Direct Methods for O_2 Imaging Using Phosphorescent Probes

The phosphorescence quenching method exploits exogenous molecular probes, which emit from their triplet spin states (as opposed to fluorescent probes, which emit from singlet spin states), to quantify O_2 content in living tissue [73]. O_2 can react with excited state molecules in the environment, quenching their luminescence. Collisional quenching is much less probable on the time scale of excited singlet states (nanoseconds) than of triplet states (microseconds to milliseconds), making phosphorescence much more sensitive than fluorescence to dynamic O_2 quenching. The phosphorescence decay time (τ) of a probe depends on the partial pressure of O_2 in the physiological pO_2 range according to the Stern–Volmer relationship:

$$\frac{\tau_o}{\tau} = 1 + k_q \tau_o pO_2 \qquad (2)$$

where τ_o and τ are the phosphorescence lifetimes in the absence of O_2 and at O_2 pressure pO_2, respectively, and k_q is the O_2 quenching constant. Parameters k_q and τ_o are characteristic of the probe molecule and the environment (solvent, pH, temperature, etc.) and do not depend on the measurement system. Phosphorescence quenching is arguably the most direct method of O_2 quantification, since the physics of quenching is well understood. In biological systems O_2 is the only small-molecule quencher present in sufficiently high concentrations: hence the specificity of the measurement.

Phosphorescence quenching as a form of biological oximetry has been developed specifically for in vivo applications. Phosphorescent probes (Box 2) are introduced directly into the medium of interest, where they report on the oxygenation in their local environments. External excitation and detection are used to retrieve the signal. Using phosphorescence, O_2 can be imaged in large areas/volumes [69, 72, 138, 139] potentially in 3D [140]; no mechanical damage is done to the tissue, and the measurement can be performed with high temporal and spatial resolution. If phosphorescence is excited by ballistic photons, as in microscopy applications [141–143], the spatial resolution is limited only by the optical diffraction. Provided that the volume contains a sufficient number of probe molecules, temporal resolution is limited only by the triplet lifetime (sub-millisecond range). However, in practice, collection of multiple decays is required to obtain a reliable measurement. Thus, phosphorescent probes with high emissivity are the key to high temporal and spatial resolution.

4 Single-Photon Excitation of O_2-Sensitive Probe

Regular phosphorescence-based pO_2 measurements by linear (one-photon) excitation can be performed with relatively simple instrumentation. For example, using a time-gated CCD camera synchronized with a pulsed- or frequency-modulated excitation source [69, 70, 138, 139, 144–147], spatially resolved oxygenation maps can be collected in tissue or vasculature with temporal resolution on the order of seconds.

Frequency-modulated illumination can be provided by an optically chopped halogen source light or an LED [144, 145]. Alternatively, in the time domain, one can use flash lamps [69, 70, 138] or a pulsed laser such as a q-switched, frequency-doubled Nd:YAG laser that delivers ~6-ns pulses at a fixed frequency (10 Hz) [139, 147]. As with OIS, the excitation light is typically coupled into a multimode fiber positioned at an angle relatively oblique to the cortical surface. The sample can also be excited with a 0° incidence angle if a dichroic mirror is used to couple the excitation and detection beams. In a wide-field microscopy system operating in one of our laboratories [139], the phosphorescence light is collected by a low-magnification infinity-corrected objective (e.g., Olympus XL Fluor 4x/340, 0.28 NA), and an image is formed on the CCD sensor by a 100-mm focal length tube lens. A long-pass filter is positioned between the objective and the tube lens to suppress the excitation light, and the signal is detected from the entire field of view at once. Higher signal-to-noise ratio (SNR) can be achieved by binning pixels on the CCD chip (at a price of lower spatial resolution). A single lifetime image (reconstruction of the phosphorescent decay) requires multiple camera exposures temporally jittered relative to the illumination. Using a pulsed laser source, each camera exposure starts at a specific programmed delay with respect to its corresponding excitation pulse (Fig. 3a). For each pixel, intensity values are plotted against the delay between the excitation pulse and CCD exposure to generate the phosphorescence decay. If frequency-modulated

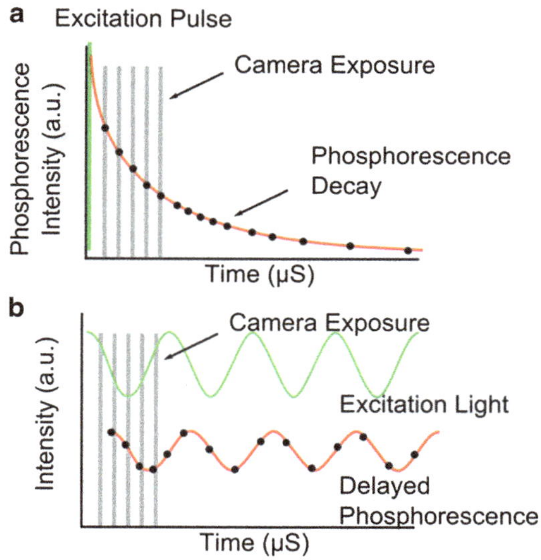

Fig. 3 Phosphorescence lifetime measurements—time-domain and frequency-domain methods. (**a**) Timing relationship for time-domain measurement of phosphorescence decay. A pulsed laser delivers light at a fixed repetition rate, with each pulse generating a phosphorescent decay. Each laser pulse triggers a corresponding brief camera exposure, delayed by a programmed period after the laser pulse. Each camera exposure captures a different portion of the phosphorescence decay. After several laser pulses, the phosphorescence decay can be measured and the lifetime is estimated by an exponential fit. (**b**) Timing relationship for frequency-domain lifetime measurement. A sinusoidally modulated excitation beam runs continuously, and a synchronized camera collects multiple brief exposures at programmed phase delays. After several exposures, the delayed sinusoidal phosphorescence waveform can be estimated, and the lifetime is estimated by $\tan(\theta)/2\pi \times$ modulation frequency, where theta is the phase delay of the phosphorescence waveform [98].

excitation is used rather than pulsed excitation, the lifetime can be computed from the phosphorescence waveform's phase delay with respect to the excitation waveform (Fig. 3b) [145]. Maps of pO_2 are computed per pixel using the Stern–Volmer relationship (Eq. 2).

An illustration of CCD-based intravascular pO_2 imaging is presented in Fig. 4. In this example, a simple laser speckle imaging system was used simultaneously with the pO_2 measurement to record relative blood flow changes (rCBF) for estimation of the cerebral metabolic rate of O_2 ($CMRO_2$) [139]. See Chapter 15 in this volume for review of laser speckle imaging.

As discussed in Sect. 2.2, CCD-based imaging (in combination with wide-field excitation of the phosphor) suffers from poor spatial resolution due to the scattering of light in tissue. Confocal microscopy can be employed to improve the spatial resolution in a single-photon excitation mode [143]. In the confocal regime, the scanning galvanometer mirrors can be programmed to direct the

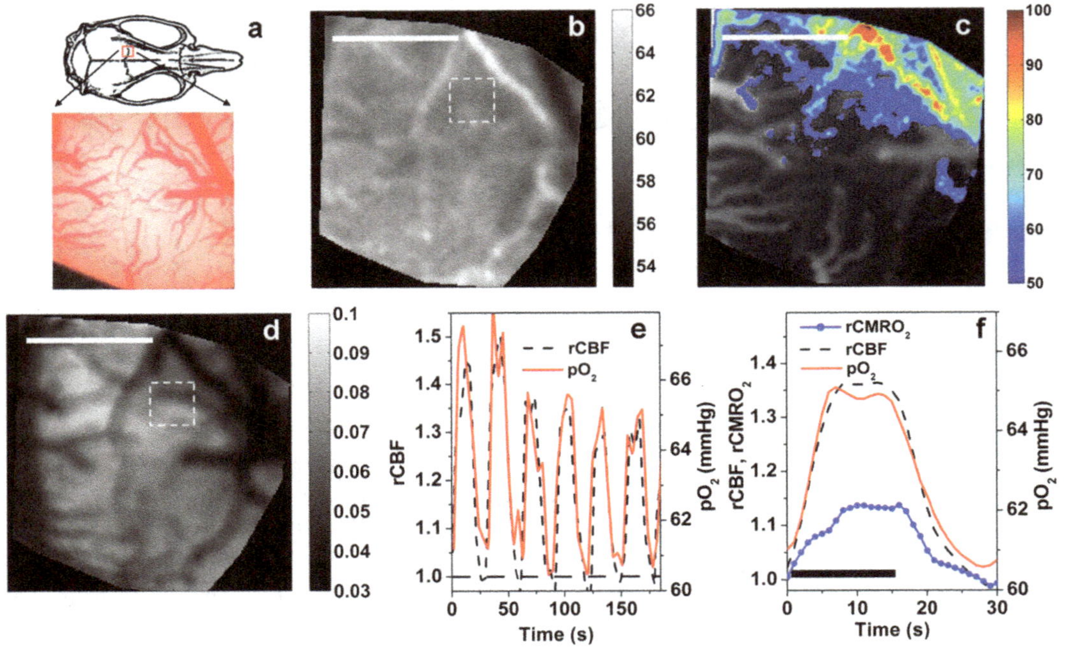

Fig. 4 CCD measurement of intravascular pO_2 with simultaneous laser speckle imaging during forepaw stimulation. (**a**) Position of the cranial window over the forepaw area of the somatosensory cortex and a photograph of the cortical vasculature. (**b**) Baseline pO_2 map. (**c**) Composite image consisting of phosphorescence intensity (*gray*) and activation area (*color*). (**d**) Baseline speckle contrast image. (**e**) Time-courses of pO_2 (*red solid curve*) and rCBF (*black dashed curve*) during several stimulation sequences extracted from the ROI marked by the *white rectangles* in (**b**) and (**d**). *Black horizontal bars* in (**e**) and (**f**) denote stimulus duration. (**f**) Trial-averaged pO_2, rCBF, and $rCMRO_2$ responses. Scale bar is 1 mm.

excitation beam to distinct locations from which pO_2 measurements are made. The detected light must be de-scanned by placing the detector after the galvanometer scanners and spatially filtered using a pinhole in front of the detector. Thus, spatial resolution is improved by (1) confining the excitation to a small volume and (2) rejecting out-of-focus phosphorescence by implementing the pinhole.

5 Two-Photon Phosphorescence Lifetime Imaging (2PLM) of O_2

In addition to single-photon excitation of O_2-sensitive probes, two-photon excitation is now being employed to achieve better depth resolution and higher spatial localization [71, 81–83, 148]. In two-photon microscopy, ultrashort laser pulses and a high NA objective are required to induce nonlinear excitation of the probe, confining the generation of the excited state to the immediate vicinity of the focal volume. In two-photon microscopes constructed in our laboratories [81, 82], phosphorescence is excited by trains of femtosecond pulses from a Ti:sapphire oscillator, gated

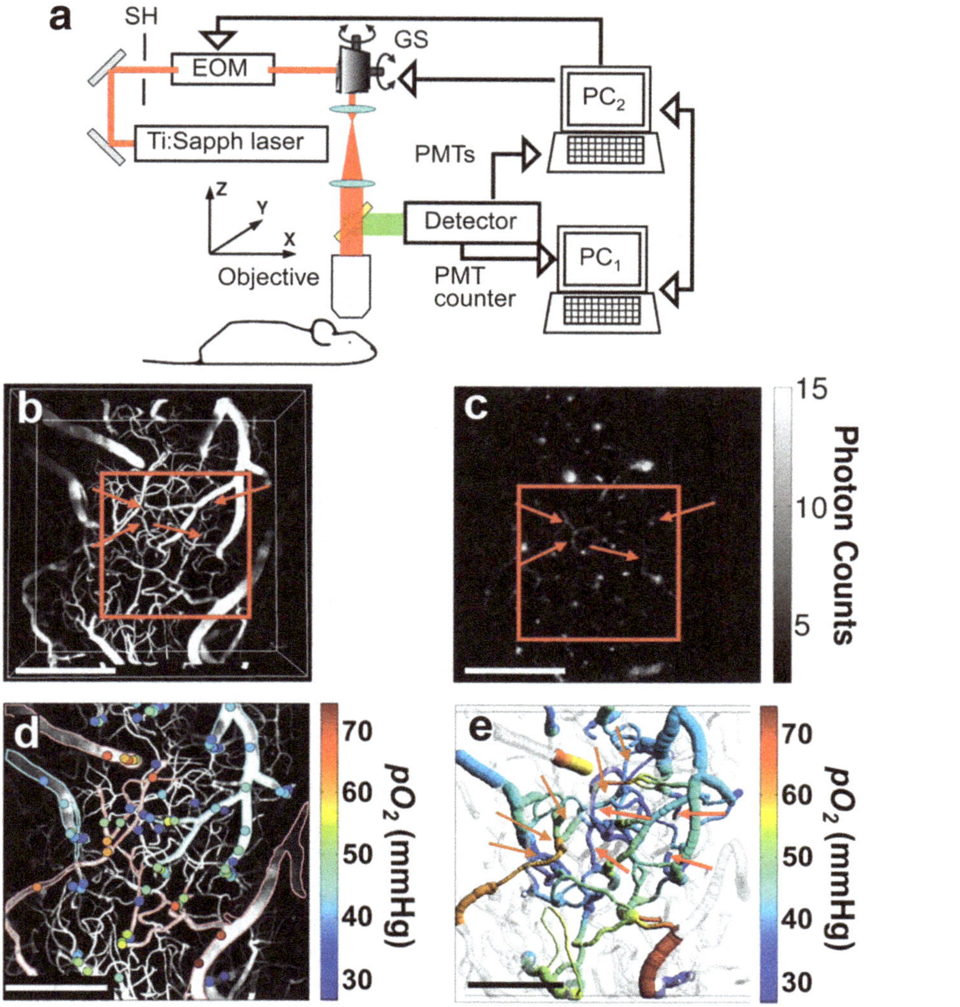

Fig. 5 Two-photon detection of phosphorescence. (**a**) Schematic of the experimental setup. *EOM* electro-optic modulator, *GS* galvanometer scanner, *PMTs* analog-mode photomultiplier tubes, *PMT counter* a Geiger-mode photomultiplier tube. (**b**) Maximum intensity projection along the *z* direction of a 250-μm-thick stack in the mouse cortex. The vasculature was labeled with dextran-conjugated fluorescein isothiocyanate (FITC). (**c**) Phosphorescence intensity image of microvasculature obtained at a depth of 240 μm below the cortical surface. The *color bar* shows the average number of photon counts in each pixel collected during a single phosphorescence decay. Four *arrows* mark the locations of the same capillary vessels in panels (**b**), (**c**), and (**e**). (**d**) Measured pO_2 values in microvasculature at various depths (*colored dots*), overlaid on the maximum intensity projection image of vasculature structure (grayscale). Digital processing was performed to remove images of the dura vessels. Edges of the major pial arterioles and venules are outlined in *red* and *blue*, respectively. (**e**) Composite image showing a projection of the imaged vasculature stack. *Red arrows* mark pO_2 measurement locations in the capillary vessels at a depth of 240 μm. *Orange arrows* point to the consecutive branches of the vascular tree, from pial arteriole (*bottom left arrow*) to the capillary and then to the connection with ascending venule (*top right arrow*). Scale bar, 200 μm.

by an electro-optic modulator. The phosphorescent signals are detected by photon-counting modules (Fig. 5a). For example, the probe can be excited by a 10-μs excitation gate, followed by a 300-μs data acquisition period. The counts are binned and subsequently used to reconstruct the decay, which is analyzed by the least-squares method to obtain the phosphorescence lifetime. The SNR of the measurement is determined solely by the number of the counted photons. Only very few photons can be collected from a near diffraction-limited volume per excitation gate at a reasonable probe concentration [81]. Therefore, decays are typically analyzed after averaging multiple excitation cycles. The size of the excitation volume is dependent on the laser power and the duration of the excitation gate [148]. These parameters can be adjusted to select the optimal combination of the excitation volume (spatial resolution) and pixel dwell time (temporal resolution). For example, to measure baseline pO_2 in the mouse cortex we usually acquire 500–2,000 excitation cycles per point using PtP-C343 (for description of this and other probes see [74, 75, 77, 80–81]) at intravascular concentration of <15 μM and with 10-μs excitation gates. The average laser power during the excitation gate is <10 mW at the focal plane. Using 2,000 cycles, a grid of 600 points can be acquired in ~6 min. We estimate that under this excitation regime the emitting volume has ~2 μm lateral (*XY*) and ~5 μm axial (*Z*) dimensions.

Using 2PLM in an upright setup [82], we can image the cortex of anesthetized mice and rats through sealed cranial windows, either injecting the phosphorescence probe into the vasculature through the femoral artery or pressure-injecting it directly into the brain tissue. Structural images of the cortical vasculature can be obtained either by imaging fluorescence emission from coumarin-343 (C343) chromophores constituting the two-photon antenna of the PtP-C343 probe or, to increase contrast, by using intravenously administered fluorescein isothiocyanate (FITC) conjugated with dextran (Fig. 5b).

O_2 imaging in the cerebral cortex is typically performed in two steps. First, we raster-scan the excitation beam over the field of view, rendering 2D survey maps of the integrated emission intensity (150 × 150 or 250 × 250 pixels, acquired in ~9 or ~25 s, respectively). A single-plane scan performed 240 μm below the brain surface in Fig. 5c demonstrates the ability of the system to resolve the structure of the microvasculature down to the capillary level. After mapping the tissue, we average multiple excitation cycles in selected locations for accurate pO_2 determination. Given the average of 500–2,000 cycles, the acquisition time corresponds to a temporal resolution of 0.16–0.76 s per single-point pO_2 measurement. Finally, using Stern–Volmer calibration plots (Eq. 2), we convert phosphorescence lifetimes, obtained by fitting, into pO_2 values (Fig. 5d–e).

5.1 Example Application: Two-Photon Imaging of Tissue pO_2 Dynamics in Rat Somatosensory Cortex with Phosphorescent Nanoprobe PtP-C343

Measurement of pO_2 based on 2PLM is uniquely capable of assessing pO_2 directly in the cortical tissue. In the example shown in Fig. 6, PtP-C343 was microinjected into the interstitial space (tissue) of the rat primary somatosensory cortex through a glass pipette attached to a pressure injector, as is usually done with calcium indicators [149]. In contrast to calcium indicators, PtP-C343 is easily dissolved in artificial cerebrospinal fluid and does not require additional solvents (such as pluronic acid or DMSO). Following microinjection, PtP-C343 quickly diffuses within ~1 mm from the injection site. Using our protocol, there is no significant decrease in the signal intensity up to 4 h after the injection. After that, the signal slowly reduces. Thus, a single injection per subject is sufficient. FITC is administered intravenously to delineate the vasculature (Fig. 6a) and allows coregistration of pO_2 measurements relative to the vascular O_2 sources. To investigate the spatial distribution of tissue pO_2 at the baseline (resting) level of neuronal activity, one can acquire grids of points (Fig. 6b). Plotted as a function of the radial distance from the arteriole (Fig. 6c), these data demonstrate a gradient indicating that O_2 diffuses into the tissue through the walls of diving arterioles [84].

When the probe is also injected intravenously, one can perform simultaneous intravascular and tissue pO_2 measurements. However, given the resolution of $\sim 2 \times 2 \times 5$ μm, capillary signals cannot be resolved if the probe is present in both intra- and extravascular compartments. Nevertheless, one can resolve intravascular pO_2 in diving arterioles, although the ontogeny of the signal at the vessel border would be uncertain (Fig. 6d). Alternatively, capillary pO_2 can be readily imaged if the probe is present only in the vasculature (but not in the tissue around it) as demonstrated in Fig. 5.

When the beam is moved from point to point, a short time period is required to stabilize the galvanometer mirrors. Therefore, the most time-efficient way to accomplish a point pO_2 measurement is to stay at the same point and collect multiple decays before moving to the next point. For functional measurements, however, the sampling rate per point must match the kinetics of the dynamic process to be imaged. Figure 7 shows an example of a dynamic pO_2 measurement performed at ~1 Hz. In this example, 50 decays were collected at each point and the number of points was adjusted to allow us to revisit the point within ~1 s (1 Hz sampling rate). Therefore, after determining the target acquisition speed, one has to compromise between the overall number of measurement points and the number of excitation cycles collected per point. Multiple stimulus trials can be averaged before fitting for the decay to increase the SNR of the measurement. For example, 20 trials were averaged in Fig. 7.

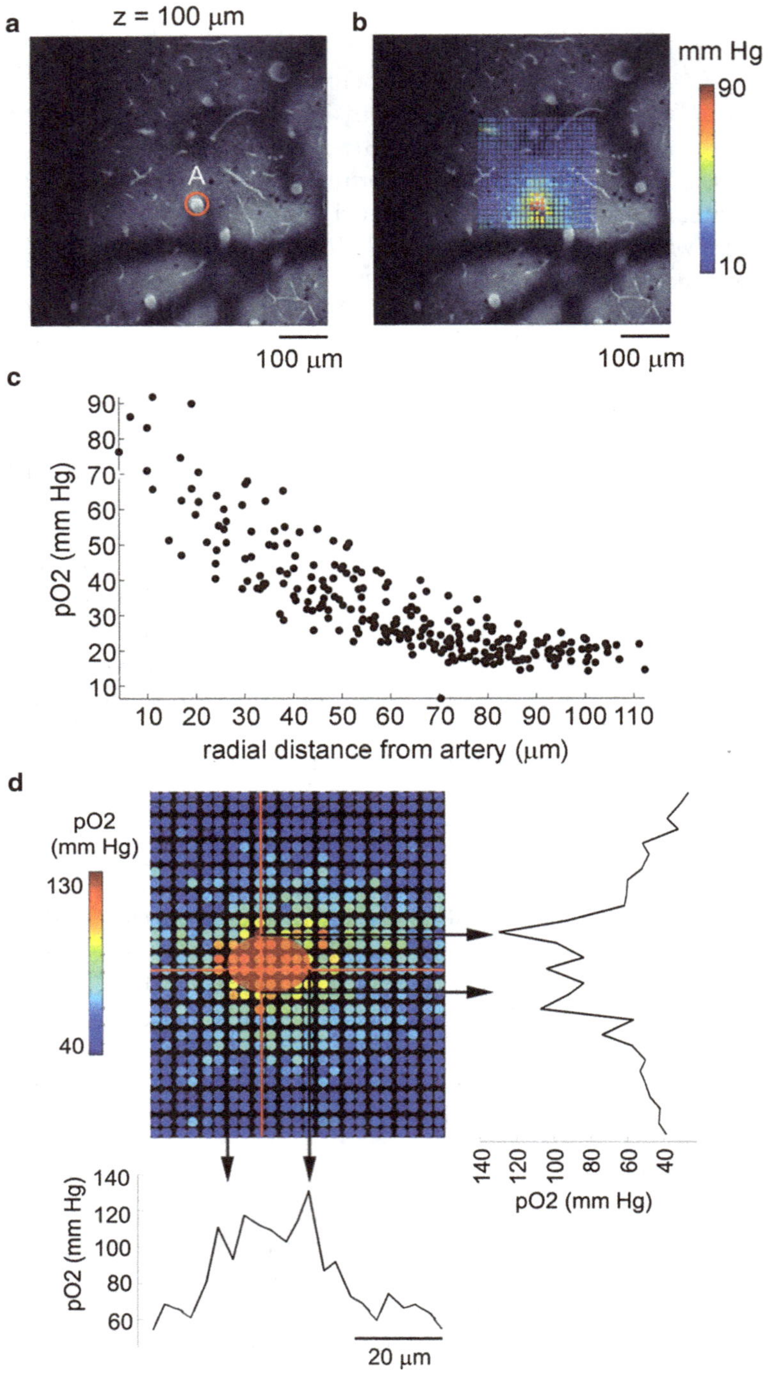

Fig. 6 Spatial distribution of baseline pO_2 values. (**a**) A reference vascular image with an arteriole ("A"). The fluorescent contrast is due to intravascular FITC. The background fluorescence is due to the excitation of coumarin molecules in PtP-C343. (**b**) The same image as in (**a**) with superimposed pO_2 values and segmented arteriolar territory. (**c**) pO_2 as a function of the radial distance from the arteriole. (**d**) The plots at the *bottom* and on the *right* show *X* and *Y* profiles cut through the *red lines* in the image. The *red circle* indicates the arteriolar boundary estimated from the FITC reference.

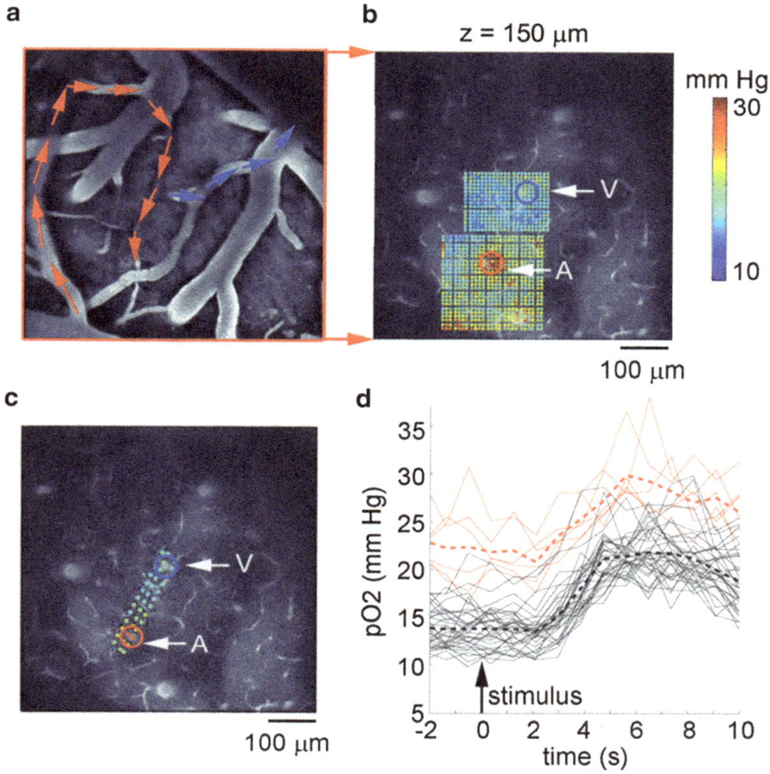

Fig. 7 Stimulus-induced tissue pO_2 changes in rat primary somatosensory cortex. (**a**) An image of the surface vasculature calculated as a maximum intensity projection (MIP) of an image stack 0–100 μm in depth using a 4× objective. Individual images were acquired every 10 μm. (**b**) The same region at the depth plane of pO_2 measurements, 150 μm below the surface. "A" and "V" indicate a diving arteriole and a surfacing venule, respectively. *Red* and *blue arrows* in (**a**) show the upstream and downstream surface arteriole and venule for "A" and "V," respectively. A grid of measured baseline pO_2 values is superimposed on the vascular reference plane. (**c**) The same vascular reference plane as in (**b**) with superimposed pO_2 measurement points used for dynamic measurements. The points are color-coded according to the baseline pO_2. (**d**) Stimulus-evoked time-courses of pO_2 change, extracted from each of the measurement points in (**c**). The traces are divided into two categories: high baseline (*red*) and low baseline (*black*). The *dotted red* and *black lines* correspond to the average for each category. The *black arrow* denotes stimulus onset.

One additional limitation of 2PLM measurements is a decrease in SNR with increasing baseline pO_2. This limitation stems directly from the O_2 sensing mechanism of the probe: quenching of the triplet state by O_2 leads to a fewer emitted photons. At higher pO_2s, more of the probe molecules are quenched, resulting in fewer photons reaching the detector. The decrease in SNR at high pO_2 is usually not a problem for baseline measurements, where it can be compensated for by averaging more excitation cycles per point.

6 Conclusions and Outlook

In this chapter we have reviewed methods for measuring intravascular and tissue oxygenation, with a particular focus on applying OIS and phosphorescence quenching in the cerebral cortex. In the context of cortical function, these methods on their own provide a somewhat one-sided view of a single (but very important) aspect of the underlying physiological processes—oxygenation. A more complete picture can be obtained by combining these imaging modalities with other measurements of neuronal, glial, vascular, and metabolic activity. As such, OIS can be readily combined with simultaneous electrophysiological recordings [9, 27, 67, 150] or fMRI [105], and can serve as a bridge between the neuronal and fMRI signals, getting around the need to filter electrical potentials recorded in the environment of changing electromagnetic fields. Likewise, 2PLM can be combined with optical imaging of neuronal activity (e.g., two-photon calcium imaging [152], at least in alternating stimulus trials; see Chapter 9 in this volume) and optical coherence tomography measurements of blood flow [107, 108]. Measurements of capillary flow—the speed of red blood cells in individual capillaries—can be extracted directly from 2PLM data when a phosphorescent O_2 probe is applied intravascularly [83].

The choice of method for measuring oxygenation would depend on the requirement for spatial resolution, compartments of interest (intravascular or tissue), allocated budget, and availability of optical engineering expertise. OIS remains the method of choice for "mesoscopic" mapping of neuronal activity on the cortical surface: It does not require delivery of extrinsic probes and is inexpensive and easy to implement in a standard neuroscience laboratory setting. The most significant drawbacks of OIS are scattering and lack of depth resolution, which result in an inherent uncertainty in the spatial origin of the imaged signals. 2PLM is situated on the opposite side of the experimental/instrumentation complexity spectrum, but offers high-resolution, deep-penetration imaging of either intra- or extravascular pO_2, depending on the location of the pO_2 probe. Confocal imaging might serve as a good compromise on the instrumentation side and does not require probes with high two-photon cross sections, offering more flexibility in the choice of probe molecules. In the two-photon regime, the most important parameter determining the temporal and spatial resolution of the measurement is the phosphorescence quantum yield of the probe. Probes based on metalloporphyrins and other chromophores exhibiting brighter phosphorescence are currently being developed. Likewise, strategies for intracellular delivery of phosphorescent O_2 probes are under investigation. In combination with transgenic methods for identification of specific neuronal cell types in the cerebral cortex, intracellular O_2 probes could be used for evaluation of cell-type specific pathology in neurodegenerative disease.

Acknowledgements

We gratefully acknowledge support from the NINDS (NS051188 and NS057198 to A.D., NS057476 and NS055104 to D.A.B.), NIBIB (EB00790 to A.M.D., EB009118 to A.D., EB2066 to B.R.R., EB007279 to S.A.V.), American Heart Association Grant No. 11SDG7600037 to S.S., and NIH SIG S10-RR022428 to D.A.B.

References

1. Grinvald A, Lieke E, Frostig RD, Gilbert CD, Wiesel TN (1986) Functional architecture of cortex revealed by optical imaging of intrinsic signals. Nature 324:361–364
2. Berwick J, Martin C, Martindale J, Jones M, Johnston D, Zheng Y, Redgrave P, Mayhew J (2002) Hemodynamic response in the unanesthetized rat: intrinsic optical imaging and spectroscopy of the barrel cortex. J Cereb Blood Flow Metab 22:670–679
3. Jones M, Berwick J, Mayhew J (2002) Changes in blood flow, oxygenation, and volume following extended stimulation of rodent barrel cortex. Neuroimage 15:474–487
4. Frostig RD, Lieke EE, Ts'o DY, Grinvald A (1990) Cortical functional architecture and local coupling between neuronal activity and the microcirculation revealed by in vivo high-resolution optical imaging of intrinsic signals. Proc Natl Acad Sci USA 87: 6082–6086
5. Malonek D, Dirnagl U, Lindauer U, Yamada K, Kanno I, Grinvald A (1997) Vascular imprints of neuronal activity: relationships between the dynamics of cortical blood flow, oxygenation, and volume changes following sensory stimulation. Proc Natl Acad Sci USA 94:14826–14831
6. Malonek D, Grinvald A (1996) Interactions between electrical activity and cortical microcirculation revealed by imaging spectroscopy: implications for functional brain mapping. Science 272:551–554
7. Martindale J, Mayhew J, Berwick J, Jones M, Martin C, Johnston D, Redgrave P, Zheng Y (2003) The hemodynamic impulse response to a single neural event. J Cereb Blood Flow Metab 23:546–555
8. Mayhew J, Johnston D, Berwick J, Jones M, Coffey P, Zheng Y (2000) Spectroscopic analysis of neural activity in brain: increased oxygen consumption following activation of barrel cortex. Neuroimage 12:664–675
9. Nemoto M, Sheth S, Guiou M, Pouratian N, Chen JW, Toga AW (2004) Functional signal- and paradigm-dependent linear relationships between synaptic activity and hemodynamic responses in rat somatosensory cortex. J Neurosci 24:3850–3861
10. Sheth S, Nemoto M, Guiou M, Walker M, Pouratian N, Toga AW (2003) Evaluation of coupling between optical intrinsic signals and neuronal activity in rat somatosensory cortex. Neuroimage 19:884–894
11. Vanzetta I, Hildesheim R, Grinvald A (2005) Compartment-resolved imaging of activity-dependent dynamics of cortical blood volume and oximetry. J Neurosci 25:2233–2244
12. Chen G, Lu HD, Roe AW (2008) A map for horizontal disparity in monkey V2. Neuron 58:442–450
13. Erinjeri JP, Woolsey TA (2002) Spatial integration of vascular changes with neural activity in mouse cortex. J Cereb Blood Flow Metab 22:353–360
14. Shtoyerman E, Arieli A, Slovin H, Vanzetta I, Grinvald A (2000) Long-term optical imaging and spectroscopy reveal mechanisms underlying the intrinsic signal and stability of cortical maps in V1 of behaving monkeys. J Neurosci 20:8111–8121
15. Sirotin YB, Das A (2009) Anticipatory haemodynamic signals in sensory cortex not predicted by local neuronal activity. Nature 457:475–479
16. Roy CS, Sherrington CS (1890) On the regulation of the blood-supply of the brain. J Physiol 11:85–158
17. Fox PT, Raichle ME (1984) Stimulus rate dependence of regional cerebral blood flow in human striate cortex, demonstrated by positron emission tomography. J Neurophysiol 51:1109–1120
18. Lassen NA, Ingvar DH (1961) The blood flow of the cerebral cortex determined by radioactive krypton. Experientia 17:42–43
19. Sokoloff L, Reivich M, Kennedy C, Des Rosiers MH, Patlak CS, Pettigrew KD,

Sakurada O, Shinohara M (1977) The [14C] deoxyglucose method for the measurement of local cerebral glucose utilization: theory, procedure, and normal values in the conscious and anesthetized albino rat. J Neurochem 28:897–916

20. Sokoloff L (1981) Localization of functional activity in the central nervous system by measurement of glucose utilization with radioactive deoxyglucose. J Cereb Blood Flow Metab 1:7–36
21. Penfield W (1933) The evidence for a cerebral vascular mechanism in epilepsy. Ann Intern Med 7:303–310
22. Jobsis FF (1977) Noninvasive, infrared monitoring of cerebral and myocardial oxygen sufficiency and circulatory parameters. Science 198:1264–1267
23. Chance B, Cohen P, Jobsis F, Schoener B (1962) Intracellular oxidation-reduction states in vivo. Science 137:499–508
24. Vanzetta I, Slovin H, Omer DB, Grinvald A (2004) Columnar resolution of blood volume and oximetry functional maps in the behaving monkey; implications for FMRI. Neuron 42:843–854
25. Vanzetta I, Grinvald A (1999) Increased cortical oxidative metabolism due to sensory stimulation: implications for functional brain imaging. Science 286:1555–1558
26. Sheth SA, Nemoto M, Guiou MW, Walker MA, Toga AW (2005) Spatiotemporal evolution of functional hemodynamic changes and their relationship to neuronal activity. J Cereb Blood Flow Metab 25:830–841
27. Sheth SA, Nemoto M, Guiou M, Walker M, Pouratian N, Toga AW (2004) Linear and nonlinear relationships between neuronal activity, oxygen metabolism, and hemodynamic responses. Neuron 42:347–355
28. Buxton RB (2001) The elusive initial dip. Neuroimage 13:953–958
29. Lindauer U, Royl G, Leithner C, Kuhl M, Gold L, Gethmann J, Kohl-Bareis M, Villringer A, Dirnagl U (2001) No evidence for early decrease in blood oxygenation in rat whisker cortex in response to functional activation. Neuroimage 13:988–1001
30. Sirotin YB, Hillman EM, Bordier C, Das A (2009) Spatiotemporal precision and hemodynamic mechanism of optical point spreads in alert primates. Proc Natl Acad Sci USA 106:18390–18395
31. Vanzetta I, Grinvald A (2001) Evidence and lack of evidence for the initial dip in the anesthetized rat: implications for human functional brain imaging. Neuroimage 13:959–967
32. Harel N, Mori N, Sawada S, Mount RJ, Harrison RV (2000) Three distinct auditory areas of cortex (AI, AII, and AAF) defined by optical imaging of intrinsic signals. Neuroimage 11:302–312
33. Spitzer MW, Calford MB, Clarey JC, Pettigrew JD, Roe AW (2001) Spontaneous and stimulus-evoked intrinsic optical signals in primary auditory cortex of the cat. J Neurophysiol 85:1283–1298
34. Versnel H, Mossop JE, Mrsic-Flogel TD, Ahmed B, Moore DR (2002) Optical imaging of intrinsic signals in ferret auditory cortex: responses to narrowband sound stimuli. J Neurophysiol 88:1545–1558
35. Chen-Bee CH, Agoncillo T, Xiong Y, Frostig RD (2007) The triphasic intrinsic signal: implications for functional imaging. J Neurosci 27:4572–4586
36. Sheth SA, Nemoto M, Guiou M, Walker M, Pouratian N, Hageman N, Toga AW (2004) Columnar specificity of microvascular oxygenation and volume responses: implications for functional brain mapping. J Neurosci 24: 634–641
37. Mc Loughlin NP, Blasdel GG (1998) Wavelength-dependent differences between optically determined functional maps from macaque striate cortex. Neuroimage 7:326–336
38. Kohl M, Lindauer U, Royl G, Kuhl M, Gold L, Villringer A, Dirnagl U (2000) Physical model for the spectroscopic analysis of cortical intrinsic optical signals. Phys Med Biol 45: 3749–3764
39. Bosking WH, Zhang Y, Schofield B, Fitzpatrick D (1997) Orientation selectivity and the arrangement of horizontal connections in tree shrew striate cortex. J Neurosci 17:2112–2127
40. Lu HD, Chen G, Tanigawa H, Roe AW (2010) A motion direction map in macaque V2. Neuron 68:1002–1013
41. Ts'o DY, Frostig RD, Lieke EE, Grinvald A (1990) Functional organization of primate visual cortex revealed by high resolution optical imaging. Science 249:417–420
42. Ts'o DY, Roe AW, Gilbert CD (2001) A hierarchy of the functional organization for color, form and disparity in primate visual area V2. Vision Res 41:1333–1349
43. Malonek D, Tootell RB, Grinvald A (1994) Optical imaging reveals the functional architecture of neurons processing shape and motion in owl monkey area MT. Proc Biol Sci 258:109–119
44. Tanigawa H, Lu HD, Roe AW (2010) Functional organization for color and orientation in macaque V4. Nat Neurosci 13: 1542–1548

45. Wang G, Tanaka K, Tanifuji M (1996) Optical imaging of functional organization in the monkey inferotemporal cortex. Science 272: 1665–1668
46. Kalatsky VA, Polley DB, Merzenich MM, Schreiner CE, Stryker MP (2005) Fine functional organization of auditory cortex revealed by Fourier optical imaging. Proc Natl Acad Sci USA 102:13325–13330
47. Bakin JS, Kwon MC, Masino SA, Weinberger NM, Frostig RD (1996) Suprathreshold auditory cortex activation visualized by intrinsic signal optical imaging. Cereb Cortex 6: 120–130
48. Masino SA, Kwon MC, Dory Y, Frostig RD (1993) Characterization of functional organization within rat barrel cortex using intrinsic signal optical imaging through a thinned skull. Proc Natl Acad Sci USA 90: 9998–10002
49. Chen LM, Friedman RM, Ramsden BM, LaMotte RH, Roe AW (2001) Fine-scale organization of SI (area 3b) in the squirrel monkey revealed with intrinsic optical imaging. J Neurophysiol 86:3011–3029
50. Chen LM, Friedman RM, Roe AW (2003) Optical imaging of a tactile illusion in area 3b of the primary somatosensory cortex. Science 302:881–885
51. Shoham D, Grinvald A (2001) The cortical representation of the hand in macaque and human area S-I: high resolution optical imaging. J Neurosci 21:6820–6835
52. Chen LM, Friedman RM, Roe AW (2009) Area-specific representation of mechanical nociceptive stimuli within SI cortex of squirrel monkeys. Pain 141:258–268
53. Grinvald A, Frostig RD, Siegel RM, Bartfeld E (1991) High-resolution optical imaging of functional brain architecture in the awake monkey. Proc Natl Acad Sci USA 88: 11559–11563
54. Vnek N, Ramsden BM, Hung CP, Goldman-Rakic PS, Roe AW (1999) Optical imaging of functional domains in the cortex of the awake and behaving monkey. Proc Natl Acad Sci USA 96:4057–4060
55. Arieli A, Grinvald A, Slovin H (2002) Dural substitute for long-term imaging of cortical activity in behaving monkeys and its clinical implications. J Neurosci Methods 114:119–133
56. Chen LM, Heider B, Williams GV, Healy FL, Ramsden BM, Roe AW (2002) A chamber and artificial dura method for long-term optical imaging in the monkey. J Neurosci Methods 113:41–49
57. Chapman B, Stryker MP, Bonhoeffer T (1996) Development of orientation preference maps in ferret primary visual cortex. J Neurosci 16:6443–6453
58. Godecke I, Kim DS, Bonhoeffer T, Singer W (1997) Development of orientation preference maps in area 18 of kitten visual cortex. Eur J Neurosci 9:1754–1762
59. Lowel S, Schmidt KE, Kim DS, Wolf F, Hoffsummer F, Singer W, Bonhoeffer T (1998) The layout of orientation and ocular dominance domains in area 17 of strabismic cats. Eur J Neurosci 10:2629–2643
60. Schmidt KE, Singer W, Galuske RA (2004) Processing deficits in primary visual cortex of amblyopic cats. J Neurophysiol 91:1661–1671
61. Schwartz TH, Bonhoeffer T (2001) In vivo optical mapping of epileptic foci and surround inhibition in ferret cerebral cortex. Nat Med 7:1063–1067
62. Zhao M, Suh M, Ma H, Perry C, Geneslaw A, Schwartz TH (2007) Focal increases in perfusion and decreases in hemoglobin oxygenation precede seizure onset in spontaneous human epilepsy. Epilepsia 48:2059–2067
63. Pouratian N, Cannestra AF, Martin NA, Toga AW (2002) Intraoperative optical intrinsic signal imaging: a clinical tool for functional brain mapping. Neurosurg Focus 13:e1
64. Toga AW, Cannestra AF, Black KL (1995) The temporal/spatial evolution of optical signals in human cortex. Cereb Cortex 5: 561–565
65. Eaton WA, Henry ER, Hofrichter J, Mozzarelli A (1999) Is cooperative oxygen binding by hemoglobin really understood? Nat Struct Biol 6:351–358
66. Dunn AK, Devor A, Dale AM, Boas DA (2005) Spatial extent of oxygen metabolism and hemodynamic changes during functional activation of the rat somatosensory cortex. Neuroimage 27:279–290
67. Devor A, Dunn AK, Andermann ML, Ulbert I, Boas DA, Dale AM (2003) Coupling of total hemoglobin concentration, oxygenation, and neural activity in rat somatosensory cortex. Neuron 39:353–359
68. Dunn AK, Devor A, Bolay H, Andermann ML, Moskowitz MA, Dale AM, Boas DA (2003) Simultaneous imaging of total cerebral hemoglobin concentration, oxygenation, and blood flow during functional activation. Opt Lett 28:28–30
69. Wilson DF, Pastuszko A, DiGiacomo JE, Pawlowski M, Schneiderman R, Delivoria-Papadopoulos M (1991) Effect of hyperventila-

tion on oxygenation of the brain cortex of newborn piglets. J Appl Physiol 70:2691–2696

70. Wilson DF, Gomi S, Pastuszko A, Greenberg JH (1993) Microvascular damage in the cortex of cat brain from middle cerebral artery occlusion and reperfusion. J Appl Physiol 74: 580–589
71. Estrada AD, Ponticorvo A, Ford TN, Dunn AK (2008) Microvascular oxygen quantification using two-photon microscopy. Opt Lett 33:1038–1040
72. Rumsey WL, Vanderkooi JM, Wilson DF (1988) Imaging of phosphorescence: a novel method for measuring oxygen distribution in perfused tissue. Science 241:1649–1651
73. Vanderkooi JM, Maniara G, Green TJ, Wilson DF (1987) An optical method for measurement of dioxygen concentration based upon quenching of phosphorescence. J Biol Chem 262:5476–5482
74. Lebedev AY, Cheprakov AV, Sakadzic S, Boas DA, Wilson DF, Vinogradov SA (2009) Dendritic phosphorescent probes for oxygen imaging in biological systems. ACS Appl Mater Interfaces 1:1292–1304
75. Vinogradov SA, Wilson DF (2011) Porphyrin-dendrimers as biological oxygen sensors. In: Campagna S, Ceroni P, Puntoriero F (eds) Designing dendrimers. Wiley, New York
76. Vinogradov SA, Wilson DF (1995) Metallotetrabenzoporphyrins. New phosphorescent probes for oxygen measurements. J Chem Soc Perkin Trans 2:103–111
77. Vinogradov SA, Lo LW, Wilson DF (1999) Dendritic polyglutamic porphyrins: probing porphyrin protection by oxygen-dependent quenching of phosphorescence. Chem Eur J 5:1338–1347
78. Rietveld IB, Kim E, Vinogradov SA (2003) Dendrimers with tetrabenzoporphyrin cores: near infrared phosphors for in vivo oxygen imaging. Tetrahedron 59:3821–3831
79. Dunphy I, Vinogradov SA, Wilson DF (2002) Oxyphor R2 and G2: phosphors for measuring oxygen by oxygen-dependent quenching of phosphorescence. Anal Biochem 310: 191–198
80. Lebedev AY, Troxler T, Vinogradov SA (2008) Design of metalloporphyrin-based dendritic nanoprobes for two-photon microscopy of oxygen. J Porphyr Phthalocyanines 12:1261–1269
81. Finikova OS, Lebedev AY, Aprelev A, Troxler T, Gao F, Garnacho C, Muro S, Hochstrasser RM, Vinogradov SA (2008) Oxygen microscopy by two-photon-excited phosphorescence. Chemphyschem 9:1673–1679
82. Sakadzic S, Roussakis E, Yaseen MA, Srinivasan VJ, Mandeville ET, Devor A, Lo EH, Vinogradov SA, Boas DA (2010) Two-photon high-resolution 3D imaging of partial pressure of oxygen in cerebral vasculature and tissue. Nat Methods 7:755–759
83. Lecoq J, Parpaleix A, Roussakis E, Ducros M, Houssen YG, Vinogradov SA, Charpak S (2011) Simultaneous two-photon imaging of oxygen and blood flow in deep cerebral vessels. Nat Med 17:893–898
84. Devor A, Sakadzic S, Saisan PA, Yaseen MA, Roussakis E, Srinivasan VJ, Vinogradov SA, Rosen BR, Buxton RB, Dale AM, Boas DA (2011) “Overshoot” of O2 is required to maintain baseline tissue oxygenation at locations distal to blood vessels. J Neurosci 31: 13676–13681
85. Tootell RB, Silverman MS, De Valois RL, Jacobs GH (1983) Functional organization of the second cortical visual area in primates. Science 220:737–739
86. Weber B, Keller AL, Reichold J, Logothetis NK (2008) The microvascular system of the striate and extrastriate visual cortex of the macaque. Cereb Cortex 18:2318–2330
87. Keller AL, Schuz A, Logothetis NK, Weber B (2011) Vascularization of cytochrome oxidase-rich blobs in the primary visual cortex of squirrel and macaque monkeys. J Neurosci 31: 1246–1253
88. Fatt I (1976) Polarographic oxygen sensors. CRC Press, Cleveland
89. Thompson JK, Peterson MR, Freeman RD (2003) Single-neuron activity and tissue oxygenation in the cerebral cortex. Science 299: 1070–1072
90. Viswanathan A, Freeman RD (2007) Neurometabolic coupling in cerebral cortex reflects synaptic more than spiking activity. Nat Neurosci 10:1308–1312
91. Offenhauser N, Thomsen K, Caesar K, Lauritzen M (2005) Activity-induced tissue oxygenation changes in rat cerebellar cortex: interplay of postsynaptic activation and blood flow. J Physiol 565:279–294
92. Erecinska M, Silver IA (2001) Tissue oxygen tension and brain sensitivity to hypoxia. Respir Physiol 128:263–276
93. Masamoto K, Takizawa N, Kobayashi H, Oka K, Tanishita K (2003) Dual responses of tissue partial pressure of oxygen after functional stimulation in rat somatosensory cortex. Brain Res 979:104–113
94. Sharan M, Vovenko EP, Vadapalli A, Popel AS, Pittman RN (2008) Experimental and theoretical studies of oxygen gradients in rat

pial microvessels. J Cereb Blood Flow Metab 28:1597–1604

95. Vazquez AL, Fukuda M, Tasker ML, Masamoto K, Kim SG (2010) Changes in cerebral arterial, tissue and venous oxygenation with evoked neural stimulation: implications for hemoglobin-based functional neuroimaging. J Cereb Blood Flow Metab 30: 428–439
96. Vovenko E (1999) Distribution of oxygen tension on the surface of arterioles, capillaries and venules of brain cortex and in tissue in normoxia: an experimental study on rats. Pflugers Arch 437:617–623
97. Koch CJ (2002) Measurement of absolute oxygen levels in cells and tissues using oxygen sensors and 2-nitroimidazole EF5. Methods Enzymol 352:3–31
98. Swartz HM, Clarkson RB (1998) The measurement of oxygen in vivo using EPR techniques. Phys Med Biol 43:1957–1975
99. Ahmad R, Kuppusamy P (2010) Theory, instrumentation, and applications of electron paramagnetic resonance oximetry. Chem Rev 110:3212–3236
100. Wang LV (2009) Multiscale photoacoustic microscopy and computed tomography. Nat Photonics 3:503–509
101. Fu D, Matthews TE, Ye T, Piletic IR, Warren WS (2008) Label-free in vivo optical imaging of microvasculature and oxygenation level. J Biomed Opt 13:040503
102. Ashkenazi S, Huang SW, Horvath T, Koo YE, Kopelman R (2008) Photoacoustic probing of fluorophore excited state lifetime with application to oxygen sensing. J Biomed Opt 13:034023
103. Geissbuehler M, Spielmann T, Formey A, Marki I, Leutenegger M, Hinz B, Johnsson K, Van De Ville D, Lasser T (2010) Triplet imaging of oxygen consumption during the contraction of a single smooth muscle cell (A7r5). Biophys J 98:339–349
104. Spielmann T, Blom H, Geissbuehler M, Lasser T, Widengren J (2010) Transient state monitoring by total internal reflection fluorescence microscopy. J Phys Chem B 114: 4035–4046
105. Mik EG, van Leeuwen TG, Raat NJ, Ince C (2004) Quantitative determination of localized tissue oxygen concentration in vivo by two-photon excitation phosphorescence lifetime measurements. J Appl Physiol 97: 1962–1969
106. Brinas RP, Troxler T, Hochstrasser RM, Vinogradov SA (2005) Phosphorescent oxygen sensor with dendritic protection and two-photon absorbing antenna. J Am Chem Soc 127:11851–11862
107. Srinivasan VJ, Sakadzic S, Gorczynska I, Ruvinskaya L, Wu W, Fujimoto JG, Boas DA (2010) Quantitative cerebral blood flow with optical coherence tomography. Opt Express 18:2477–2494
108. Srinivasan VJ, Sakadzic S, Gorczynska I, Ruvinskaya S, Wu W, Fujimoto JG, Boas DA (2009) Depth-resolved microscopy of cortical hemodynamics with optical coherence tomography. Opt Lett 34:3086–3088
109. Deneux T, Faugeras O, Takerkart S, Masson G, Vanzetta I (2011) A new variational method for erythrocyte velocity estimation in wide-field imaging in-vivo. IEEE Trans Med Imaging 30:1527–1545
110. Turner DA, Foster KA, Galeffi F, Somjen GG (2007) Differences in O2 availability resolve the apparent discrepancies in metabolic intrinsic optical signals in vivo and in vitro. Trends Neurosci 30:390–398
111. Nemoto M, Nomura Y, Sato C, Tamura M, Houkin K, Koyanagi I, Abe H (1999) Analysis of optical signals evoked by peripheral nerve stimulation in rat somatosensory cortex: dynamic changes in hemoglobin concentration and oxygenation. J Cereb Blood Flow Metab 19:246–259
112. Narayan SM, Esfahani P, Blood AJ, Sikkens L, Toga AW (1995) Functional increases in cerebral blood volume over somatosensory cortex. J Cereb Blood Flow Metab 15: 754–765
113. Narayan SM, Santori EM, Blood AJ, Burton JS, Toga AW (1994) Imaging optical reflectance in rodent barrel and forelimb sensory cortex. Neuroimage 1:181–190
114. Grinvald A (1992) Optical imaging of architecture and function in the living brain sheds new light on cortical mechanisms underlying visual perception. Brain Topogr 5: 71–75
115. Huang D, Swanson EA, Lin CP, Schuman JS, Stinson WG, Chang W, Hee MR, Flotte T, Gregory K, Puliafito CA et al (1991) Optical coherence tomography. Science 254: 1178–1181
116. Wojtkowski M, Bajraszewski T, Targowski P, Kowalczyk A (2003) Real-time in vivo imaging by high-speed spectral optical coherence tomography. Opt Lett 28:1745–1747
117. Harbig K, Chance B, Kovach AG, Reivich M (1976) In vivo measurement of pyridine nucleotide fluorescence from cat brain cortex. J Appl Physiol 41:480–488
118. Lothman E, Lamanna J, Cordingley G, Rosenthal M, Somjen G (1975) Responses of

electrical potential, potassium levels, and oxidative metabolic activity of the cerebral neocortex of cats. Brain Res 88:15–36

119. Mayevsky A (1984) Brain NADH redox state monitored in vivo by fiber optic surface fluorometry. Brain Res 319:49–68

120. Baraghis E, Devor A, Fang Q, Srinivasan VJ, Yaseen MA, Wu W, Lesage F, Ayata C, Kasischke KA, Boas DA, Sakadzic S (2011) Two-photon microscopy of cortical NADH fluorescence intensity changes: correcting contamination from the hemodynamic response. J Biomed Opt 16:106003

121. Kramer RS, Pearlstein RD (1979) Cerebral cortical microfluorometry at isosbestic wavelengths for correction of vascular artifact. Science 205:693–696

122. Boas DA, Jones SR, Devor A, Huppert TJ, Dale AM (2008) A vascular anatomical network model of the spatio-temporal response to brain activation. Neuroimage 40:1116–1129

123. Devor A, Ulbert I, Dunn AK, Narayanan SN, Jones SR, Andermann ML, Boas DA, Dale AM (2005) Coupling of the cortical hemodynamic response to cortical and thalamic neuronal activity. Proc Natl Acad Sci USA 102: 3822–3827

124. Hillman EM (2007) Optical brain imaging in vivo: techniques and applications from animal to man. J Biomed Opt 12:051402

125. Berwick J, Johnston D, Jones M, Martindale J, Martin C, Kennerley AJ, Redgrave P, Mayhew JE (2008) Fine detail of neurovascular coupling revealed by spatiotemporal analysis of the hemodynamic response to single whisker stimulation in rat barrel cortex. J Neurophysiol 99:787–798

126. Devor A, Hillman EM, Tian P, Waeber C, Teng IC, Ruvinskaya L, Shalinsky MH, Zhu H, Haslinger RH, Narayanan SN, Ulbert I, Dunn AK, Lo EH, Rosen BR, Dale AM, Kleinfeld D, Boas DA (2008) Stimulus-induced changes in blood flow and 2-deoxyglucose uptake dissociate in ipsilateral somatosensory cortex. J Neurosci 28:14347–14357

127. Devor A, Tian P, Nishimura N, Teng IC, Hillman EM, Narayanan SN, Ulbert I, Boas DA, Kleinfeld D, Dale AM (2007) Suppressed neuronal activity and concurrent arteriolar vasoconstriction may explain negative blood oxygenation level-dependent signal. J Neurosci 27:4452–4459

128. Bouchard MB, Chen BR, Burgess SA, Hillman EM (2009) Ultra-fast multispectral optical imaging of cortical oxygenation, blood flow, and intracellular calcium dynamics. Opt Express 17:15670–15678

129. Calcinaghi N, Jolivet R, Wyss MT, Ametamey SM, Gasparini F, Buck A, Weber B (2011) Metabotropic glutamate receptor mGluR5 is not involved in the early hemodynamic response. J Cereb Blood Flow Metab 31: e1–e10

130. Jones PB, Shin HK, Boas DA, Hyman BT, Moskowitz MA, Ayata C, Dunn AK (2008) Simultaneous multispectral reflectance imaging and laser speckle flowmetry of cerebral blood flow and oxygen metabolism in focal cerebral ischemia. J Biomed Opt 13:044007

131. Polimeni JR, Granquist-Fraser D, Wood RJ, Schwartz EL (2005) Physical limits to spatial resolution of optical recording: clarifying the spatial structure of cortical hypercolumns. Proc Natl Acad Sci USA 102:4158–4163

132. Tian P, Teng IC, May LD, Kurz R, Lu K, Scadeng M, Hillman EM, De Crespigny AJ, D'Arceuil HE, Mandeville JB, Marota JJ, Rosen BR, Liu TT, Boas DA, Buxton RB, Dale AM, Devor A (2010) Cortical depth-specific microvascular dilation underlies laminar differences in blood oxygenation level-dependent functional MRI signal. Proc Natl Acad Sci USA 107:15246–15251

133. Boas D, Culver J, Stott J, Dunn A (2002) Three dimensional Monte Carlo code for photon migration through complex heterogeneous media including the adult human head. Opt Express 10:159–170

134. Tian P, Devor A, Sakadzic S, Dale AM, Boas DA (2010) Monte Carlo simulation of the spatial resolution and depth sensitivity of two-dimensional optical imaging of the brain. J Biomed Opt 16:016006

135. Wilson BC, Adam G (1983) A Monte Carlo model for the absorption and flux distributions of light in tissue. Med Phys 10:24–830

136. Chen BR, Bouchard MB, McCaslin AF, Burgess SA, Hillman EM (2011) High-speed vascular dynamics of the hemodynamic response. Neuroimage 54:1021–1030

137. Tian P, Teng IC, Dunn AK, Boas DA, Dale AM, Devor A (2007) Hemodynamic augmentation in response to a paired whisker stimulus as a function of cortical distance. In: Program No. 87.6. Neuroscience 2007 abstracts, Society for Neuroscience, San Diego. Online

138. Vinogradov SA, Lo LW, Jenkins WT, Evans SM, Koch C, Wilson DF (1996) Noninvasive imaging of the distribution in oxygen in tissue in vivo using near-infrared phosphors. Biophys J 70:1609–1617

139. Sakadzic S, Yuan S, Dilekoz E, Ruvinskaya S, Vinogradov SA, Ayata C, Boas DA (2009)

Simultaneous imaging of cerebral partial pressure of oxygen and blood flow during functional activation and cortical spreading depression. Appl Opt 48:D169–D177
140. Apreleva SV, Wilson DF, Vinogradov SA (2006) Tomographic imaging of oxygen by phosphorescence lifetime. Appl Opt 45:8547–8559
141. Golub AS, Barker MC, Pittman RN (2007) PO2 profiles near arterioles and tissue oxygen consumption in rat mesentery. Am J Physiol Heart Circ Physiol 293:H1097–H1106
142. Golub AS, Pittman RN (2008) PO2 measurements in the microcirculation using phosphorescence quenching microscopy at high magnification. Am J Physiol Heart Circ Physiol 294:H2905–H2916
143. Yaseen MA, Srinivasan VJ, Sakadzic S, Wu W, Ruvinskaya S, Vinogradov SA, Boas DA (2009) Optical monitoring of oxygen tension in cortical microvessels with confocal microscopy. Opt Express 17:22341–22350
144. Wilson DF, Vinogradov SA, Grosul P, Vaccarezza MN, Kuroki A, Bennett J (2005) Oxygen distribution and vascular injury in the mouse eye measured by phosphorescence-lifetime imaging. Appl Opt 44:5239–5248
145. Shonat RD, Kight AC (2003) Oxygen tension imaging in the mouse retina. Ann Biomed Eng 31:1084–1096
146. Shonat RD, Wilson DF, Riva CE, Pawlowski M (1992) Oxygen distribution in the retinal and choroidal vessels of the cat as measured by a new phosphorescence imaging method. Appl Opt 31:3711–3718
147. Ponticorvo A, Dunn AK (2010) Simultaneous imaging of oxygen tension and blood flow in animals using a digital micromirror device. Opt Express 18:8160–8170
148. Sinks LE, Robbins GP, Roussakis E, Troxler T, Hammer DA, Vinogradov SA (2010) Two-photon microscopy of oxygen: polymersomes as probe carrier vehicles. J Phys Chem B 114:14373–14382
149. Garaschuk O, Milos RI, Konnerth A (2006) Targeted bulk-loading of fluorescent indicators for two-photon brain imaging in vivo. Nat Protoc 1:380–386
150. Boorman L, Kennerley AJ, Johnston D, Jones M, Zheng Y, Redgrave P, Berwick J (2010) Negative blood oxygen level dependence in the rat: a model for investigating the role of suppression in neurovascular coupling. J Neurosci 30:4285–4294
151. Kennerley AJ, Berwick J, Martindale J, Johnston D, Papadakis N, Mayhew JE (2005) Concurrent fMRI and optical measures for the investigation of the hemodynamic response function. Magn Reson Med 54:354–365
152. Stosiek C, Garaschuk O, Holthoff K, Konnerth A (2003) In vivo two-photon calcium imaging of neuronal networks. Proc Natl Acad Sci USA 100:7319–7324

Chapter 15

Laser Speckle Imaging of Cerebral Blood Flow

Pavel Zakharov, Frank Scheffold, and Bruno Weber

Abstract

Optical imaging of cerebral hemodynamics has been used extensively to investigate the complex interplay between neurons and the blood-borne supply of oxygen and energy substrates. Among several techniques, the so-called laser speckle imaging (LSI) has found widespread applications for studying (patho-)physiological aspects of neurovascular coupling. LSI is a rather simple and cost-effective method to acquire two-dimensional maps of cerebral blood flow from the cortex of experimental animals. In this chapter, we introduce the physical underpinnings as well as the most important technical and practical aspects of its implementation.

Key words Optical imaging, Cerebral blood flow, Laser speckle, Hemodynamics, Laser, In vivo, Somatosensory system, Barrel cortex

1 Introduction

Early on, high-resolution functional neuroimaging with optical tools was tightly linked to cerebral blood flow (CBF). Indeed, pioneering work of Chance et al. [1] and Jöbsis [2] demonstrated that the intrinsic properties of the chromophores and their relationship to hemodynamics could be applied for monitoring neuronal activity. This work paved the way for the first high-resolution optical mapping of hemodynamic signals by the group of Amiram Grinvald using charge-coupled device (CCD) technology [3, 4]. Since then, optical imaging of intrinsic signals (OIS) has been extensively applied for studying cortical processing. Apart from using OIS as a surrogate for neuronal activity and thereby for mapping cortical processing, optical imaging became an important tool for investigating neurovascular coupling. This research field has grown considerably during the last 2 decades, as the processes governing the regulation of CBF are still far from being understood (for a recent review, see Attwell et al. [5]). Many of the widely used noninvasive imaging tools to study the human brain in action—such as blood oxygenation level-dependent (BOLD) functional

Bruno Weber and Fritjof Helmchen (eds.), *Optical Imaging of Neocortical Dynamics*, Neuromethods, vol. 85,
DOI 10.1007/978-1-62703-785-3_15, © Springer Science+Business Media New York 2014

magnetic resonance imaging—rely on hemodynamic signals. Furthermore, a number of the most relevant neurological diseases (such as stroke or Alzheimer's disease) show abnormalities in vascular structure and function [6]. Although OIS is based on the hemodynamic response that follows an increase in neuronal activity (see also Chapter 14 in this volume), it is important to note that blood oxygenation and CBF are by no means identical. Interestingly, changes in blood oxygenation and CBF can be uncoupled in certain (patho-)physiological conditions. For example, Lindauer et al. [7] could show that increased baseline blood oxygenation caused by hyperbaric hyperoxia can completely abolish the deoxy-Hb response, while the CBF response remains unchanged. For many questions related to neurovascular coupling and pathophysiological states, direct measurements of CBF are thus mandatory. This is the main reason for today's usage of complementary optical methods that directly monitor CBF. In this field laser speckle imaging (LSI) has found widespread application, as it allows imaging of local changes in CBF with rather high spatial and temporal resolution [8–11].

LSI is an implementation of the family of methods based on quasi-elastic light scattering (QELS). The method employs the statistical analysis of temporal fluctuations of light scattered by microscale dynamic scattering heterogeneities in order to extract quantitative information about their dynamics. The motion of scattering objects, such as red blood cells (RBCs), results in Doppler shifts, which are minuscule as compared to the frequency of light. Thus the scattering is coined "quasi-elastic" meaning that the light energy is essentially conserved throughout the whole process.

The frequency-shifted light is further optically mixed with the light scattered by the static tissues. If within the volume under observation the scatterers move in random directions with fluctuating velocity the Doppler shifts add up and a Doppler broadening is observed rather than a shift. Such random motion can be due to the intrinsic Brownian motion but can also be mimicked by the complex blood flow pattern in vascularized tissue. For the case of the cortical microvascular system which lacks defined flow directions the light reaching the photodetector produces the noise-like fluctuations of the photocurrent with the spectral properties defined by the scattering geometry, number of scattering particles (typically RBCs), and their velocity. It can be shown [12] that the first moment of the power spectrum is proportional to product of scatterer concentration and average velocity, while the zero-th moment is proportional to the concentration. The variations of photocurrent are typically processed with the spectrum analyzer, which provides the estimation of the spectral moments. The technique is usually referred to as Laser Doppler Velocimetry (LDV) or flowmetry (LDF).

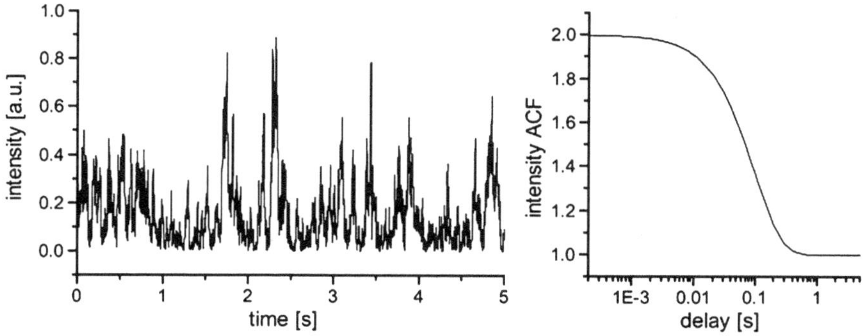

Fig. 1 Illustration of the intensity fluctuations caused by the movement of scattering particles (*left*) and corresponding intensity correlation function (*right*). From [15].

For this case the narrow line-width of incident laser light is slightly broadened due to motion, which can be directly detected by (Doppler-) spectroscopy but can also be regarded as time fluctuating random diffraction pattern. The latter is known as *speckle* and it is fluctuating due to changes in position of the scatterers within the scattering volume. Such fluctuations can be analyzed in the time domain with the help of the intensity correlation function, which is related to the frequency power spectrum via the Fourier transform (Wiener–Khinchin theorem) [13]. A hardware or software correlator can be used for the analysis of these fluctuations. The resulting temporal fluctuations of light intensities in the millisecond range are easily analyzed while the corresponding spectral broadening of the order of $\Delta f = 1/\tau \sim 1$ kHz is often more difficult to measure. The formalism outlined in the next section can be further applied to single or multiple scattering alike. The former is often called "dynamic light scattering" (DLS) while the latter is know under the name of "diffusing wave spectroscopy" (DWS) [14].

In general, in QELS the temporal fluctuations of scattered light intensity $I(t)$ are measured (see Fig. 1) and the intensity correlation function (ICF) is calculated:

$$g_2(\tau) = \frac{\langle I(t) I(t+\tau) \rangle_t}{\langle I(t) \rangle_t^2}, \tag{1}$$

where $\langle \cdots \rangle_t$ denotes time averaging. The physical meaning is attributed to the field $E(t)$ correlation function $g_1(\tau)$:

$$g_1(\tau) = \frac{\langle E(t) E^*(t+\tau) \rangle_t}{\langle |E(t)| \rangle_t^2}, \tag{2}$$

which is related to the measurable ICF via the Siegert relation [16]:

$$g_2(\tau) = 1 + \beta |g_1(\tau)|^2, \tag{3}$$

where β is usually referred to as coherence factor and depends on the detection conditions. In an ideal case when single speckle mode is detected it is approaching the value of 1. In the case of complete loss of coherence $\beta = 0$.

While the approaches to data analysis of Doppler and DLS/DWS methods are somewhat different, the underlying physical principles are equal [17] and thus it makes sense to refer to these methods using the most general term of QELS.

2 Method Description

2.1 Theoretical Basis of the Laser Speckle Methods

Among the dynamic processes in the body, one of the most prominent and best studied is blood circulation. QELS methods are suitable for the characterization of blood flow, where moving RBCs act as dynamic scatterers. In 1972 Riva et al. have demonstrated the applicability of the QELS for the measurements of blood flow in the retinal arteries, as well as whole blood flowing through capillary tubes [18]. Even in the presence of multiple scattering by RBCs, the authors were able to obtain a linear relation of the flow rate and frequency of the peak in the intensity fluctuations spectrum. Three years later Stern performed measurements of the microcirculation in the finger [19]. Due to the plurality of the flow directions and speeds of flow in the microvascular network no peak has been observed in the spectrum of the intensity fluctuations; however, energy contained in the low-frequency part can be used to characterize perfusion flow. Further on, Stern et al. reported observations of the pulsatile component in the Doppler system and studied the blood perfusion of the outer cortex of the rat kidney in response to the application of several drugs [20]. Later on Bonner and Nassal [21] have developed a model, which laid the theoretical ground for the application of LDF for the microcirculation analysis.

In the beginning of the 1980s, Fercher and Briers [22, 23] suggested to use the amount of speckle image blurring for the estimation of the correlation time of intensity fluctuations. The dynamics of the speckle pattern affects the obtained image if the decorrelation time is comparable to the exposure time T of the camera since the speckle pattern evolves during the exposure. This leads to the blurring of the speckle and decreases the contrast of speckle image as detected by the camera. This approach has been coined Laser Speckle Contrast Analysis (LASCA) since it employs the spatial contrast (standard deviation of the time-integrated intensity pattern divided by the mean intensity) as a measure of the blurring:

$$K(T) = \frac{\sigma(I_T)}{\langle I_T \rangle} = \sqrt{\frac{\langle I_T^2 \rangle}{\langle I_T \rangle^2} - 1} \qquad (4)$$

where I_T denotes the intensity integrated over the time period T, $\sigma(I_T)$ is its standard deviation, and $\langle \ldots \rangle$ refers to means obtained over the spatial pixels.

With faster dynamics, and thus shorter correlation time of the intensity fluctuations, the effect of blurring becomes larger and the contrast K decreases. This relation between contrast and correlation function can be quantified in the following way [24–26] (in order to keep the notation simple we will discuss the properties of the square of contrast K^2):

$$K^2(T) = \frac{2}{T}\int_0^T \left[g_2(\tau) - 1\right]\left(1 - \frac{\tau}{T}\right) d\tau = \frac{2\beta}{T}\int_0^T \left|g_1(\tau)\right|^2 \left(1 - \frac{\tau}{T}\right) d\tau \quad (5)$$

For some specific simple forms of the correlation function analytic solutions can be derived for the correlation time dependence K [25].

For example, for a single exponential decay of the correlation function $g_1{}^s(\tau) = \exp(-\tau/\tau_c)$, where τ_c is called correlation time, the contrast is given by the following expression [25]:

$$K^2(T) = \beta\left[exp(-2x) - 1 + 2x\right] / 2x^2, \quad (6)$$

where $x = T/\tau_c$.

For the case of DWS backscattering from a turbid medium undergoing simple Brownian motion, the field correlation function is given by the following equation $g_1^m(\tau) = exp\left(-\gamma\sqrt{6\tau/\tau_0}\right)$ and one finds for the contrast

$$K^2(T) = \beta\left[\left(3 + 6\sqrt{x} + 4x\right)exp\left(-2\sqrt{x}\right) - 3 + 2x\right] / 2x^2 \quad (7)$$

where $x = 6\gamma^2 T/\tau_0$.

Both expressions essentially assume a single diffusive relaxation process such as provided by Brownian motion of monosized micron scale objects dispersed uniformly in a medium. Within that framework the former result of (6) is exact in the case of single scattering while the latter (7) is a very good approximation for the case of multiple scattering. One should keep in mind that the assumptions made are certainly much too simple given the complexity of real tissue structure and dynamics. Nevertheless, given the simplicity of the results and the lack of more sophisticated models, both equations have been used in the past to analyze LSI on experimental animals.

The experimental arrangement for the speckle contrast imaging is relatively simple. A quantitative interpretation of the contrast values requires some assumptions concerning the particular shape of the correlation function and the impact of a static component in the detected signal [27]. In addition, the contrast has the maximal sensitivity (with respect to the underlying relaxation processes) if the correlation time is of the order of the exposure time [28] and it is significantly reduced if the exposure time is much smaller or much larger than the correlation time.

In this limit a conversion of the contrast to the correlation values can be performed only with significant error.

Since the original work by Riva et al. [18] hundreds of scientific papers have been published on various applications of light scattering methods to resolve physiological questions. A number of companies have successfully commercialized the QELS systems for monitoring the microcirculation, e.g., Moor Instruments Inc. (UK), Perimed Instruments AB (Sweden), Aimago SA (Switzerland). The majority of these devices are based on the laser Doppler paradigm; however, the LSI system has also been commercialized recently.

Nevertheless, the DLS methods are still suffering from several important drawbacks, which impede their penetration to the regular clinical practice. The major problem is the absence of a quantitative link with the real physiological parameters of microvascular flow, expressed in common units. Instead most of the devices provide relative units. Another drawback is the significant temporal and spatial variability of the measurements. The temporal variability is usually related to the intrinsic non-stationarity of the flow, which is changing during a heartbeat cycle, breath, and the dynamic processes in the tissue itself, which is also called biological zero. The spatial variability is usually related to the small sampling volume of the fiber system (within 1 mm^3), which is affected by the local structure of the vasculature and the relative amount of Doppler-shifted photons in the signal [29].

The so-called non-ergodicity problem might arise if the detected light consists of components with different dynamics, e.g., light scattered by moving RBCs and light only reflected from the static tissues or bone. The non-ergodicity of the signal is not addressed in the proper way if the impact of the non-shifted photons is ignored in the analysis, which is often the case. This in turn results in the incorrect estimate of blood flow.

A correlation-based analysis of QELS signals can potentially address some of the problems mentioned above since an extensive theory and various experimental techniques have been developed for the quantitative interpretation of data, also addressing the problem of non-ergodicity [30–32]. For example, it is essential to analyze the speckle statistics not only in temporal but also in spatial domain, which is the basis of the so-called multi-speckle technique [33–35]. The original LASCA method based on the spatial speckle statistics ensures ergodic averaging, since the speckle size on the detector is adjusted to the pixel size and calculation within LASCA window $N \times N$ performs averaging over up to N^2 statistically independent speckles. Unfortunately, simplifications and assumptions of LASCA are also sources of drawbacks, e.g., the original method is not able to properly analyze the speckle field with multiple dynamic components, for example, consisting of a dynamic contribution from moving RBCs, a slow evolving component caused by vasomotion, and the static contribution from the rigid tissues or bone.

However, it has been demonstrated that with the help of the additional temporal analysis [36] or using images obtained with different exposure times [37] the static component can be separated from the contrast signal. This is the main idea underpinning the approach of dynamic laser speckle imaging (dLSI), which is discussed in the next section.

2.2 Dynamic Laser Speckle Imaging

There exist a number of practically important experimental situations, when the scattering fields of interest are mixed with the nonfluctuating electric fields scattered from static regions. A prominent example is the measurement with a strong static background or dynamical imaging of a volume covered by a static scattering surface. This is the case, for example, for a layered medium with dynamics localized in a deeper layer (e.g., when imaging through the skull).

Although the removal of the dura mater to access the brain cortex is still required in most cases, some groups perform imaging through the thinned skull of rats or through the intact skull of mice [36, 38–41]. This has considerable advantages compared to the open skull experiments, as surgery is simplified and undesired side effects, such as cortical swelling, are avoided.

The influence of the skull on the laser speckle signal has been quantitatively addressed recently [27, 37]. The presence of a static component in the speckle image affects the shape of the field correlation function by introducing a nonzero baseline which results in the overestimation of the contrast [27]. One of the efficient approaches to separate dynamic and static components is the multi-exposure technique [37, 42–44]. Essentially, this technique performs the contrast analysis of the speckle fields integrated with different exposure times instead of a single one, used in the standard LASCA. The images obtained with longer exposure times provide information on the slow dynamics, while shorter times—on faster dynamics. The contrast of the image sampled with large exposure time gives an indication of the contrast baseline introduced by the static or pseudo-static components that can be subsequently subtracted from the contrast of images with shorter exposure in order to provide the quantitatively correct estimation of the blood flow dynamics. In order to obtain multi-exposed images the gating of the laser beam can be performed using, e.g., an acousto-optic modulator, while the camera is kept with the constant exposure [37]. This allows obtaining the images with constant intensity level while changing the exposure during the measurements.

We have recently proposed [36] a somewhat simpler processing scheme using a single exposure time which we called dLSI. dLSI allows accounting for static contribution to the optical signal and thus facilitates correct qualitative and quantitative interpretation of CBF. At the same time it does not require modifications of the typical technically simple LASCA experimental settings and image acquisition procedure, which is a major advantage of this approach.

Contributions of the static and dynamic components to the contrast cannot be distinguished in a traditional LASCA approach since one obtains the weighted integral of the field correlation function [22, 23]. It is however possible to derive a rather simple formalism to deal with the dynamics of intensity fluctuations in the presence of a static component. We can define the static contribution as the light that has been scattered by rigid structures only (such as bone or a skin layer which is not highly perfused with blood). The detected intensity in this case is composed of a non-fluctuating static part I_s and a dynamic part I_d with a relative static contribution to the total intensity: $\rho = \langle I_s \rangle / (\langle I_s \rangle + \langle I_d \rangle)$. In this case the correlation function of the mixed field can be written as a composition of the correlation function of the pure dynamic part $g_{1d}(\tau)$ and the time-independent baseline ρ from static light [27, 45]:

$$g_1(\tau) = (1-\rho)|g_{1d}(\tau)| + \rho \tag{8}$$

The spatial contrast of the superposition of static and dynamic speckles can be found by substituting (8) into (5):

$$\begin{aligned} K^2(T) &= \frac{2\beta}{T}\int_0^T \left[(1-\rho)|g_{1d}(\tau)| + \rho\right]^2 \left(1-\frac{\tau}{T}\right) d\tau \\ &= \beta\left[(1-\rho)^2 K_{2d}^2(T) + 2\rho(1-\rho)K_{1d}^2(T) + \rho^2\right] \end{aligned} \tag{9}$$

where the K^2_{1d} and K^2_{2d} represent the contrast of the dynamic field and dynamic intensity, respectively:

$$\begin{aligned} K_{1d}^2(T) &= \frac{2}{T}\int_0^T |g_{1d}(\tau)|\left(1-\frac{\tau}{T}\right) d\tau \\ K_{2d}^2(T) &= \frac{2}{T}\int_0^T |g_{1d}(\tau)|^2\left(1-\frac{\tau}{T}\right) d\tau \end{aligned} \tag{10}$$

Together K^2_{1d} and K^2_{2d} define the mixed dynamic contrast:

$$K_{12d}^2(T) = \beta\left[(1-\rho)^2 K_{2d}^2(T) + 2\rho(1-\rho)K_{1d}^2(T)\right] \tag{11}$$

Since $K^2_{12d}(T)$ is dependent on ρ it is not a pure measure of dynamics. However, it decays to zero for the infinitely short correlation time and reaches the maximum of $\beta(1-\rho^2)$ in the limit of infinitely long correlation time. The total measured contrast can be also written as following:

$$K^2(T) = K_{12d}^2(T) + \beta\rho^2 \tag{12}$$

In the absence of a static component ($\rho = 0$) (9) reduces to (5) and the mixed dynamic contrast equals dynamic intensity contrast K^2_{2d}.

In an LSI experiment the exposure time T (typically from 1 to 20 ms, [28]) is chosen to be larger than the intensity correlation time τ_c (in the range 0.001–0.1 ms for CBF). The time interval between successive frames Δt is for technical reasons larger than T and we find $\Delta t > T >> \tau_c$. We can than assume that $g_{1d}(\Delta t) \approx 0$.

Hence two speckle patterns detected in two sequential frames are correlated only due to the presence of static speckles. We can thus find the amount of static light in the detected signal by comparison of sequential frames and by making use of a multi-speckle averaging scheme [27, 46].

With the correlation of dynamic intensity close to zero in (8) at lag time Δt we can experimentally obtain the static contribution from the correlation of speckle patterns recorded in sequential camera frames:

$$\rho^2 = \left[g_2(\Delta t) - 1 \right] / \beta \tag{13}$$

Since the contribution of the static light can be position-dependent, ρ must be computed locally as well.

One can introduce the obtained ρ into (9) in order to relate the experimentally obtained contrast K^2 to the correlation time τ_c or relaxation time τ_0 of the dynamic part of the correlation function defined by DLS or DWS models. This can be done numerically by calculating the function $K^2(\tau_c)$ or $K^2(\tau_0)$ and taking ρ into account. Instead of integrating the correlation function in (9), one can use the analytical expressions for the dependence of dynamic contrasts $K_{1d}{}^2$ and $K_{2d}{}^2$ on corresponding relaxation times, available for the simplest functional forms of field correlation function (such as single or stretched exponential) [25]. Box 1 provides a step-by-step guide for implementation of the dLSI method.

Box 1 dLSI Algorithm

Further we present a step-by-step guide for implementation of the dLSI method to the laser speckle data.

Step 1: Mixed dynamic contrast

The local mixed dynamic contrast K^2_{12d} can be found for the set of pixels $i = 1 \cdots N$ as following:

$$K^2_{12d} = \frac{\left\langle (I_2 - I_1)^2 \right\rangle}{2\langle I_1 \rangle \langle I_2 \rangle} \tag{14}$$

where $\langle \cdots \rangle$ denotes at least the spatial averaging over a selected set of pixels which can be extended with temporal statistics collected over the same pixels. Speckle intensities $I_1(x_i)$ and $I_2(x_i)$ with $i = 1 \ldots N$ defining the same set of pixels represent two time separated speckle images (for example, sequential images). The mean intensities are found from averaging over the corresponding set of pixels:

$$\langle I \rangle = \frac{1}{N} \sum_{i=1}^{N} I(x_i)$$

and

$$\left\langle (I_2 - I_1)^2 \right\rangle = \frac{1}{N} \sum_{i=1}^{N} \left[I_2(x_i) - I_1(x_i) \right]^2$$

(continued)

Box 1 (continued)

Equation (14) implies that the square of the mixed dynamic contrast is equal to half of the mean squared difference of the sequential image frames normalized with the product of the mean intensities of those frames or the so-called second-order structure function [47] at the inter-frame time. The mixed contrast itself provides a good measure of the overall dynamics in the presence of the static component, since it eliminates the pure static baseline in (9). It can be used to generate the maps of dynamics if the quantitative interpretation is less crucial.

Step 2: Static contribution

As can be seen from (11) the dynamic contrast K_{12d}^2 still includes the unknown static contribution ρ, which is required for the further quantitative conversion to the correlation time. The static contribution can be found with an additional processing step using the same speckle patterns I_1 and I_2 similar to the step 1:

$$\rho = \frac{1}{\beta^{1/2}}\left[\frac{\langle I_1 I_2\rangle}{\langle I_1\rangle\langle I_2\rangle} - 1\right]^{1/2} \tag{15}$$

where

$$\langle I_1 I_2\rangle = \frac{1}{N}\sum_{i=1}^{N} I_1(x_i)\, I_2(x_i)$$

is the mean product of the intensities at the pixels on sequential images. Equation (15) is essentially the multi-speckle calculation of the correlation of images I_1 and I_2.

The coherence factor β used in this calculation has to be obtained separately. It can be calibrated using a solid white medium such as a block of Teflon with $\rho = 1$.

Step 3: Conversion to correlation times

Using the previously obtained static contribution ρ and assuming a certain shape of the dynamic field correlation function $g_{1d}(\tau)$ (e.g., single or stretched exponent), we can find theoretical dependence of dynamic contrast K^2_{12d} on the corresponding correlation or relaxation time. Using the value of mixed dynamic contrast obtained on step 1 we can numerically invert this relation to obtain the correlation time.

These steps should be repeated for all local pixel subsets composing the images in order to obtain the maps of the correlation time.

It has been demonstrated [36] that with the dLSI approach it is possible to perform functional LSI through the thinned or intact skull of rats with higher accuracy.

2.3 Experimental Setup for Laser Speckle Imaging

2.3.1 Animal Preparation

Surgery is preferentially performed under isoflurane anesthesia, whereas during the actual experiment the rodents should be anesthetized with α-chloralose (s.c.). Catheters are placed in a femoral artery and vein for the continuous monitoring of the arterial blood pressure (90–110 mmHg) and for the administration of pharmacological agents. In our lab, animals are normally tracheotomized and mechanically ventilated to maintain physiological arterial levels of pH (7.35–7.45), pCO_2 (34–38 mmHg), and pO_2 (>100 mmHg). The body temperature is held constant at 37 °C using a regulated heating pad. During the measurements, animals are placed in a stereotactic frame (David Kopf Instruments, Tujunga, USA). In our hands, more reproducible and stable results are obtained with the thinned bone preparation. For this, the skull covering the cortex is carefully thinned to translucency using a surgical drill (Osseodoc, Bien-Air Surgery S.A. Le Noirmont, Switzerland). The thinned bone is then covered with 1 % agar and a round cover glass (approximately 5 mm diameter for the use in rats, smaller for mice). The cover slip prevents drying of the agar. To avoid reflections, it should be placed parallel to the imaging plane.

For single whisker stimulation, we insert one vibrissa into a cannula, which is attached to a piezoelectric ceramic wafer (Piezo Systems Inc., Cambridge, MA, USA). A computer-triggered pulse (50 ms) sent to the wafer produces a 1 mm rostral to caudal deflection of the vibrissa. The stimulation lasts 4 s with a frequency of 4 Hz. Before stimulation a baseline phase of at least 2 s should be acquired. A 60 s inter-trial interval is recommended to allow the hemodynamic signal to return to baseline.

2.3.2 Optical Imaging

Cortical images can be acquired using a CCD camera (we use a 12-bit Pixelfly VGA camera, PCO Imaging, Kelheim, Germany). Either a custom-made tandem lens system or a stereomicroscope (e.g., Leica MZ16 FA, Leica Microsystems, Heerbrugg, Switzerland) can be used for imaging. We focus 0.5 mm below the cortical surface. Collimated 785 nm laser light is shown onto the imaged cortical area (e.g., TuiOptics, Munich, Germany). We normally use free beam optics (mirrors and lenses) to collimate and steer the light (see Fig. 2). The images (640 × 480 pixels) are

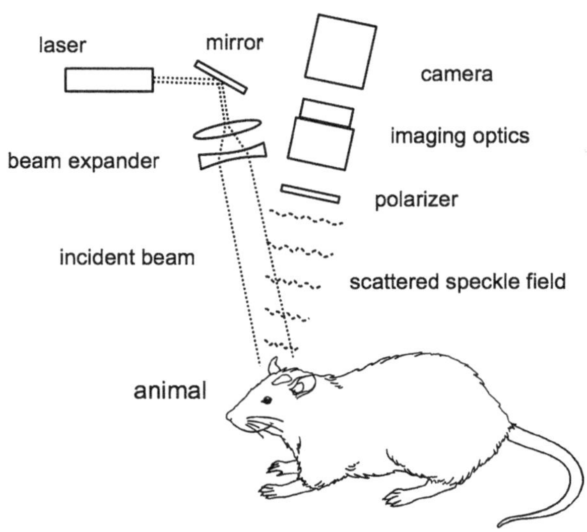

Fig. 2 Typical experimental setup for the laser speckle imaging. Coherent light from the laser is expanded and used to illuminate the tissue. Diffusively scattered light forms a speckle field, which produces an image on the camera through the polarizer and imaging optics.

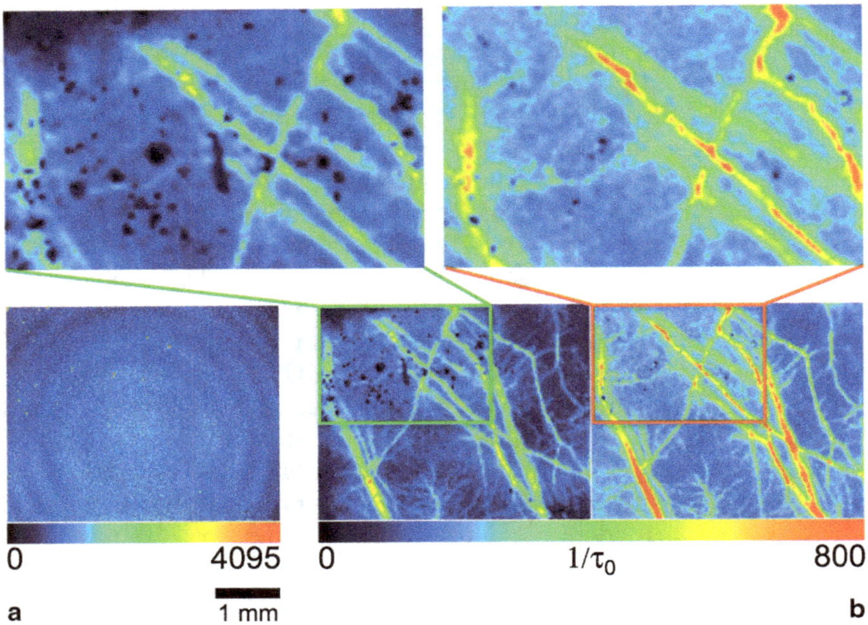

Fig. 3 (**a**) Raw speckle reflection recorded with 10 ms exposure time, measured through thinned skull. (**b**) *Left*: Reciprocal correlation time (RCT) derived from classical LASCA. (**b**) *Right*: RCT computed with dynamic laser speckle imaging. Note the significant reduction of specular reflections and effects of uneven illumination in the *right* panel. From [36].

acquired with a frame rate of around 50 Hz. For most of the applications, our camera specifications are sufficient; however, a larger number of pixels as well as a higher frame rate could be advantageous in certain situations. An exposure time of 10 ms is ideal, although other groups proposed slightly different values [28]. The f-stop of the imaging optics should be adjusted to find a best compromise between speckle contrast and number of speckles [48]. This is accomplished when the size of the speckles is approximately identical to the pixel size of the image sensor [10].

2.4 Laser Speckle Imaging in the Rat Somatosensory Cortex

The raw reflection signal from the somatosensory cortex is shown in Fig. 3a. Applying either the classical LASCA or dLSI algorithm yields the flow map depicted in Fig. 3b, where the surface vessels are easy to identify. The relative signal change following the 4-s stimulation is shown in Fig. 4, where a circumscribed region shows a marked increase in CBF reflected by the increase in change in the reciprocal correlation time ($\Delta\tau_0^{-1}$), estimated from DWS approximation of (7). The time–activity curve shows an increase in the hemodynamic signal with a delay of approximately 0.5 s. The peak amplitude is reached shortly before the end of the stimulation period. It is also noteworthy that the area of increased blood flow extends over a region that is considerably larger than the known barrel size, which is in the order of 0.4 mm for the principal whisker C1.

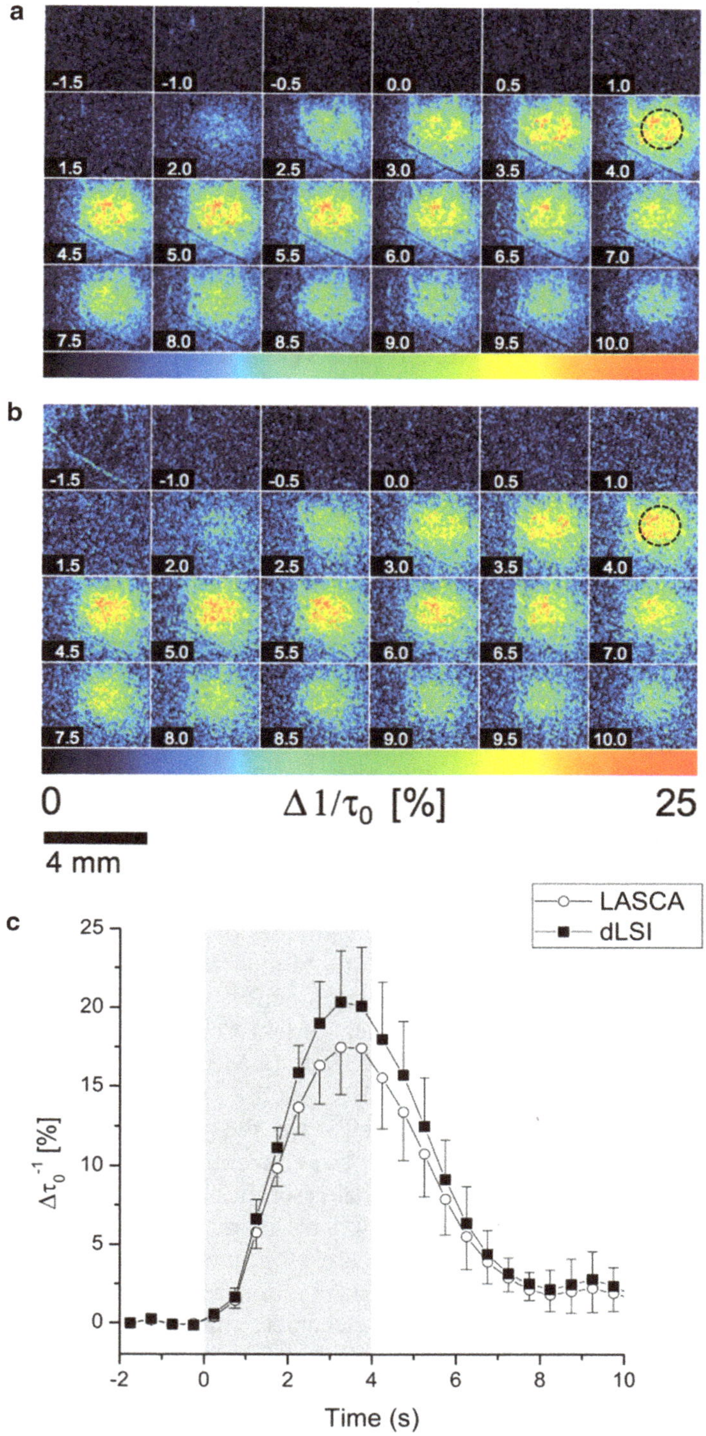

Fig. 4 Dynamic laser speckle imaging in the somatosensory cortex during single vibrissa (C1) stimulation. Maps represent relative changes of the RCT obtained from LASCA (**a**) and dLSI (**b**) signal upon single whisker stimulation (time step 0.5 s). (**c**) Time–activity curves averaged over 5 animals (10 trials each) from activated area (lines show average; error bars represent standard error of the mean). *Gray* area represents the duration of the whisker stimulation. Modified from [36].

3 Discussion

Much of what we know today about the cerebral hemodynamics was generated with the use of functional neuroimaging tools. For the acquisition of data with high spatial resolution, animal experiments are a prerequisite and optical approaches have been very successful. LSI was introduced in the field of neuroimaging more than a decade ago [10] and has since been subject to many technical developments and is nowadays used extensively in many research areas.

The technical implementation and data analysis for LSI have significantly progressed over the last years. To a large extent this was facilitated by the rapid development of digital image registration technologies as well as by advances in affordable computational power, which made acquisition and processing of the speckle images a straightforward task. One of the major advantages of the time-integrated speckle imaging is that it can cope with some of the limitations of digital cameras: limited frame rate and dynamic range. For a typical DLS/DWS measurement such drawback poses a significant challenge, since high temporal resolution is required to monitor speckle intensity fluctuations. This leads to the decreased signal to noise ratio due to the limited light detected under such conditions. On the contrary, for the speckle contrast analysis a higher exposure time is desirable which typically yields sufficient intensity signal.

The relatively simple and cost-effective implementation is one of the advantages of LSI. From an instrumentation point of view, the setup is similar to OIS. As a matter of fact, many groups use LSI in conjunction with OIS or multi-wavelength spectroscopy. The latter combination has the advantage that the cerebral metabolic rate of oxygen ($CMRO_2$) can be extracted. This is possible because the $CMRO_2 = CBF \times OEF$ and $OEF \approx HbR_V/HbT_V$, where HbR_V and HbT_V are the venous deoxyhemoglobin and venous total hemoglobin concentrations [11]. The size of the field of view and the spatial resolution of LSI are sufficient for most of the applications used to image the cortical hemodynamics in rodents. The temporal resolution is normally in the range of 0.5–10 Hz, which is also good enough to capture the essentials of the hemodynamic response characteristics.

There are also disadvantages of LSI that should be addressed. Above all, LSI is not able to produce quantitative CBF data. In most cases LSI data is presented as relative hemodynamic changes (e.g., $\Delta\tau_0^{-1}$ [%]; see Fig. 4), whereas a unit such as ml/min/g would be preferred in applications in which absolute values are needed. Furthermore, LSI is a two-dimensional imaging technique and to our knowledge a tomographic approach that would yield depth information has not yet been developed. Depth information is needed for laminar analysis and also for the identification of single cortical vessels, which is easily possible using CBF measurements with two-photon imaging (see Chapter 16 in this volume).

Increased performance of the detectors and processors opened new possibilities for further developments of the LSI methods. In addition to the original method based on the spatial statistics the temporal analysis has been proposed [49] as well as combined spatiotemporal statistics [50, 51]. Analysis in the temporal domain allows increasing the spatial resolution at the expense of the temporal resolution, since in the case of system ergodicity, measurements can be performed at every single pixel of the camera matrix. Spatiotemporal analysis allows the optimization of a combination of spatial and temporal resolution for each particular application. However, one has to use such methods with care since under certain conditions spatial speckle statistics—which is equivalent to ensemble averaging—cannot be substituted by time averaging. This is typically the case in the presence of strong static components, such as bone. In this case the static speckle field formed by bone is always present in the signal and cannot be properly averaged by the observation with a single spatial detector even over infinite time. In these situations spatial averaging is required for the quantitative assessment of the results. It has been shown that LSI can be further developed to address the case of heterogeneous dynamics. Two techniques have been introduced for this purpose: dLSI and multi-exposure speckle imaging [37]. Both are using the additional speckle statistics to get a better estimation of the correlation function on the different temporal scales. In a case of dLSI the standard speckle imaging equipment can be used. This demonstrates that the LSI methods can be further improved without losing the benefits of a simple experimental arrangement and straightforward processing steps.

References

1. Chance B, Cohen P, Jobsis F, Schoener B (1962) Intracellular oxidation-reduction states in vivo. Science 137:499–508
2. Jobsis FF (1977) Noninvasive, infrared monitoring of cerebral and myocardial oxygen sufficiency and circulatory parameters. Science 198:1264–1267
3. Grinvald A, Lieke E, Frostig RD, Gilbert CD, Wiesel TN (1986) Functional architecture of cortex revealed by optical imaging of intrinsic signals. Nature 324:361–364
4. Frostig RD, Lieke EE, Ts'o DY, Grinvald A (1990) Cortical functional architecture and local coupling between neuronal activity and the microcirculation revealed by in vivo high-resolution optical imaging of intrinsic signals. Proc Natl Acad Sci U S A 87:6082–6086
5. Attwell D, Buchan AN, Charpak S, Lauritzen M, MacVicar BA, Newman EA (2010) Glial and neuronal control of brain blood flow. Nature 468
6. Zlokovic BV (2011) Neurovascular pathways to neurodegeneration in Alzheimer's disease and other disorders. Nat Rev Neurosci 12:723–738
7. Lindauer U, Leithner C, Kaasch H, Rohrer B, Foddis M, Fuchtemeier M, Offenhauser N, Steinbrink J, Royl G, Kohl-Bareis M, Dirnagl U (2010) Neurovascular coupling in rat brain operates independent of hemoglobin deoxygenation. J Cereb Blood Flow Metab 30: 757–768
8. Weber B, Burger C, Wyss MT, von Schulthess GK, Scheffold F, Buck A (2004) Optical imaging

of the spatiotemporal dynamics of cerebral blood flow and oxidative metabolism in the rat barrel cortex. Eur J Neurosci 20:2664–2670
9. Boas DA, Dunn AK (2010) Laser speckle contrast imaging in biomedical optics. J Biomed Opt 15:011109
10. Dunn AK, Bolay H, Moskowitz MA, Boas DA (2001) Dynamic imaging of cerebral blood flow using laser speckle. J Cereb Blood Flow Metab 21:195–201
11. Dunn A, Devor A, Andermann M, Bolay H, Moskowitz M, Dale A, Boas D (2003) Simultaneous imaging of total cerebral hemoglobin concentration, oxygenation and blood flow during functional activation. Opt Lett 28:28–30
12. Nilsson GE, Salerud EG, Strömberg TK (2003) Laser Doppler perfusion monitoring and imaging. In: Vo-Dinh T (ed) Biomedical photonics handbook. CRC Press, Boca Raton, FL
13. Pike ER (1977) Introduction to photon correlation spectroscopy and velocimetry. In: Cummins HZ, Pike ER (eds) Photon correlation spectroscopy and velocimetry. Plenum Press, New York
14. Maret G, Wolf P-E (1987) Multiple light scattering from disordered media. The effect of brownian motion of scatterers. Z Phys B 65:409–413
15. Zakharov P, Scheffold F (2009) Advances in dynamic light scattering techniques. In: Kokhanovsky AA (ed) Light scattering reviews 4. Springer, Berlin, pp 433–467
16. Berne BJ, Pecora R (2000) Dynamic light scattering. With applications to chemistry, biology, and physics. Dover, New York
17. Briers JD (1996) Laser Doppler and time-varying speckle: a reconciliation. J Opt Soc Am A 13:345
18. Riva CE, Ross B, Benedek GB (1972) Laser Doppler measurements of blood flow in capillary tubes and retinal arteries. Invest Ophthalmol 11:936–944
19. Stern MD (1975) In vivo evaluation of microcirculation by coherent light scattering. Nature 254:56–58
20. Stern MD, Lappe DL, Bowen PD, Chimosky JE, Holloway GA, Keiser HR, Bowman RL (1977) Continuous measurement of tissue blood flow by laser-Doppler spectroscopy. Am J Physiol 232:H441–H448
21. Bonner R, Nossal R (1981) Model for laser Doppler measurements of blood flow in tissue. Appl Opt 20:2097–2107
22. Fercher AF, Briers JD (1981) Flow visualization by means of single-exposure speckle photography. Opt Commun 37:326–329
23. Briers JD, Fercher AF (1982) Retinal blood-flow visualization by means of laser speckle photography. Invest Ophthalmol Vis Sci 22:255–259
24. Schätzel K (1990) Noise on photon correlation data. I. Autocorrelation functions. Quantum Opt J European Opt Soc Part B 2:287–305
25. Bandyopadhyay R, Gittings AS, Suh SS, Dixon PK, Durian DJ (2005) Speckle-visibility spectroscopy: a tool to study time-varying dynamics. Rev Sci Instrum 76:93110
26. Wang Z, Hughes S, Dayasundara S, Menon RS (2007) Theoretical and experimental optimization of laser speckle contrast imaging for high specificity to brain microcirculation. J Cereb Blood Flow Metab 27:258–269
27. Zakharov P, Völker A, Buck A, Weber B, Scheffold F (2006) Quantitative modeling of laser speckle imaging. Opt Lett 31: 3465–3467
28. Yuan S, Devor A, Boas DA, Dunn AK (2005) Determination of optimal exposure time for imaging of blood flow changes with laser speckle contrast imaging. Appl Opt 44: 1823–1830
29. Braverman IM (1995) Anatomy and physiology of the cutaneous microcirculation. In: Berardesca E, Elsner P, Maibach HI (eds) Bioengineering of the skin: cutaneous blood flow and erythema. CRC Press, Boca Raton, FL
30. Pusey PN, Megen WV (1989) Dynamic light scattering by non-ergodic media. Physica A 57:705–742
31. Xue JZ, Pine D, Milner S, Wu XL, Chaikin P (1992) Nonergodicity and light scattering from polymer gels. Phys Rev A 46:6550–6563
32. Scheffold F, Skipetrov SE, Romer S, Schurtenberger P (2001) Diffusing-wave spectroscopy of nonergodic media. Phys Rev E 63:61404–61411
33. Kirsch S, Frenz V, Schärtl W, Bartsch E, Sillescu H (1996) Multispeckle autocorrelation spectroscopy and its application to the investigation of ultraslow dynamical processes. J Chem Phys 104:1758–1761
34. Viasnoff V, Lequeux F, Pine DJ (2002) Multispeckle diffusing-wave spectroscopy: a tool to study slow relaxation and time dependent dynamics. Rev Sci Instrum 73: 2336–2344
35. Zakharov P, Bhat S, Schurtenberger P, Scheffold F (2006) Multiple-scattering suppression in dynamic light scattering based on a digital camera detection scheme. Appl Optics 45:1756
36. Zakharov P, Völker AC, Wyss MT, Haiss F, Calcinaghi N, Zunzunegui C, Buck A, Scheffold F, Weber B (2009) Dynamic laser speckle imaging of cerebral blood flow. Opt Express 17:13904–13917
37. Parthasarathy AB, Tom WJ, Gopal A, Zhang X, Dunn AK (2008) Robust flow measurement with multi-exposure speckle imaging. Opt Express 16:1975–1989

38. Berwick J, Johnston D, Jones M, Martindale J, Redgrave P, McLoughlin N, Schiessl I, Mayhew JE (2005) Neurovascular coupling investigated with two-dimensional optical imaging spectroscopy in rat whisker barrel cortex. Eur J Neurosci 22:1655–1666
39. Calcinaghi N, Jolivet R, Wyss MT, Ametamey SM, Gasparini F, Buck A, Weber B (2011) Metabotropic glutamate receptor mGluR5 is not involved in the early hemodynamic response. J Cereb Blood Flow Metab 31:e1–e10
40. Devor A, Dunn AK, Andermann ML, Ulbert I, Boas DA, Dale AM (2003) Coupling of total hemoglobin concentration, oxygenation, and neural activity in rat somatosensory cortex. Neuron 39:353–359
41. Lindauer U, Villringer A, Dirnagl U (1993) Characterization of CBF response to somatosensory stimulation: model and influence of anesthetics. Am J Physiol 264:H1223–H1228
42. Parthasarathy AB, Kazmi SMS, Dunn AK (2010) Quantitative imaging of ischemic stroke through thinned skull in mice with Multi Exposure Speckle Imaging. Opt Express 1:246–259
43. Smausz T, Zölei D, Hopp B (2009) Real correlation time measurement in laser speckle contrast analysis using wide exposure time range images. Appl Optics 48:1425
44. Thompson OB, Andrews MK (2010) Tissue perfusion measurements: multiple-exposure laser speckle analysis generates laser Doppler-like spectra. J Biomed Opt 15:027015
45. Boas DA, Yodh AG (1997) Spatially varying dynamical properties of turbid media probed with diffusing temporal light correlation. JOSA 14:192–215
46. Apollo P, Wong Y, Wiltzius P (1993) Dynamic light scattering with a ccd camera. Rev Sci Instrum 64:2547–2549
47. Schätzel K (1983) Noise in photon correlation and photon structure functions. J Mod Opt 30:155–166
48. Skipetrov SE, Peuser J, Cerbino R, Zakharov P, Weber B, Scheffold F (2010) Noise in laser speckle correlation and imaging techniques. Opt Express 18:14519–14534
49. Cheng H, Luo Q, Zeng S, Chen S, Cen J, Gong H (2003) Modified laser speckle imaging method with improved spatial resolution. J Biomed Opt 8:559–564
50. Qiu J, Li P, Luo W, Wang J, Zhang H, Luo Q (2010) Spatiotemporal laser speckle contrast analysis for blood flow imaging with maximized speckle contrast. J Biomed Opt 15:016003
51. Peuser J, Belhaj-Saif A, Hamadjida A, Schmidlin E, Gindrat A-D, Völker AC, Zakharov P, Hoogewoud H-M, Rouiller EM, Scheffold F (2011) Follow-up of cortical activity and structure after lesion with laser speckle imaging and magnetic resonance imaging in nonhuman primates. J Biomed Opt 16:096011

Chapter 16

Two-Photon Microscopy to Measure Blood Flow and Concurrent Brain Cell Activity

Andy Y. Shih, Jonathan D. Driscoll, Michael J. Pesavento, and David Kleinfeld

Abstract

The cerebral vascular system services the constant demand for energy during neuronal activity in the brain. Attempts to delineate the logic of neurovascular coupling have been greatly aided by the advent of two-photon laser scanning microscopy to concurrently image blood flow and the activity of individual neurons and astrocytes involved in the control of the flow. Here we review the procedures to generate optical access to the cortex for both rats and mice, determine the receptive fields of the exposed cortical areas, and use two-photon microscopy to accurately measure blood flow in individual cortical vessels concurrent with local cellular activity. We illustrate the techniques with acute recordings from rats and chronic recordings from mice.

Key words Astrocytes, Flux, Neurons, Scanning, Vasculature, Vasomotion

1 Introduction

Blood is a vital and limited resource in the brain. All aspects of neuronal and non-neuronal activity require a supply of oxygen and glucose—a need that constantly evolves with changes in brain activity [1, 2]. How is the distribution of blood controlled relative to these changing needs? Delimiting this phenomenon, commonly termed functional hyperemia or neurovascular coupling, remains an active area of research [3]. However, recent studies have highlighted important conditions under which neural activity and blood flow become decoupled [4–6], and thus raise basic questions about neurovascular coupling [7]. One set of questions concerns the patterns of neuronal signals that lead to vasoactivity. A second set concerns the astrocytes that ensheath the vasculature and their role as intermediary cells that deliver signals from neurons to blood vessels.

The answers to the above questions depend on the ability to image blood flow and cells throughout the depth of cortex,

Bruno Weber and Fritjof Helmchen (eds.), *Optical Imaging of Neocortical Dynamics*, Neuromethods, vol. 85, DOI 10.1007/978-1-62703-785-3_16, © Springer Science+Business Media New York 2014

1.2–1.5 mm in rat and 1.0–1.2 mm in mouse. Imaging at this depth with subcellular resolution is facilitated by two-photon laser scanning microscopy [8], an optical sectioning technique in which absorption of light to excite fluorescent molecules occurs only at the laser focus. Past studies have made use of two-photon microscopy to examine vascular dynamics and blood flow in multiple brain regions, including somatosensory cortex [4, 9–27] and the olfactory bulb [6, 28–31], down to depths of 600 μm, which is sufficient to resolve vessels and neurons in layer 4. Recent advancements show that two-photon microscopy can achieve imaging depths that allow single microvessels to be studied throughout the full depth of cortex [32], and neuronal dynamics down to layer 5b [33], which is an important issue since vascular regulation appears to initiate in middle and deeper layers of cortex [9]. A further advantage is the concomitant use of exogenous and endogenous fluorescence-based functional reporters to observe cellular activity, such as changes in intracellular Ca^{2+} concentration [22, 28, 31] and the ratio of NADP to NAD^{+} [34, 35], concurrent with blood flow and vessel diameter changes.

Here we discuss basic procedures of single and multi-vessel two-photon imaging of blood flow dynamics, concurrent with cellular activity, in the somatosensory cortex of anesthetized and awake rodents. We further provide case studies. The equipment and algorithms used in these studies have been summarized elsewhere, including comprehensive reviews of basic methodology of vascular imaging [36], hardware [37, 38] and software [39, 40] for two-photon microscopy, and algorithms for data analysis [41–44]. Additional work has addressed the use of two-photon microscopy to image histological tissue with labeled vasculature [7, 45–48].

2 Optical Access and Mapping of Somatosensory Cortex

2.1 Brain Window Preparation

Both rats and mice have their place in cerebral blood flow imaging studies. The relatively large size of rats allows them to tolerate anesthesia better than mice and makes them the animal of choice for complex surgical procedures. Extended cranial windows can be fabricated to permit access to multiple regions of cortex, blood samples may be obtained at multiple time points in a procedure, and physiological parameters can be readily controlled. However, a current disadvantage of cranial windows, and thus rats, is that the imaging quality degrades within days. The use of mice has two advantages. First, they allow researchers to exploit the wide range of vascular-related transgenic animals. Second, transcranial windows with a thinned skull may be fabricated. The use of transcranial windows obviates potential problems with inflammation and changes in cranial volume and is excellent for repeated imaging studies over many months. On the down side, blood sampling and physiological control is limited with mice compared with rats.

Cranial windows. The generation of a cranial window for optical access in rats and mice differs on a number of levels. In rats, the overlying bone must be completely removed. Further, the dura mater must be carefully resected to the edge of the imaging window for optical access [49–51]. The window must then be resealed to restore intracranial pressure and to minimize motion artifacts caused by heart beat and breathing. Very large windows can be generated, i.e., 4 × 6 mm, to facilitate easy positioning of electrodes and cannulas. The clarity through such a cranial window is initially optimal as the materials overlying the pial surface cause minimal scattering. However, dural regrowth degrades the imaging quality such that repeated imaging is limited, in our hands, to about 4 days. As a result, chronically implanted windows for repeated imaging in rats are rarely reported and the use of pharmacological agents to suppress inflammation may also affect the phenomenon under study, such as the magnitude of injury in experimental stroke models [52]. Longer lasting windows may, in principle, be achieved by using inert substances such as Kwiksil silicone for movement suppression [53].

Cranial windows in mice are surgically less demanding, as the dura is thin and does not need to be removed for optical access. Detailed methods have been described for cranial windows [54, 55] and transcranial, thinned skull windows [56].

Transcranial windows. An alternate method that is suitable for mice is to generate a stable transcranial window, where the skull is thinned, polished, and reinforced with a thin layer of glue and cover slip [57]. These windows, which may be as large as 2 × 2 mm, greatly minimize disruption of the intracranial milieu, reduce inflammation, and prevent bone regrowth. While the imaging depth and clarity are somewhat reduced compared to windows with complete bone removal, polished and reinforced cranial windows in mice have been proven to give excellent clarity for two-photon imaging months after the initial surgery. This procedure has so far failed with rats because of the lower clarity of the dura and skull.

2.2 Localization of Active Areas

Standard brain atlases provide approximate coordinates for different brain regions as a means to locate the vasculature in relation to loci of neuronal activity. Yet somatotopically refined maps are often of considerable value. The lissencephalic structure of the rodent brain permits different regions in cortex to be further mapped to determine receptive fields using a variety of classical tools. For sensory areas, these include surface electrodes and optical imaging of intrinsic signals (OIS). The latter technique avoids any contact with the brain and uses changes in the intensity of reflected light to report a change in the ratio of oxy- to deoxyhemoglobin that occurs secondary to changes in neuronal activity [58, 59]. Motor areas may also be mapped in one of the three ways: (1) measuring limb

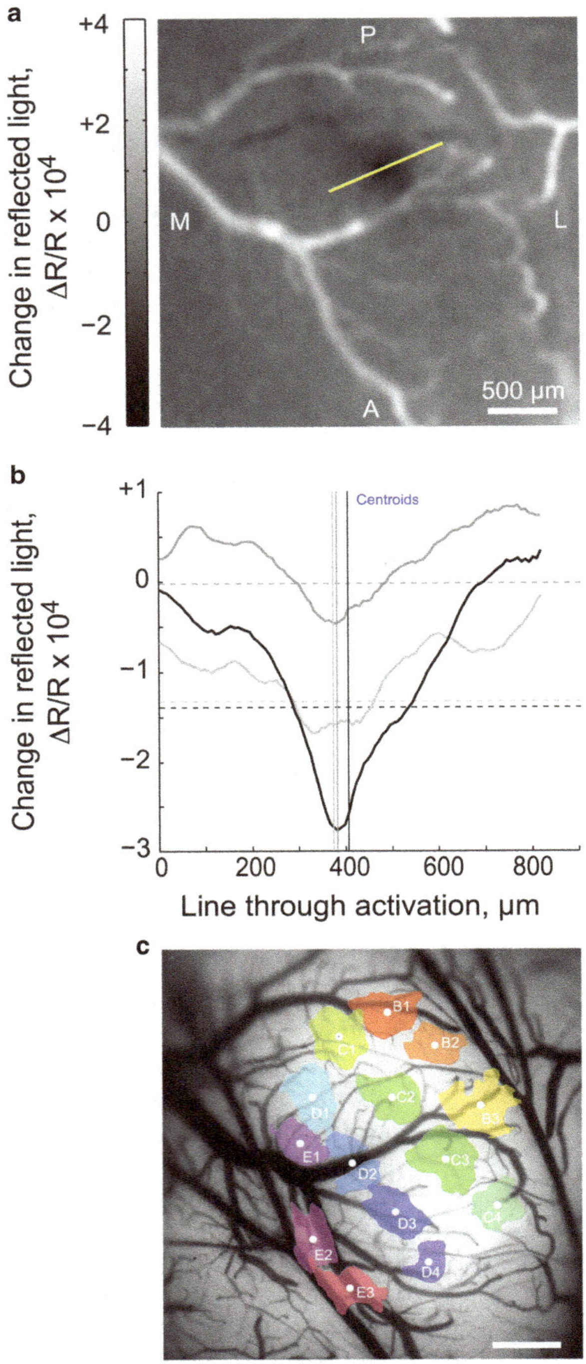

Fig. 1 Optical imaging of intrinsic signals for functional region targeting. A female Long Evans adult (2 month) rat was anesthetized with isoflurane (2 % in O_2 for induction and <1 % sustained) and a 4 mm × 4 mm closed craniotomy performed over the vibrissa area of primary somatosensory cortex. The OIS was obtained as the reflectance at 633 nm as a function of time and trial. The sampling

or vibrissa movement in response to intracortical microstimulation with bipolar microelectrodes; (2) measuring electromyogenic activity in response to intracortical microstimulation; or (3) through the use of focal illumination in conjunction with mice that express channelrhodopsin in projection neurons [60].

We illustrate the mapping process with optical imaging of intrinsic signals (OIS) for the case of vibrissa primary sensory cortex (Fig. 1). A craniotomy was prepared, as described previously [49], and individual vibrissae on the face of the animals were mechanically stimulated. Stimulation of a single vibrissa leads to a net decrease in reflectance at 633 nm within a localized region of vibrissa primary sensory cortex (Fig. 1a). This corresponds to a local decrease in blood oxygenation. The reflectance, denoted $R_i(x, y, t)$ where i labels the trial, was quantified as a function of time and trial. We computed the smoothed, trial-averaged value as $R(x, y, t) \equiv Gaussian(\sigma)_{x,y} \{<R_i(x, y, t)>_i\} - Median_{All\ x,\ y}\{<R_i(x, y, t)>_i\}$ where the Gaussian filter had a width of $\sigma = 4$ pixels. For the data of Fig. 1, the initial 4 s of the average signal constitutes baseline, i.e., $R_{base}(x, y) \equiv <R(x, y, t)> 0\ s \leq t \leq 4\ s$, the second 4 s were used to deflect individual vibrissae, and a period after the onset of stimulation was used to form $R_{response}(x, y) \equiv <R(x, y, t)> 4.5\ s \leq t \leq 9\ s$. Finally, the IOS is expressed as $\Delta R/R \equiv [R_{response}(x, y) - R_{base}(x, y)]/R_{base}(x, y)$. The centroid of this signal and the overall shape of the active region remain roughly constant between blocks of trials (Fig. 1b), although the amplitude varies greatly between blocks. Nonetheless, IOS imaging serves as a useful tool for mapping the center of activation in cortex for the different vibrissa (Fig. 1c). The branching arrangement of pial arterioles and venules, which is unique to each animal, serves as fiducials to relocate the active region between OIS and two-photon measurements.

Fig. 1 (continued) period is 50 ms, the pixel width is 8 μm, the period of each trial was 18 s, and we obtained 30 trials. (**a**) An example of the IOS for deflection of the C2 vibrissa, realized as a ± 12° movement by 10 Hz square wave filtered with a sixth order 100 Hz Bessel low pass filter. The *dark* declivity indicates reduced reflectance of red light, suggesting an increase of deoxygenated hemoglobin and thus increased neural activity in that region. (**b**) Intrinsic signal for a profile line through the centroid of the activated region in panel a for three separate blocks of trials. The *horizontal dashed lines* are a 50 % decrease of the intrinsic signal; the *solid vertical lines* indicate the centroids. (**c**) The thresholded IOS image, e.g., *dashed line* in panel b, for 14 different vibrissae across the cortical surface. The surface map was obtained by reflectance at 475 nm. Thresholds values were B1 = −1, B2 = −2.5, B3 = 0, C1 = −2, C2 = −1, C3 = 0, C4 = −1.5, D1 = −2, D2 = −1, D3 = −1.5, D4 = −3, E1 = −0.5, E2 = −3, E3 = 0; all × 10^{-4}. In some areas the vessels were masked from the calculation, e.g., vibrissa D2.

3 Blood Flow Measurements

3.1 Measurement of Blood Flow Dynamics in Single Cortical Vessels

We consider first the measurement of the flux of blood flow. This involves a user-defined scan profile for two-photon microscopy to simultaneously measure both the speed of red blood cells (RBCs) and the diameter of vessels as a means to compute the volume flux of RBCs. The vasculature is labeled with an intravenous bolus of fluorescein- or rhodamine-conjugated dextran for green and red light emission, respectively. The cerebral vasculature within the window is first imaged at low resolution with a 5× objective (Fig. 2a); this image can be directly compared with that made with OIS (Fig. 1c). High resolution imaging of surface pial vessels, penetrating vessels, and subsurface capillaries can then be performed in smaller regions with a high numerical aperture water dipping objective, such as a 20×–40× objective with a numerical aperture ranging from 0.8 to 1.0 (Fig. 2b).

When the serum is labeled, RBCs exclude the dextran dye and appear as dark objects moving against a bright fluorescent background. We used custom software to direct the imaging laser beam in a user-defined path within the imaging plane (Box 1) [42, 44]. Linear segments of constant scan speed traverse along the length of the center of the vessel and across the width of the vessel to measure RBC speed and lumen diameter, respectively. These linear scan segments are connected by polynomial splines, where connecting portions of the scan are accelerated to allow for rapid data collection across multiple vessels [44] (Fig. 2c).

The resulting line scan data forms a space-time image, typically displayed with the individual scan lines stacked on each other (Fig. 2d). In principle, many vessels that lie in the same plane can be measured simultaneously. Portions of the scan path along the centerline of the vessel lumen reveal angled streaks within the cascade image. The centerline velocity is proportional to the slope of the RBC streaks, measured from vertical (Fig. 2e). Determining this slope is most efficiently and robustly performed with a Radon transform of the data [41].

A velocity time series is calculated by transforming successive time-windowed portions of the line scans (right panel in Fig. 2f). The temporal spacing of successive windows must be close enough to resolve the highest velocity modulation frequency, the heart rate, which is ~6 Hz for rats and ~10 Hz for mice. In addition, the window size must be large enough to capture enough streak lines so that the Radon transform has sufficient data to calculate an accurate velocity value. We find that a window size of 40 ms with a spacing of 10 ms is a good compromise. In addition to heart rate, other physiological signals detected in the RBC velocity time series include breathing, at ~1 Hz for rats and ~2 Hz for mice, and vasomotion at ~0.1 Hz for rat and 0.1–1 Hz for mice [43, 57, 61].

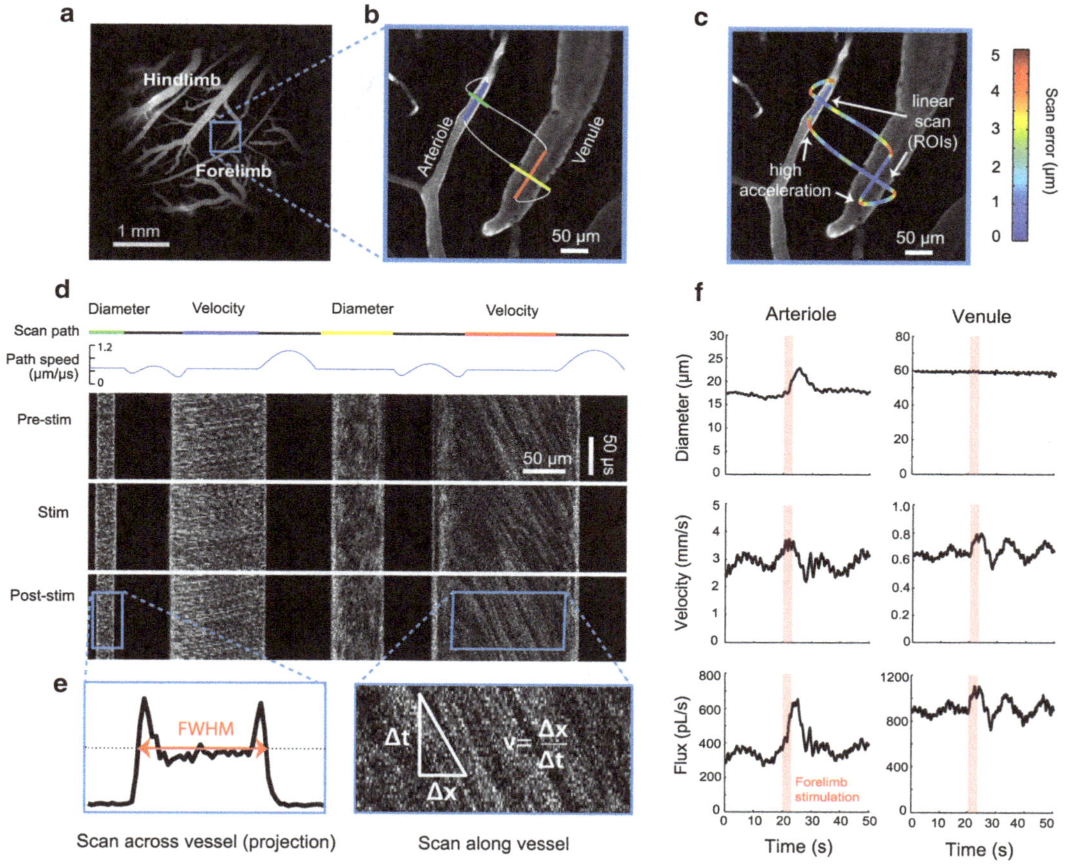

Fig. 2 Simultaneous measurement of diameter and velocity in two vessels using spatially optimized line scans. (**a**) Image of fluorescently stained vessels in somatosensory cortex of a Sprague Dawley rat. The forelimb and hindlimb representations across cortex were mapped using OIS, similar to that in Fig. 1. (**b**) Image of a surface arteriole and venule, with scan pattern superimposed. Portions of the scan path along the length are used to calculate RBC velocity, while portions across the diameter of the vessels are used to calculate diameter. Scans were acquired at a rate of 735 lines per second. (**c**) Scan path, colored to show the error between the desired scan path and the actual path the mirrors traversed. The error along linear portions of the image is about 1 μm and increases when the mirrors undergo rapid acceleration. The error between successive scans of the same path is less than 0.15 μm, several times lower than the point spread function for two-photon microscopy. (**d**) Scan mirror speed as a function of time (*top*). Note that portions used to acquire diameter and velocity data are constant speed (*top*). The line scans generated from the path can be stacked sequentially as a function of time to produce a raw cascade image (*bottom*). (**e**) Vessel diameter is calculated as the full width at half maximum of a time average of several scans across the width of a vessel (*left*). Red blood cell velocity calculated from the angle of the streaks caused by the flow of RBCs. (**f**) Data traces of diameter, velocity, and flux for the arteriole and venule, processed to remove heart rate and smoothed with a running window. Both vessels show an increase in flux in response to forelimb stimulation. In the arteriole, this flux increase is due to simultaneous increases of lumen diameter and RBC velocity. In contrast, flux increase in the venule is due only to an increase in RBC velocity, as diameter is unchanged by stimulation. All panels adapted from Driscoll et al. [44].

As with the velocity calculation, the diameter calculation is taken from a time-windowed portion of the data (Fig. 2d). The same window size and spacing used for velocity is also used for diameter so that both parameters are concurrently calculated. Vessel diameter is defined as full width at half maximum of the vessel profile for each window (left panel in Fig. 2e); the intensity profile tends to increase near the edges due to the exclusion of RBCs from the glycocalyx and endothelial surface layer. The two outermost half maximal points of these peaks are used to calculate the vessel boundary. Linear interpolation is used to add subpixel accuracy to the diameter measurement.

When the diameter of the vessel is much greater than that of the RBC, the flow is laminar and nearly parabolic [14, 62]. The two vascular parameters, RBC velocity and lumen diameter, are combined to calculate the volume flux, RBCs and plasma, for each vessel. The volume flux through the vessel is given by

$$\vec{\mathrm{F}} = \langle \vec{\mathrm{v}} \rangle A = \frac{\pi}{8} \vec{\mathrm{v}}(0)\, d^2$$

where $\vec{v}(0)$ is the time averaged RBC velocity at the center line of the vessel, A is the cross sectional area of the vessel lumen, and d is the lumen diameter. This formula underestimates the flux as the nonzero spatial extent of the RBC flattens the parabola of Poiseuille flow.

As an example, we measure the change in flux at the level of single penetrating vessels in response to somatosensory stimulation. These vessels deliver blood from the communicating vessels on the surface of cortex to the subsurface microvessels [15]. Both diameter and RBC velocity in the arteriole respond to stimulation (left column in Fig. 2f). The flux through the arteriole increases to a peak of 86 % over baseline, compared with a much smaller peak increase of 29 and 24 % for diameter and velocity measurements alone, respectively. The increase in RBC velocity is partially masked by a peak in the underlying vasomotor fluctuation, but remains a significant increase over an average 1 min period of basal activity. In contrast to the arteriole, a neighboring venule exhibits no change in lumen diameter, but a 23 % change in RBC velocity. As a result, the flux increases in the venule by 23 % as well (right column in Fig. 2f). This increase in venous flux would not be detected with methods that measure diameter only.

3.2 Simultaneous Imaging of Blood Flow and Local Cellular Activity

An important goal is to identify the cell types and signaling pathways that regulate neurovascular coupling. The activity of various neuronal cell types and astrocytes can be monitored in superficial cortex by bulk loading the tissue with functional dyes, in this case Oregon Green Bapta-1 AM (OGB-1 AM), to detect changes in intracellular

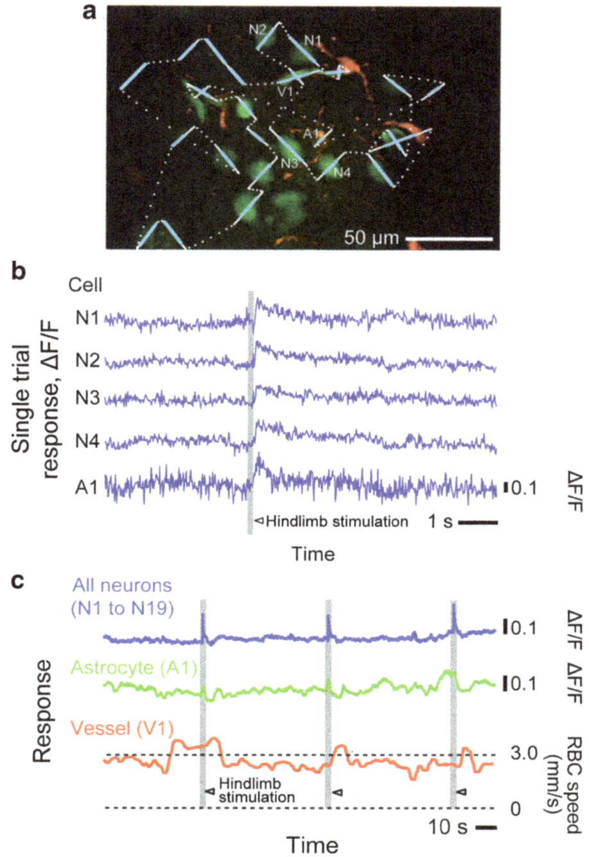

Fig. 3 Example of automated cell segmentation and fast scanning for functional imaging in rat parietal cortex. (**a**) A full field image of 19 neurons (N), 1 astrocyte (A), and three blood vessels, obtained at 4 frames/s, with a scan path superimposed on cells determined by our machine learning algorithm. All cells and vessels are scanned at 110 Hz. The *green channel* shows the fluorescence from OGB-1 and fluorescein while the *red channel* shows fluorescence from SR101. The scan path is superimposed on the image; *solid lines* are sections of the scan used to record responses while *dashed lines* are section of high scan acceleration. (**b**) Activity of neurons and an astrocyte, indicated in panel a, in response to a single weak electrical shock to the forelimb. (**c**) The Ca^{2+} response of the astrocyte (A1), the average neuronal response (N1–N19), and the speed of red blood cells in one capillary (V1). All panels adapted from Valmianski et al. [42].

Ca^{2+} concentration [63, 64] (Fig. 3a) (see Chapters 4 and 9 in this volume). The Ca^{2+} indicators may be injected by pipette into the active region through a vent in the cranial window under two-photon guidance, or prior to sealing the window during the initial craniotomy procedure. Astrocytes are selectively labeled with the astrocytic marker sulforhodamine 101 (SR101) in the same tissue to distinguish them from neurons in a second imaging channel [65].

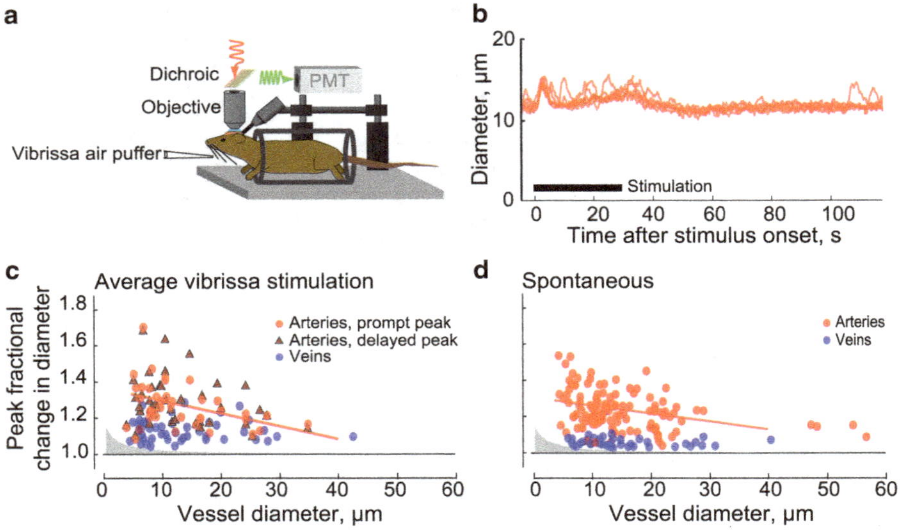

Fig. 4 Spontaneous and stimulus-induced vascular dynamics in the cortex of awake mouse. (**a**) Schematic of the experimental setup. The awake mouse is head fixed by means of a bolt and sits passively in an acrylic cylinder beneath the two-photon microscope. Air puffers for sensory stimulation are aimed at the vibrissa and as a control at the tail. (**b**) Individual dilation responses to 30 s vibrissae stimulation. (**c**) Plot of peak averaged dilation responses to 30 s vibrissae stimulation. Early arterial peaks, in the 0–10 s interval after stimulation, are denoted by *red circles*; regression slope = 0.007 μm^{-1} ($r^2 = 0.15$, $p < 0.02$). Late arterial peaks, greater than 10 s after onset, are denoted by *red triangles*; the linear regression (not shown) is not significant. Venules are denoted by *blue dots*; the linear regression is not significant. (**d**) Plot of peak value of the spontaneous dilations for arteries, in red, and veins, in blue. *Grey* area shows the 0.2 μm resolution limit of detectable changes. *Lines* show linear regressions; slope = −0.004 μm^{-1} for arterioles is significant ($r^2 = 0.13$, $p < 0.001$), while that for veins (not shown) is not significantly different from zero. All panels adapted from Drew et al. [16].

We consider a particular field of view that contains 20 identified cells, i.e., 19 neurons and one astrocyte, along with three blood vessels. Even at this level it is tedious to localize each labeled cell and guide the laser scan path manually. We thus used a machine learning algorithm to locate each soma [42] (Box 1) in conjunction with full-field images from the OGB-1 emission channel. Cells that were colabeled with SR101 were automatically labeled as astrocytes; the coordinates of selected blood vessels were also marked. We then calculated the fastest scan path through each soma along with selected microvessels (Fig. 3a) (Box 1). The typical signal-to-RMS-noise ratio, which we define as the ratio of the peak of the response to the RMS noise during the baseline, for a change in intracellular Ca^{2+} induced by a single sensory stimulus is ~10 (Fig. 3b). Further, the resulting line scan image contains stereotypical angled streaks for calculation of RBC speed, as well as intensity traces from selected neurons and astrocytes, similar to that seen in Fig. 2c. The composite data permits comparison of changes in astrocytic Ca^{2+} levels, together with changes in the speed of RBCs in a nearby microvessel, with the composite neuronal activity (Fig. 3c).

3.3 Stimulated and Basal Hemodynamics in Awake Mice

We consider the quantification of blood flow both at and below the cortical surface in awake mice. Mice were habituated to head fixation and blood flow was measured in single vessels through a chronically implanted transcranial window (Fig. 4a). Robust arteriole dilations could be evoked by prolonged contralateral whisker stimulation (Fig. 4b, c, red). Small arterioles dilated proportionally more than larger arterioles (Fig. 4c). Pial venules, typically thought to be static in terms of diameter, show a delayed and weak dilation in the awake state that had no dependence on initial diameter (Fig. 4c). These data suggest that functional hyperemia changes detected by BOLD fMRI contain a significant and possibly dominant contribution from large changes in arteriole volume, in concurrence with recent studies [66], rather than in venules [67].

Individual traces of the arteriole diameter show a high degree of variation and spontaneous dilation (Fig. 4b). Similar to the case of stimulus-induced changes, small arterioles dilated proportionally more than larger arterioles for both spontaneous events (Fig. 4d). The critical point raised by this data is that the magnitude of both prompt and delayed arterial responses induced by stimulation are similar in magnitude to spontaneous arterial dilations, with typical dilations of 30 % and maximum dilations near 50 % for both cases (Fig. 4b–d). Thus, from the point of control, stimulus-induced changes in blood flow are small, i.e., on the order of the noise level. The corollary is that individual stimulus events cannot be distinguished from spontaneous dilations based on magnitude alone [20, 68].

3.4 Imaging of Blood Flow in Deep Cortical Layers

An important development in two-photon microscopy is the ability to image deep cortical layers that are responsible for the output of neuronal processing [69]. Further, different cell types predominate in different layers, so that deep imaging permits neurovascular coupling to be studied in changing environments. Longer wavelengths of light penetrate deeper into tissue as a result of reduced scattering, absorption, and optical aberration by the tissue. In the example of Fig. 5, a cranial window was prepared in a mouse and the dye Alexa 680-conjugated dextran was used to label the blood serum. The dye was excited with 1,280 nm wavelength light from a Ti:sapphire pumped optical parametric oscillator and was found to enable imaging through the entire depth of cortex, i.e., 1 mm deep (Fig. 5a, b) [32]. RBC velocities in capillaries could be resolved as deep as 900 μm into cortex (Fig. 5c, d).

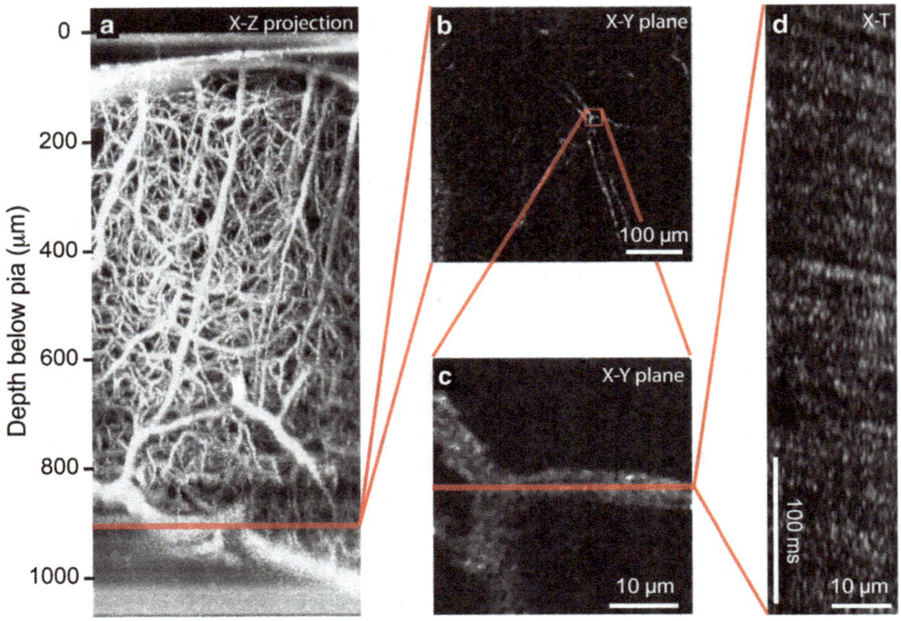

Fig. 5 Deep imaging of cortical angioarchitecture and blood flow. (**a**) Maximum intensity projections of mouse cortical vasculature in the coronal orientation. The image stack was collected over the entire depth of cortex through a cranial window with the dura intact. To reduce scattering and improve imaging depth, a long wavelength of excitation, i.e., 1,280 nm, was used and the blood plasma was labeled by intravenous injection of Alexa 680 conjugated to dextran. (**b**) Single planar image taken from panel a at the depth of the *red line*. (**c**) Magnified image taken from a vessel from the region in the *red box* in panel b. (**d**) Magnified image taken from a vessel from the region in the *red box* in panel b. All panels adapted from Kobat et al. [32].

4 Summary

Two-photon microscopy provides a number of advantages that will aid the study of the mechanisms underlying neurovascular coupling and cerebrovascular disease in animal models, including (1) the resolution needed to visualize single cortical vessels and their surrounding cells, (2) penetration depths of 250 μm through a transcranial window and 500 μm with craniotomies at 800 nm excitation, and imaging to 1,000 μm depth with longer excitation wavelengths, (3) reduced photodamage, (4) high speed user-defined line scans for near simultaneous measurement of RBC velocity, lumen diameter, and local cellular activity, and (5) the opportunity to image vascular dynamics deep in the cortex of awake mice. Advances in two-photon microscopy that enable rapid line scans in all three dimensions [70], as opposed to just within a plane, will enable real-time estimates of global flow within a region.

A wide variety of functional fluorescent dyes can be exploited for studies of neurovascular control. Recent focus has been on Ca^{2+} sensors to study cellular activity. Exogenous sensors have the advantage of labeling both neurons and astrocytes, along with a meshwork of processes that intervene between cell bodies. However, the cellular source of a signal may be difficult to resolve when focusing on fine processes. Transgenic mice with different neuronal subtypes labeled with fluorescent proteins can aid the separation of responses from different cell populations [71]. Genetically encoded Ca^{2+} sensors exclusively expressed in a specific cell type can be essential in separating signals from individual cells [72]. Finally, cyclic adenosine monophosphate sensors would also be of value since Ca^{2+}-independent pathways may also be involved in neurovascular control.

Box 1 Arbitrary Scan Patterns Through Automatically Selected Targets

In order to generate a standard two-dimensional two-photon microscopy image, the laser focal point is scanned systematically across the entire field of view in a raster pattern. Repeating the scan pattern generates a time series of frames and results in an easy to interpret movie of the entire field of view or a zoom of a smaller region. However, in many cases the relevant optical signal is spread through small portions of the field of view, such as when a sparse network of labeled neurons fire in response to stimuli. In such cases, it is advantageous to scan primarily across the active regions of interest, to maximize the number of signal photons acquired per unit of time. This not only maximizes the signal to noise ratio, but allows imaging on much faster timescales than the typical raster scan frame rates.

Frequently, the field of view contains only a few obvious neurons, or regions of interest (ROIs), which can be selected by the user during the experiment. In cases where a large number of ROIs are present, such as a sparse network of labeled neurons which are responsive to stimuli, machine learning can be used to automatically detect the regions and create an optimized scan path which passes through all of them in a minimum amount of time (Fig. 3). Typically, a computer algorithm is trained on experimental data which has been annotated by a human to mark relevant areas that the algorithm learns to recognize. During an experiment, the algorithm can then be used to quickly and automatically identify the ROIs. A typical workflow is [42]:

- A region of cortex is labeled with a Ca^{2+} indicator. An initial, full frame movie is collected while the animal is stimulated so that a subset of the cells responds with increased fluorescence. Multiple training and validation datasets can be created by taking multiple movies over the same or similar regions.
- After the experiment, the movies are marked by a human expert. Regions are labeled as a "cell" or "possibly a cell." Unmarked regions are considered to be "not a cell." This training data set is used to create a pixel classifier, which is used to determine if a given pixel is or is not a cell as well as a "blob" classifier, which uses the shapes of pixels to determine actual cell regions. The data is used to generate a set of mathematical rules, which can be implemented by the computer, to form a basis for the classifiers [75]. The classifiers operate on eight features in the dataset: the mean, variance, covariance, and correlation contained in the images, as well as the mean, variance, covariance, and correlation normalized by the standard deviation (*see* Fig. 6 and Table 1). After identifying pixels which are likely to belong to a cell, a morphological classifier using features such as size and eccentricity are used to delineate intact cells.
- During subsequent experiments, a movie is taken of a region similar to that used in the original experiment. This movie, along with the classifiers created previously, is fed into the software, which automatically generates a set of ROIs along, with an initial laser scan path across them. The ROIs can then be modified by the user, typically by adding or removing ROIs.
- The time it takes to scan across all the ROIs will depend on the order in which they are scanned. Optimizing the order is the classic "traveling salesman" problem, which has been extensively studied in computer science and mathematics. In general, it is not possible to search through all possible paths, and some computational shortcuts are taken. We use the ANT optimization algorithm to search rapidly through a subset of all possible paths as a means to find the shortest path that travels through all regions [76].

(continued)

Box 1 (continued)

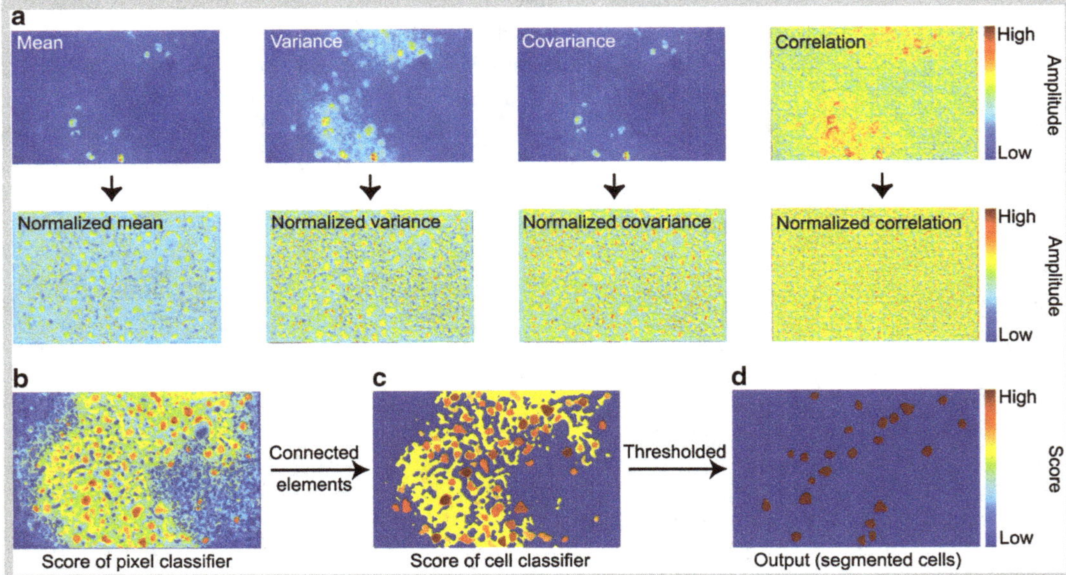

Fig. 6 Example of cell segmentation for functional imaging of neurons. (**a**) The four unnormalized filtered version of the raw data (*see* Table 1, formula indices 1 to 4). The color corresponds to amplitude of the filtered output. The normalized versions of filtered images from panel a (*see* Table 1, formula indices 5 to 8). (**b**) The output of the first classifier. The color corresponds to the likelihood that a given pixel is a cell. (**c**) The output of the second step classifier, with isolated pixels removed with a 5 × 5 pixel median filter, along with the output values then thresholded to form clusters of pixels that are candidate cells. We chose six levels, which correspond to pixels lying in the top 5, 10, 15, 20, 25, and 30 percent of the maximum amplitude. (**d**) Final classification made by thresholding the output shown in panel c. All panels adapted from Valmianski et al. [42].

Table 1
Feature map of time-series image data

Index	Feature map
1	$I^{Mean}_{x,y} = \left\langle I_{x,y}(t) \right\rangle$ [a,b]
2	$I^{Var}_{x,y} = \left\langle \left[I_{x,y}(t) - I^{Mean}_{x,y} \right]^2 \right\rangle$
3	$I^{Cov}_{x,y} = \sqrt{\left\langle \left[I_{x,y}(t) - I^{Mean}_{x,y} \right]\left[I_{x+1,y}(t) - I^{Mean}_{x+1,y} \right]^2 \right\rangle + \left\langle \left[I_{x,y}(t) - I^{Mean}_{x,y} \right]\left[I_{x,y+1}(t) - I^{Mean}_{x,y+1} \right]^2 \right\rangle}$
4	$I^{Corr}_{x,y} \equiv \sqrt{\frac{\left\langle \left[I_{x,y}(t) - I^{Mean}_{x,y} \right]\left[I_{x+1,y}(t) - I^{Mean}_{x+1,y} \right]^2 \right\rangle}{I^{Var}_{x,y}\, I^{Var}_{x+1,y}} + \frac{\left\langle \left[I_{x,y}(t) - I^{Mean}_{x,y} \right]\left[I_{x,y+1}(t) - I^{Mean}_{x,y+1} \right]^2 \right\rangle}{I^{Var}_{x,y}\, I^{Var}_{x,y+1}}}$

(continued)

Box 1 (continued)

Table 1 (continued)

Index	Feature map
5	$I_{x,y}^{NormMean} = \frac{I_{x,y}^{Mean} - I_{x,y}^{Mean}}{\sigma_{x,y}^{Mean}}$ [c]
6	$I_{x,y}^{NormVar} = \frac{I_{x,y}^{Var} - I_{x,y}^{Var}}{\sigma_{x,y}^{Var}}$
7	$I_{x,y}^{NormCov} = \frac{I_{x,y}^{Cov} - I_{x,y}^{Cov}}{\sigma_{x,y}^{Cov}}$
8	$I_{x,y}^{NormCorr} = \frac{I_{x,y}^{Corr} - I_{x,y}^{Corr}}{\sigma_{x,y}^{Corr}}$

[a] $I(t) \equiv \frac{1}{N}\sum_{t=1}^{N} I(t)$

[b] $I(t) = I(t) \otimes W_5$, where W_n is a uniform filter of n pixels

[c] $\hat{I}(t) = I(t) \otimes W_{21}$ and $\hat{\sigma}_{x,y} = \frac{1}{\sqrt{(21)^2 - 1}}\sqrt{\sum_{x'=x-10}^{x+10}\sum_{y'=y-10}^{y+10}\left[I_{x',y'}(t) - \hat{I}_{x,y}\right]^2}$

The intermixing of cell types within a small tissue volume also hinders electrophysiological and pharmacological approaches to query the role of cell types in neurovascular coupling. New advances with light-activated opsins [73] or engineered receptors with unnatural affinities for exogenous chemicals [74] may help to unravel issues with cellular specificity [7]. However, caveats should also be considered, as activation of one cell type does not preclude activation of nontargeted cells linked within the same circuitry. The manipulation of cell-specific vasoactive signaling cascades will be an important step in dissecting the chemical basis of neurovascular coupling, but will also be challenging as new tools to knock-down gene expression will need to be developed [7].

References

1. Fox PT, Raichle ME (1986) Focal physiological uncoupling of cerebral blood flow and oxidative metabolism during domatosensory stimulation in human subjects. Proc Natl Acad Sci U S A 83:1140–1144
2. Leybaert L (2005) Neurobarrier coupling in the brain: a partner of neurovascular and neurometabolic coupling? J Cereb Blood Flow Metab 25:2–16
3. Attwell D, Buchan AM, Charpak S, Lauritzen M, MacVicar BA, Newman EA (2010) Glial and neuronal control of brain blood flow. Nature 468:232–243
4. Devor A, Hillman EM, Tian P, Waeber C, Teng IC, Ruvinskaya L, Shalinsky MH, Zhu H, Haslinger RH, Narayanan SN, Ulbert I, Dunn AK, Lo EH, Rosen BR, Dale AM, Kleinfeld D, Boas DA (2008) Stimulus-induced changes in

blood flow and 2-deoxyglucose uptake dissociate in ipsilateral somatosensory cortex. J Neurosci 28:14347–14357

5. Sirotin YB, Das A (2008) Anticipatory haemodynamic signals in sensory cortex not predicted by local neuronal activity. Nature 457:475–479
6. Jukovskaya N, Tiret P, Lecoq J, Charpak S (2011) What does local functional hyperemia tell about local neuronal activation? J Neurosci 31:1579–1582
7. Kleinfeld D, Blinder P, Drew PJ, Driscoll JD, Muller A, Tsai PS, Shih AY (2011) A guide to delineate the logic of neurovascular signaling in the brain. Front Neuroenergetics 1:1–9
8. Denk W, Strickler JH, Webb WW (1990) Two-photon laser scanning fluorescence microscopy. Science 248:73–76
9. Tian P, Teng I, May LD, Kurz R, Lu K, Scadeng M, Hillman EM, De Crespigny AJ, D'Arceuil HE, Mandeville JB, Marota JJ, Rosen BR, Lui TT, Boas DA, Buxton RB, Dale AM, Devor A (2010) Cortical depth-specific microvascular dilation underlies laminar differences in blood oxygenation level-dependendent functional MRI signal. Proc Natl Acad Sci U S A 107:15246–15251
10. Sigler A, Mohajerani MH, Murphy TH (2009) Imaging rapid redistribution of sensory-evoked depolarization through existing cortical pathways after targeted stroke in mice. Proc Natl Acad Sci U S A 106:11758–11764
11. Zhang S, Murphy TH (2007) Imaging the impact of cortical microcirculation on synaptic structure and sensory-evoked hemodynamic responses *in vivo*. Public Libr Sci Biol 5:e119
12. Brown CE, Li P, Boyd JD, Delaney KR, Murphy TH (2007) Extensive turnover of dendritic spines and vascular remodeling in cortical tissues recovering from stroke. J Neurosci 27:4101–4109
13. Zhang S, Boyd J, Delaney KR, Murphy TH (2005) Rapid reversible changes in dendritic spine structure *in vivo* gated by the degree of ischemia. J Neurosci 25:5228–5333
14. Schaffer CB, Friedman B, Nishimura N, Schroeder LF, Tsai PS, Ebner FF, Lyden PD, Kleinfeld D (2006) Two-photon imaging of cortical surface microvessels reveals a robust redistribution in blood flow after vascular occlusion. Public Libr Sci Biol 4:258–270
15. Nishimura N, Schaffer CB, Friedman B, Lyden PD, Kleinfeld D (2007) Penetrating arterioles are a bottleneck in the perfusion of neocortex. Proc Natl Acad Sci U S A 104:365–370
16. Drew PJ, Shih AY, Kleinfeld D (2011) Fluctuating and sensory-induced vasodynamics in rodent cortex extends arteriole capacity. Proc Natl Acad Sci U S A 108:8473–8478
17. Blinder P, Shih AY, Rafie CA, Kleinfeld D (2010) Topological basis for the robust distribution of blood to rodent neocortex. Proc Natl Acad Sci U S A 107:12670–12675
18. Shih AY, Friedman B, Drew PJ, Tsai PS, Lyden PD, Kleinfeld D (2009) Active dilation of penetrating arterioles restores red blood cell flux to penumbral neocortex after focal stroke. J Cereb Blood Flow Metab 29:738–751
19. Winship IR, Plaa N, Murphy TH (2007) Rapid astrocyte calcium signals correlate with neuronal activity and onset of the hemodynamic response *in vivo*. J Neurosci 27:6268–6272
20. Kleinfeld D, Mitra PP, Helmchen F, Denk W (1998) Fluctuations and stimulus-induced changes in blood flow observed in individual capillaries in layers 2 through 4 of rat neocortex. Proc Natl Acad Sci U S A 95:15741–15746
21. Devor A, Tian P, Nishimura N, Teng IC, Hillman EM, Narayanan SN, Ulbert I, Boas DA, Kleinfeld D, Dale AM (2007) Suppressed neuronal activity and concurrent arteriolar vasoconstriction may explain negative blood oxygenation level-dependent signaling. J Neurosci 27:4452–4459
22. Wang X, Lou N, Xu Q, Tian GF, Peng WG, Han X, Kang J, Takano T, Nedergaard M (2006) Astrocytic Ca^{2+} signaling evoked by sensory stimulation *in vivo*. Nat Neurosci 9:816–823
23. Fernández-Klett F, Offenhauser N, Dirnagl U, Priller J, Lindauer U (2010) Pericytes in capillaries are contractile *in vivo*, but arterioles mediate functional hyperemia in the mouse brain. Proc Natl Acad Sci U S A 107: 22290–22295
24. McCaslin AF, Chen BR, Radosevich AJ, Cauli B, Hillman EM (2010) *In vivo* 3D morphology of astrocyte-vasculature interactions in the somatosensory cortex: implications for neurovascular coupling. J Cereb Blood Flow Metab 31:795–806
25. Nishimura N, Rosidi NL, Iadecola C, Schaffer CB (2010) Limitations of collateral flow after occlusion of a single cortical penetrating arteriole. J Cereb Blood Flow Metab 30:1914–1927
26. Stefanovic B, Hutchinson E, Yakovleva V, Schram V, Russell JT, Belluscio L, Koretsky AP, Silva AC (2007) Functional reactivity of cerebral capillaries. J Cereb Blood Flow Metab 28:961–972
27. Hutchinson EB, Stefanovic B, Koretsky AP, Silva AC (2006) Spatial flow-volume dissociation of the cerebral microcirculatory response to mild hypercapnia. Neuroimage 32:520–530
28. Petzold GC, Albeanu DF, Sato TF, Murthy VN (2008) Coupling of neural activity to blood flow in olfactory glomeruli is mediated by astrocytic pathways. Neuron 58:879–910
29. Lecoq JL, Tiret P, Najac M, Sheperd GM, Greer CA, Charpak S (2009) Odor-evoked oxygen consumption by action potential and synaptic transmission in the olfactory bulb. J Neurosci 29:1424–1433

30. Chaigneau E, Oheim M, Audinat E, Charpak S (2003) Two-photon imaging of capillary blood flow in olfactory bulb glomeruli. Proc Natl Acad Sci U S A 100:13081–13086
31. Chaigneau E, Tiret P, Lecoq J, Ducros M, Knöpfel T, Charpak S (2007) The relationship between blood flow and neuronal activity in the rodent olfactory bulb. J Neurosci 27:6452–6460
32. Kobat D, Durst ME, Nishimura N, Wong AW, Schaffer CB, Xu C (2009) Deep tissue multi-photon microscopy using longer wavelength excitation. Opt Express 17:13354–13364
33. Mittmann W, Wallace DJ, Czubayko U, Herb JT, Schaefer AT, Looger LL, Denk W, Kerr JN (2011) Two-photon calcium imaging of evoked activity from L5 somatosensory neurons *in vivo*. Nat Neurosci 14:1089–1093
34. Kasischke KA, Lambert EM, Panepento B, Sun A, Gelbard HA, Burgess RW, Foster TH, Nedergaard M (2011) Two-photon NADH imaging exposes boundaries of oxygen diffusion in cortical vascular supply regions. J Cereb Blood Flow Metab 31:68–81
35. Murphy TH, Li P, Betts K, Liu R (2008) Two-photon imaging of stroke onset *in vivo* reveals that NMDA-receptor independent ischemic depolarization is the major cause of rapid reversible damage to dendrites and spines. J Neurosci 28:756–772
36. Shih AY, Driscoll JD, Drew PJ, Nishimura N, Schaffer CB, Kleinfeld D (2012) Two-photon microscopy as a tool to study blood flow and neurovascular coupling in the rodent brain. J Cereb Blood Flow Metab 32(7):1277–1309
37. Tsai PS, Kleinfeld D (2009) *In vivo* two-photon laser scanning microscopy with concurrent plasma-mediated ablation: principles and hardware realization. In: Frostig RD (ed) Methods for in vivo optical imaging, 2nd edn. CRC Press, Boca Raton, pp 59–115
38. Driscoll JD, Shih AY, Iyengar S, Field JJ, White GA, Squier JA, Cauwenberghs G, Kleinfeld D (2011) Photon counting, censor corrections, and lifetime imaging for improved detection in two-photon microscopy. J Neurophysiol 104: 1803–1811
39. Nguyen Q-T, Dolnick EM, Driscoll J, Kleinfeld D (2009) MPScope 2.0: a computer system for two-photon laser scanning microscopy with concurrent plasma-mediated ablation and electrophysiology. In: Frostig RD (ed) Methods for in vivo optical imaging, 2nd edn. CRC Press, Boca Raton, pp 117–142
40. Nguyen Q-T, Tsai PS, Kleinfeld D (2006) MPScope: a versatile software suite for multiphoton microscopy. J Neurosci Methods 156:351–359
41. Drew PJ, Blinder P, Cauwenberghs G, Shih AY, Kleinfeld D (2010) Rapid determination of particle velocity from space-time images using the Radon transform. J Comput Neurosci 29:5–11
42. Valmianski I, Shih AY, Driscoll J, Matthews DM, Freund Y, Kleinfeld D (2010) Automatic identification of fluorescently labeled brain cells for rapid functional imaging. J Neurophysiol 104:1803–1811
43. Kleinfeld D, Mitra PP (2011) Applications of spectral methods in functional brain imaging. In: Yuste R (ed) Imaging: a laboratory manual. Cold Spring Harbor Laboratory Press, New York, pp 12.11–12.17
44. Driscoll JD, Shih AY, Drew PJ, Cauwenberghs G, Kleinfeld D (2011) Two-photon imaging of blood flow in cortex. In: Helmchen F, Konnerth A, Yuste R (eds) Imaging in neuroscience: a laboratory manual. Cold Spring Harbor Laboratory Press, New York, pp 927–938
45. Tsai PS, Blinder P, Kaufhold JP, Squier JD, Kleinfeld D (2011) All-optical, in situ histology of brain tissue with femtosecond laser pulses. In: Helmchen F, Konnerth A, Yuste R (eds) Imaging in neuroscience: a laboratory manual. Cold Spring Harbor Laboratory Press, New York, pp 437–446
46. Tsai PS, Friedman B, Ifarraguerri AI, Thompson BD, Lev-Ram V, Schaffer CB, Xiong Q, Tsien RY, Squier JA, Kleinfeld D (2003) All-optical histology using ultrashort laser pulses. Neuron 39:27–41
47. Tsai PS, Kaufhold J, Blinder P, Friedman B, Drew P, Karten HJ, Lyden PD, Kleinfeld D (2009) Correlations of neuronal and microvascular densities in murine cortex revealed by direct counting and colocalization of cell nuclei and microvessels. J Neurosci 18: 14553–14570
48. Ragan T, Sylvan JD, Kim KH, Huang H, Bahlmann K, Lee RT, So PT (2007) High-resolution whole organ imaging using two-photon tissue cytometry. J Biomed Opt 12:014015
49. Kleinfeld D, Delaney KR (1996) Distributed representation of vibrissa movement in the upper layers of somatosensory cortex revealed with voltage sensitive dyes. J Comp Neurol 375:89–108
50. Levasseur JE, Wei EP, Raper AJ, Kontos AA, Patterson JL (1975) Detailed description of a cranial window technique for acute and chronic experiments. Stroke 6:308–317
51. Morii S, Ngai AC, Winn HR (1986) Reactivity of rat pial arterioles and venules to adenosine and carbon dioxide: with detailed description of the closed cranial window technique in rats. J Cereb Blood Flow Metab 6:34–41
52. Tuor UI, Simone CS, Barks JD, Post M (1993) Dexamethasone prevents cerebral infarction without affecting cerebral blood flow in neonatal rats. Stroke 24:452–457

53. Dombeck DA, Graziano MS, Tank DW (2009) Functional clustering of neurons in motor cortex determined by cellular resolution imaging in awake behaving mice. J Neurosci 29:13751–13760
54. Holtmaat A, Bonhoeffer T, Chow DK, Chuckowree J, De Paola V, Hofer SB, Hübener M, Keck T, Knott G, Lee WC, Mostany R, Mrsic-Flogel TD, Nedivi E, Portera-Cailliau C, Svoboda K, Trachtenberg JT, Wilbrecht L (2009) Long-term, high-resolution imaging in the mouse neocortex through a chronic cranial window. Nat Protoc 4:1128–1144
55. Mostany R, Portera-Cailliau C (2008) A method for 2-photon imaging of blood flow in the neocortex through a cranial window. J Vis Exp 12:678
56. Yang G, Pan F, Parkhurst CN, Grutzendler J, Gan WB (2010) Thinned-skull cranial window technique for long-term imaging of the cortex in live mice. Nat Protoc 5:201–208
57. Drew PJ, Shih AY, Driscoll JD, Knutsen PM, Davalos D, Blinder P, Akassoglou K, Tsai PS, Kleinfeld D (2010) Chronic optical access through a polished and reinforced thinned skull. Nat Methods 7:981–984
58. Frostig RD, Lieke EE, Ts'o DY, Grinvald A (1990) Cortical functional architecture and local coupling between neuronal activity and the microcirculation revealed by *in vivo* high-resolution optical imaging of intrinsic signals. Proc Natl Acad Sci U S A 87:6082–6086
59. Grinvald A, Lieke EE, Frostig RD, Gilbert CD, Wiesel TN (1986) Functional architecture of cortex revealed by optical imaging of intrinsic signals. Nature 324:361–364
60. Ayling OG, Harrison TC, Boyd JD, Goroshkov A, Murphy TH (2009) Automated light-based mapping of motor cortex by photoactivation of channelrhodopsin-2 transgenic mice. Nat Methods 6:219–224
61. Mayhew JEW, Askew S, Zeng Y, Porrill J, Westby GWM, Redgrave P, Rector DM, Harper RM (1996) Cerebral vasomotion: 0.1 Hz oscillation in reflectance imaging of neural activity. Neuroimage 4:183–193
62. Rovainen CM, Woolsey TA, Blocher NC, Wang D-B, Robinson OF (1993) Blood flow in single surface arterioles and venules on the mouse somatosensory cortex measured with videomicroscopy, fluorescent dextrans, nonoccluding fluorescent beads, and computer-assisted image analysis. J Cereb Blood Flow Metab 13: 359–371
63. Stosiek C, Garaschuk O, Holthoff K, Konnerth A (2003) *In vivo* two-photon calcium imaging of neuronal networks. Proc Natl Acad Sci U S A 100:7319–7324
64. Garaschuk O, Milos RI, Konnerth A (2006) Targeted bulk-loading of fluorescent indicators for two-photon brain imaging *in vivo*. Nat Protoc 1:380–386
65. Nimmerjahn A, Kirchhoff F, Kerr JN, Helmchen F (2004) Sulforhodamine 101 as a specific marker of astroglia in the neocortex *in vivo*. Nat Methods 29:31–37
66. Kim T, Kim SG (2011) Temporal dynamics and spatial specificity of aterial and venous blood volume changes during visual stimulation: Implication for BOLD quantification. J Cereb Blood Flow Metab 31:1211–1222
67. Buxton RB, Wong EC, Frank LR (1998) Dynamics of blood flow and oxygenation changes during brain activation: the balloon model. Magn Reson Med 39:855–864
68. Drew PJ, Duyn JH, Galanov E, Kleinfeld D (2008) Finding coherence in spontaneous oscillations. Nat Neurosci 11:991–993
69. Helmchen F, Denk W (2005) Deep tissue two-photon microscopy. Nat Methods 2:932–940
70. Botcherby EJ, Smith CW, Kohl MM, Débarre D, Booth MJ, Juškaitis R, Paulsen O, Wilson T (2012) Aberration-free three-dimensional multiphoton imaging of neuronal activity at kHz rates. Proc Natl Acad Sci U S A 109: 2919–2924
71. Sohya K, Kameyama K, Yanagawa Y, Obata K, Tsumoto T (2007) GABAergic neurons are less selective to stimulus orientation than excitatory neurons in layer II/III of visual cortex, as revealed by in vivo functional Ca^{2+} imaging in transgenic mice. J Neurosci 27:2145–2149
72. Tian L, Hires SA, Mao T, Huber D, Chiappe ME, Chalasani SH, Petreanu L, Akerboom J, McKinney SA, Schreiter ER, Bargmann CI, Jayaraman V, Svoboda K, Looger LL (2009) Imaging neural activity in worms, flies and mice with improved GCaMP calcium indicators. Nat Methods 6:875–881
73. Lee JH, Durand R, Gradinaru V, Zhang F, Goshen I, Kim DS, Fenno LE, Ramakrishnan C, Deisseroth K (2010) Global and local fMRI signals driven by neurons defined optogenetically by type and wiring. Nature 465:788–792
74. Alexander GM, Rogan SC, Abbas AI, Armbruster BN, Pei Y, Allen JA, Nonneman RJ, Hartmann J, Moy SS, Nicolelis MA, McNamara JO, Roth BL (2009) Remote control of neuronal activity in transgenic mice expressing evolved G protein-coupled receptors. Neuron 63:27–39
75. Freund Y (2009) A more robust boosting algorithm, *arXive*, Arxiv/0905.2138
76. Di Caro G, Dorigo M (1998) AntNet: distributed stigmergetic control for communications networks. J Artif Intell Res 9:317–365

INDEX

Bruno Weber and Fritjof Helmchen (eds.), *Optical Imaging of Neocortical Dynamics*, Neuromethods, vol. 85, DOI 10.1007/978-1-62703-785-3, © Springer Science+Business Media New York 2014

Q

R

S

T

U

V

W

Y

www.ingramcontent.com/pod-product-compliance
Lightning Source LLC
LaVergne TN
LVHW080030210926
848481LV00011B/3144
* 9 7 8 1 4 9 3 9 5 9 7 0 9 *